Evolutionary Psychology

Series Editors
Todd K. Shackelford
Viviana A. Weekes-Shackelford

The Springer Series in Evolutionary Psychology is the first series of volumes dedicated to this increasingly important discipline within psychology. The series will reflect the multidisciplinary nature of this field encompassing evolutionary theory, biology, cognitive psychology, anthropology, economics, computer science, and paleoarchaeology. It will explore the underlying psychological mechanisms, and information processing systems housed in the brain as well as the various triggers for their activation. Its scientific assumptions rest on The concept that evolution is the only known causal process capable of creating complex organic mechanisms as are exhibited in human and animal life. Further, it seeks to show how information processing is adaptively influenced by input from the environment. Overall, the series will address the range of functionally specialized evolved mechanisms, mediated by contextual circumstances, that become combined and integrated to produce manifest behavior. The Series will address key areas of research within the field, including sexual behavior; conflict between the sexes; parenting; kinship; cooperation and altruism; aggression; warfare; and status and prestige. A premier resource in evolutionary psychology for researchers, academics and students, this series will provide. the field continuing and comprehensive coverage of this high profile area.

More information about this series at http://www.springer.com/series/10583

Todd K. Shackelford • Ranald D. Hansen
Editors

The Evolution of Sexuality

Springer

Editors
Todd K. Shackelford
Department of Psychology
Oakland University
Rochester
Michigan
USA

Ranald D. Hansen
Department of Psychology
Oakland University
Rochester
Michigan
USA

ISSN 2197-9898 ISSN 2197-9901 (electronic)
ISBN 978-3-319-09383-3 ISBN 978-3-319-09384-0 (eBook)
DOI 10.1007/978-3-319-09384-0
Springer Cham Heidelberg New York Dordrecht London

Library of Congress Control Number: 2014946299

Printed on acid-free paper

Springer is part of Springer Science+Business Media (www.springer.com)

Foreword

Those buzzes… Forget the short-sighted funding agencies, snail-paced build-up of data, not-quite-significant results, hung-over students, and empty conference theaters. Remember instead the thrill of a new angle, a new field, an inspired student, a successful protégé, a new collaborator, new funding, a big breakthrough—Eureka! These are the moments that keep us going. These are the ones that make us glad we are scientists and, in the case of the contributors to this volume, that make us glad we work on the Evolution of Sexuality. Such buzz-moments are magical. Every one of us has a list and most need little encouragement to reminisce their way through that list until somebody shuts us up. But as the invited writer of the Foreword to this volume I have carte-blanche—so for the next 100 pages…

Sometimes I feel that I have lived through the whole evolution of the Evolution of Sexuality; that I was there at the big bang. I was not of course. Darwin had found a few words to say on the matter long before my time. Even so, as a Zoology undergraduate in the early sixties I found no such subject being taught or even envisaged. At Bristol as in most places, evolution meant phylogeny, genetics meant what Mendel did with his peas, sex was the structure of reproductive systems plus maybe eggs, and behavior was not quite nice and was an art-subject taught by psychologists. But even before I graduated something was stirring—and luckily for me there was somebody in my year-class at Bristol who shared my excitement.

Geoff Parker and I were very different characters, yet around 1964 we both found a lifetime's inspiration from a single controversy. Outside of Bristol a war was raging in the world of evolutionary theory, triggered by V. C. Wynne-Edwards' 1962 book extolling the power of group selection to bring about evolutionary change. People were taking sides, even in Bristol, with Geoff and I aligning with what seemed to be the minority at the time, the Individual Selectionists. In a bar or a refectory we would defend our cause against all-comers. We would even go to the then-extreme of claiming individual selection to be a force in the evolution of behavior. It was our first experience of the buzz to be gained from the defending of an academic idea that, apparently, was outrageous. At a time when few believed that behavior of any sort was even heritable, our backs were against the wall for much of the time.

Over the course of the postgraduate years that followed, Geoff was destined to spend much of each summer sleeping in the back of a van in the English countryside, all the better to watch the first cowpats dropped each day in nearby fields. But before that, during the first long winter of 1965–1966, we shared a freezing apartment on the outskirts of Bristol. There, on those nights that we were both in, we would sit bent over the single gas fire, singeing gas-tainted toast on long forks and talking about sex—as postgraduates do. Unlike most postgraduates, our conversations revolved around dung-flies and Geoff's ground-breaking notion of sperm competition. It is fun to think that for those few months that fire-lit room in Bristol could have been the only place in the world where sperm competition was being discussed. But even that early in the subject's existence we did from time to time dare to ask… We did wonder… We did talk about how… But our conclusion was always the same: "It can't be done, can it?" And we would go back to talking about dung-flies. Or—another favorite sexual topic because it seemed so impenetrable—we would agonise over why there were two sexes, and only two. That question really stumped us, maybe even shook our faith a little. How could such a question possibly be answered by individual selection?

Soon we discovered that we were not alone in our excitement over evolutionary questions about behavior. Outside of Bristol it seemed that increasingly, and reassuringly, even quite rational people were becoming just as excited. Traditional zoologists still did not consider such questions quite scientific or even quite decent. They still considered them primarily the domain of psychologists. But everywhere new generations of academics were forging cross-discipline links—and following a deadly bite from a virulent Patas Monkey that "everywhere" came to include Bristol.

Until 1965 the Head of the Department of Psychology at Bristol had been Professor Ronald Hall. But on his return from a field-trip to Africa collecting data on wild primates, he was bitten by a member of a troop he had established in captivity. Whatever the infective agent, it was obscure and lethal and Ronald Hall sadly died from the bite, leaving a large section of his department in disarray. To reorient themselves, his orphaned group began holding weekly seminars to discuss each other's interests and to find a new direction. And hearing that Geoff and I might be receptive to an evolutionary approach, they invited us to attend.

Those meetings were a revelation to us all, a first faltering taste of the discipline that was to become "Evolutionary Psychology" in Psychology and "Behavioral Ecology" or "Sociobiology" in Zoology. The atmosphere at the seminars was amazing, and we all quickly became zealots. It really did feel as though we were on the verge of a new and exciting field. Some of the subject's pioneers came to attend and talk. Hans Kruuk, for example, who took us through the observational and experimental methods he was developing to ask evolutionary questions about lion behavior in the Serengeti. The main focus of the group was social behavior, mainly in birds and mammals, but that inevitably involved sexual behavior too. And whenever possible we swung the talk round to the question that interested us all: could we really apply the same evolutionary approach to humans? Few thought it possible or even, in a career sense, desirable, and at the time it seemed just a dream. It was an exciting dream though—and in 1967 that dream took a major step nearer to reality. *The Naked Ape* appeared.

Desmond Morris' book was dynamite, blasting its way into people's consciousness. Love it or hate it, everybody had an opinion on it. And for those of us desperately seeking reliable ways to apply evolutionary principles to behavior the book served two great purposes. First, it jettisoned into academic and public arenas alike how much could be gained from discussing *Homo sapiens* as just another animal. And secondly, it demonstrated clearly how not to do it. Pitfalls existed, and Desmond, with whom I was to collaborate on projects more than once in the years to come, had fallen into many. His book was a brilliant catalyst, but also a stark warning of the mistakes that could be made.

After *The Naked Ape*, what budding zoologist/psychologist would not be interested in the evolution of human sexuality? I never even contemplated the possibility as I moved from Bristol to my first temporary teaching post at the University of Newcastle upon Tyne—but I was in for a shock. From the euphoria of the evolutionary ethos at Bristol, I felt I had stepped back into the Dark Ages. My nemesis came in the form of the Department's resident palaeontologist who insisted that the only truly scientific way to study evolution was via fossils. And as fossils patently did not behave his stance was that there was no point even contemplating the matter—and he suggested quite forcefully to our mutual students that they should ignore everything I said on the subject. Perhaps because he was established and respectable and I apparently was neither, he generally held sway. I could see the suspicion in my tutees eyes as armed with copies of Desmond's book I tried to get them as excited about the questions raised as I was. It was hard work, leaving me feeling more like a smutty miscreant than a motivator.

Fortunately, not all of my students at Newcastle were so easily swayed by enemy propaganda. Officially, I was being employed as an entomologist and among the undergraduates in one series of my Entomology lectures in the early 1970s was a man destined to pioneer the study of a whole field in the evolution of sexuality. Professor Tim Birkhead of Sheffield University has kindly said or written more than once that it was my lectures on Geoff Parker's sperm competition work on dung flies that inspired his choice of career and research direction. Admittedly, after well over a decade of friendship, Tim and I did later cross swords and harsh words have been written and said. But none of that conflict can erase the pleasure and pride I took from the early years of seeing him opening up the whole study of sperm competition in birds.

Newcastle brought another buzz moment. In a single memorable phone conversation with Geoff Parker, by then at Liverpool, I discovered that we had both dreamed-up a solution to our long-standing niggle at the conundrum of two sexes. The answer, we had both decided, had little to do with maleness and femaleness and everything to do with eggs and sperm. Explain the evolution of anisogamy, we encouraged each other, and everything else would fall into place. Computer modelling was the obvious tool and Geoff was just the person to do it—and he did. He has since improved on that original 1972 *Journal of Theoretical Biology* paper of ours—but as far as I know the main principle still stands. And as that question was one of the first that he and I had jointly battered our brains against while toasting bread in a freezing flat, I still get a surge of excitement when I remember the phone call that initiated our laying it to rest.

In 1975, I moved from Newcastle to a permanent position in Zoology at the University of Manchester. Many more buzz-moments lay ahead though not all involved the evolution of sexuality—but those that did more or less defined the final phase of my academic life and made those years by far the most exciting and rewarding of all. Although still partly an entomologist and by then also an ornithologist I was taken on at Manchester mainly to teach Behavioural Ecology. The subject was still very much in its infancy. It would be another 6 years before John Krebs and Nick Davies published their now-classic introduction to the subject. Even Richard Dawkins' *The Selfish Gene* was still a year from appearing. So I had considerable freedom over what to teach and how to teach it. I did dutifully cover everything considered mainstream at the time, but I also indulged myself in a lecture or two by asking "what if" we could study human behavioral ecology in the same way as for other animals.

I made no secret during those lectures of the fact that I thought human sperm competition would be a fantastic subject area. "Totally impossible to study, of course," I would say, "unless one of you…" and I would wave my hand around the class "…can think of how it could be done." I had probably made that gesture and said those words 6 or 7 years in succession before, by chance, there was a man present who had been a medical student for a while before deciding he preferred zoology. He thought he did know how it could be done and came hot-footing to my room soon after to tell me so. His name was Mark Bellis.

I shudder to think how many curries and units of alcohol Mark and I consumed over the next few months and years as we discussed possibilities, ways and means, and where it all could lead. Maybe we also discussed safety and ethics—but I doubt it. Our investigations began in an age when "just be sensible" was the only mantra for safety, and if ethical committees existed they had no real muscle—which was just as well if a particular External Examiner to our Department a year or two into our work was typical. This man refused even to discuss one of our student's projects in her viva and reduced her to floods of tears by calling her a "painted whore". Her crime? Collecting copulatory ejaculates from pairs of volunteers who although cohabiting were unmarried. If Mark and I had tried to start our work even just 10 years later… Well, we probably would not have been allowed to start at all.

Those early years were a whirl of discussions, arguments, and inspirations; so many eureka moments crammed into such a short time that who-first-said-what was soon lost in leaky memories. We cannot even remember with certainty which of us dreamed up our Kamikaze Sperm hypothesis. My main contribution to the cauldron from which we fished the idea stemmed from a cocktail party thrown in London in 1981 by the adorably eccentric Dame Dr Miriam Rothschild. An expert on fleas and an enthusiastic lepidopterist she had invited all 44 of the speakers from a special *Biology of Butterflies* symposium organized by the Royal Entomological Society of London. Difficult though I find it to believe now, I was one of those speakers—talking not about sex but migration. At that party, over a glass or two of unidentifiable beverages, I had my one and only conversation with a young American lepidopterist, Bob Silberglied. Our paths would surely have crossed many more times if Bob had not died so tragically in the icy waters of the Potomac River just a few months later as a passenger aboard the ill-fated Air Florida Flight 90, but even just that single meeting left a lasting legacy with me. Among the things we discussed was his exciting off-the-wall idea that the then so-called eunuch morph of butterfly sperm might actually have evolved to serve some adaptive function in sperm competition. So when, a few years later, Mark Bellis, who knew a lot more about human sperm than I did at the time, began to describe all the different infertile morphs known to exist in perfectly normal human ejaculates, everything fell into place. It matters not a jot to anybody else, but we would both love to know which of us first suggested: "So maybe these infertile human morphs serve some adaptive purpose too."

Everywhere that Mark and I went, uninhibited conversations would erupt about the evolution of sperm shape, penis shape, testis size, thrusting, masturbation, mucus, orgasm and homosexuality. We must have emptied many a genteel bar and cozy dining room as we hammered out our ideas—and to our list of room-emptiers we eventually added the topic of the "flowback". We had been musing over the female orgasm from the beginning, firmly believing it to be part of the overall story of sperm competition but unable to see how. The stumbling block was our suspicion that the majority of female climaxes had nothing to do with copulation, something for which we later obtained data. So how then could those climaxes have anything to do with sperm competition?

The eventual game-changer was a chance remark by a girlfriend of mine. At the time I was a single parent looking after three young sons but had just met the woman, Elizabeth Oram, who later became the mother of my next three children. Mark and I were well into our work showing that the number and types of sperm that men ejaculate vary exquisitely and apparently adaptively from one occasion to the next. So when Liz said that she couldn't see why such fine adjustments were important when so much of the ejaculate comes back out of the woman within the hour, her remark triggered a whole new series of over-drink conversations. Liz could easily have regretted her comment because soon afterwards she had been press-ganged into pioneering—with more than a little indignity—a technique for the collection of flowback that was standard enough to pass on to other female volunteers. A whole new study had been born which had never been attempted before, has not been repeated since, and may never be allowed again, not least for ethical reasons.

Ethical reasons… Mark and I are fully aware how lucky we were to have started our studies when we did. The 10 years we spent working freely on human sperm competition could not have been more exciting and I would not have swapped them for anything. Each day seemed to bring some new twist, development, or idea. But towards the end of that decade, ethical committee dictate and health-and-safety restrictions driven by the specter of AIDS were beginning to bite. We probably would not have been allowed to continue our work in the cavalier way that suited us anyway, but in the end it did not really matter. We were both beginning to feel it was time to move on to new and different challenges—and we did.

For my part, I had always fantasized about being a full-time writer, particularly a novelist, preferably near the warmer shores of the Mediterranean, and in 1996, it seemed that there would never be a better moment to make the leap. It did not mean that my fascination with the Evolution of Sexuality had lessened in any sense, just that I could now explore new ways to indulge that fascination. As for Mark… He rapidly became a Professor and Director of the Centre for Public Health at John Moores University, Liverpool. Then in 2009, he was awarded an Order of the British Empire (OBE) for services to Healthcare, and in 2013 moved into politics as Director of the Policy, Research and Development division of Public Health in Wales; another successful protégé, another real buzz.

To an extent, while Mark and I were active participants the study of human sperm competition was nestling somewhere within biology, medicine, anthropology, and psychology. Our invitations to lecture came from all of these disciplines, and we were never quite sure how to describe ourselves and our approach. But on our departure, the subject moved quickly into the single arena of evolutionary psychology. Early workers such as Randy Hepburn and Dev Singh found ingenious new ways to study the subject without actually collecting sperm, and in recent years, this approach has been hugely and successfully expanded in all sorts of ways by Todd Shackelford and the team he has assembled, though I am thrilled to hear that he is now collecting sperm as well.

Human sperm competition is still as exciting a field for me as it ever was—but it is only one small part of the mushrooming whole that is the modern discipline of the Evolution of Sexuality. If somebody had shown me the Table of Contents from this volume during my undergraduate years early in the 1960s when Geoff and I were daily having to justify the study of the evolution of any form of behavior as a valid subject… If somebody had told me that these were the things respected scientists would be discussing 50 years into the future… I would have been totally incredulous, and incredibly excited. It really would all have seemed a brilliant dream. But…

Would I actually trade all my buzz-moments from the narrow pioneering past for those being generated across such a wide and fascinating range of subject areas in this field now?

That's a tough one.

January 2014 Robin Baker

Preface

In March 2013, we welcomed dozens of scholars from the USA and Europe to join us at Oakland University in Rochester, Michigan, for a day-long interdisciplinary conference on "The Evolution of Sexuality." We invited as panelists some of the leading scholars in sexual science from many different disciplines, including psychology, criminology, biology, anthropology, archeology, law, philosophy, and medicine. Each of these scholars had conducted and published substantial work addressing sexuality from an evolutionary perspective. This volume showcases the groundbreaking empirical and theoretical work from several of these panelists and other distinguished conference guests.

Biologist Robin Baker provides a thoughtful Foreword to the volume, in many ways setting the stage for the remaining chapters. The first five chapters present summaries of research on the evolution of sexuality from several different disciplinary perspectives. In the first chapter, biologist Tracey Chapman presents a masterful review of sexual conflict. The potential for sexual conflict is pervasive, especially in outbreeding, nonmonogamous species. Sexual conflict results from divergence between the sexes over reproductive resources. Chapman's focus is on our current understanding of sexual conflict from the perspective of evolutionary biology, and she draws skillfully upon studies across diverse species. Chapman reviews a rich literature demonstrating that sexual conflict can occur over a range of different reproductive traits and behaviors, from who to mate with, to how much parental care to give. The intensity of sexual conflict over the level of expression of a reproductive trait or behavior has been assessed by measuring its costs and benefits, in terms of reproductive output, for individuals of each sex. Chapman argues persuasively that outcomes of sexual interactions between males and females can be viewed in terms of Hamilton's quartet of social behaviors: mutual benefit (co-operation), selfishness, altruism, and spite. Chapman showcases recent work that has focused on the mechanisms used by individuals to calibrate their responses to perceived threat levels from sexual competitors.

Literary scholar Judith Saunders presents an overview of the new field called "Darwinian literary analysis", which she helped to create. In a wonderfully original contribution, Saunders applies an evolutionary lens to sexuality in literatures of the past and present, showcasing the value of Darwin's insights for securing a richer appreciation of core themes and characters in literature. Next, anthropologist Henry Harpending and polymath Gregory Cochran develop a quantitative genetic model of positive assortative mating for a neutral trait. Harpending and Cochran argue that even though a trait is selectively neutral, the mating system can mimic strong selection both for and against that trait, depending on the group membership of an individual. As a consequence, the mating system can generate large group differences rapidly, and if the system persists, arbitrary groups can be transformed into hereditary castes.

In their chapter, psychologists Yael Sela and her colleagues review recent research investigating whether oral sex might function as "mate retention." Men and women perform mate retention behaviors to reduce the likelihood of their long-term partner's infidelity. One mate retention strategy used by both sexes is to increase their partner's relationship satisfaction by provisioning her or him with benefits. Sela and colleagues review recent work indicating that men who report performing more mate retention behaviors, in general, and more benefit-provisioning mate retention behaviors, in particular, also report greater interest in, and more time spent, performing oral sex on their female partner. Likewise, women who report performing more benefit-provisioning mate retention behaviors also report greater interest in, and more time spent, performing oral sex on their male partner—but this relationship is stronger for men. Sela and colleagues highlight sex similarities and differences, discuss explanations for the results, address limitations of this research, and suggest future directions for research investigating oral sex as a mate retention behavior.

Anthropologist and historian Laura Betzig opens her chapter with reference to a cold day in January of 1649 when, after 7 years of wars, the king of England was taken to Whitehall Palace and beheaded. In the months that followed, the parliament passed an Act Abolishing the Office of King, an Act Abolishing the House of Peers, and an Act Declaring England to be a Commonwealth. As Betzig recounts, arguments have been presented that the English Revolution was caused by the rise of Puritanism, the rise of the middle class, and the difficulty of fighting domestic wars on three fronts. Betzig presents a compelling argument in this chapter for another cause. Betzig argues that Charles I was executed—and his son eventually restored—because his wife was the daughter of a king of France. According to Betzig, Queens of England had always struggled on behalf of their sons. Betzig's thesis is that this war was an effect, at least partly, of parent-offspring conflict.

The next three chapters present summaries of different areas of recent research investigating female sexual psychology and behavior. First, psychologist James Roney reviews evidence for the functional roles of hormonal signals in the regulation of women's sexual motivation. Hormone production fluctuates over time with events leading up to and following ovulation, and evolved mechanisms can use circulating hormone concentrations as information about current reproductive states. Roney reviews work documenting that, in most mammalian females, current fecundity is positively signaled by the combination of high estrogen and low progesterone. According to Roney, given the costs of sexual behavior, we might predict that sexual motivation will be higher when fecundity is higher. And in fact, estradiol positively predicts female sexual motivation across a wide range of mammals, whereas progesterone negatively predicts female sexual motivation. If women's sexual motivation is similarly calibrated to fluctuations in fecundity, Roney argues, we can likewise expect women's libido to be positively and negatively predicted by fluctuations in estradiol and progesterone, respectively. Roney reviews some of the fascinating results from a recent study conducted in his lab which demonstrated that estradiol concentrations were, in fact, positive predictors of within-cycle fluctuations in women's daily reports of sexual desire, whereas progesterone concentrations were strongly negative predictors. Roney concludes the chapter by arguing that these findings suggest that phylogenetically conserved brain mechanisms use hormonal signals to partially calibrate women's sexual motivation to fluctuations in fecundity.

The difficulty of inducing orgasm in women, the variability of orgasm between women, and the lack of an established relationship of orgasm with conception have led some researchers to conclude that female orgasm is a nonfunctional byproduct of male orgasm. As anthropologists John Wheatley and David Puts argue in their chapter, however, other researchers have presented evidence that orgasm is an adaptation in women. Wheatley and Puts review the evidence for these opposing hypotheses and present a convincing argument in favor of the adaptationist hypothesis. First, the authors discuss the phenomenological, anatomical, and neurological correlates of women's orgasm, which are inconsistent with the hypothesis that female orgasm is a byproduct. Wheatley and Puts then present powerful evidence that female orgasm enhances the likelihood of conception, and they summarize evidence that female orgasm functions as a mechanism for choosing mates of high genetic quality, investment potential, or both.

Biologist Randy Thornhill and psychologist Steve Gangestad address the functional design and phylogeny of women's sexuality. Thornhill and Gangestad note that in the past two decades, substantial research has documented estrus in women. Estrus in women is accompanied by a suite of sexual preferences, manifested in the fertile phase of the menstrual cycle, for partners with traits that indicate male genetic and phenotypic quality. Thornhill and Gangestad hypothesize that women's estrus is an adaptation to obtain genes, including conditionally via extra-pair copulation, that enhance the reproductive value of resulting offspring. The authors present a compelling argument that women's estrus is ancient phylogenetically, and has homology and functional similarity with estrus throughout vertebrates. Women's sexuality at infertile cycle points and other infertile times is referred to as "extended sexuality." As Thornhill and Gangestad note, extended sexuality is common in Old World primates and may also be common in pair-bonding, socially monogamous birds. According to the authors, the kinds of preferences associated with women's extended sexuality corroborate the hypothesis that its function is to obtain nongenetic material benefits and services from men in exchange for granting sexual access. Concealed estrus is present in women as evidenced by men's limited ability (compared to other male mammals) to detect estrus, women's limited behavioral changes (compared to other female mammals) during estrus, and estrous women's efforts to limit male mate guarding. The authors conclude the chapter with a superb review and discussion of concealed estrus in women, marshalling evidence to support the hypothesis that concealed estrus functions in extra-pair copulation to cuckold a regular partner in service of better genes for offspring, while maintaining a regular partner's material benefits.

The next two chapters provide reviews of recent empirical and theoretical work on the evolution of male sexual psychology and behavior, with a focus on homosexual psychology and behavior. Psychologists Paul Vasey and Doug VanderLaan address transgendered male androphilia in the human ancestral environment. *Androphilia* refers to predominant sexual attraction and arousal to adult males, whereas *gynephilia* refers to predominant sexual attraction and arousal to adult females. According to Vasey and VanderLaan, the manner in which male androphilia is expressed varies cross-culturally. *Sex-gender congruent male androphiles* occupy the gender role typical of their sex, behave in a relatively masculine manner, and identify as "men." In contrast, *transgendered male androphiles* often behave in a highly effeminate manner and identify as neither "men," nor "women." Instead, they often identify as members of a third gender. The authors review work indicating that, despite exhibiting different gender role presentations and gender identities, both forms of male androphilia are characterized by the same biodemographic and developmental correlates, indicating that they share a common etiological basis. As discussed by Vasey and VanderLaan, male androphilia presents an evolutionary paradox because it appears to have a genetic component, yet it compromises reproduction and archaeological evidence suggests that it has persisted for many thousands of years. Vasey and VanderLaan argue that the ancestral form of male androphilia was the transgendered form. The authors review some of their own stunningly creative research on a population of transgendered male androphiles in Samoa designed to test hypotheses addressing the evolution of male androphilia. The results of this research indicate that the mothers, paternal grandmothers and maternal grandmothers of *fa'afafine* produce more offspring than those of male gynephiles, which is consistent with the Sexually Antagonistic Gene hypothesis. The results of this research also indicate that *fa'afafine* display elevated avuncular tendencies and behavior compared to women and gynephilic men, which is consistent with the Kin Selection Hypothesis. Vasey and VanderLaan also highlight the results of recent research indicating that the *fa'afafine's* avuncular cognition displays elements of adaptive design.

All else equal, men who are less interested in having reproductive sex will produce fewer offspring and thus selection will eliminate the genes that contribute to their sexual orientation. However, according to psychologist Austin Jeffery, all else is not equal, as sexual orientation corresponds with a constellation of traits that may inform reproductive success. Jeffery presents two original hypotheses regarding the adaptive logic of reduced interest in the opposite sex. The first hypothesis addresses the tradeoff between offspring quantity and quality, proposing that men who abstain from sex with women make more effective parents. The second hypothesis invokes sperm competition and suggests that sneak copulating men can benefit from reduced arousal towards women. The question of exclusive homosexuality is addressed in the final section. Jeffery contends that self-identification as an exclusive homosexual is the product of a culture that promotes exclusive sexualities to isolate and remove non-heterosexuals from the reproductive arena. According to Jeffery, "coming out" as a homosexual man serves to distinguish oneself as a non-competitor to local sexual rivals, alleviating the severity of one's victimization at their hands. Jeffery argues that the modern notion of sexual identity has corrupted our understanding of sexuality as a fluid and functional product of evolved cognitive mechanisms. According to Jeffery, reliance on categorical sexual archetypes subverts our ability to characterize sexual variance, not only by limiting the depth of our measures, but also by limiting the depth of our theoretical thinking.

The final two chapters are broad in scope, addressing the evolution of both female and male sexual psychology and behavior. Psychologist David Schmitt opens his wide-ranging chapter noting that psychologists have identified myriad ways men and women differ in emotion, behavior, and cognition. Social role theorists assume that men's and women's psychological differences are the result exclusively of sex role socialization processes and sociopolitical power differentials. These theorists assume psychological sex differences will be smaller in cultures with more egalitarian sex role socialization and greater sociopolitical gender equity. In this chapter, Schmitt presents evidence across 21 data sources that directly challenges this assumption of social role theory. Schmitt reports that sex differences in most psychological traits—and even in many physical traits, including height, obesity, and blood pressure—are much larger in cultures with more egalitarian sex role socialization and greater sociopolitical gender equity. Schmitt argues that three alternative evolutionary perspectives on psychological sex differences—obligate sex differences, facultatively mediated sex differences, and emergently moderated sex differences—better explain the universal and culturally variable sex differences reliably observed across human cultures.

In the final chapter, psychologists Michael Pham and Todd Shackelford begin with a brief introduction to sperm competition theory. Sperm competition occurs when a female copulates with two or more males within a sufficiently brief time period, resulting in sperm of the different males competing to fertilize ova. Sperm competition has been documented or inferred to occur in many species. Pham and Shackelford review the evidence for sperm competition in humans. Specifically, the authors review literature indicating apparently convergent adaptations to sperm competition in humans and nonhumans. Pham and Shackelford discuss future research directions, and conclude that the research that documents anatomical, biological, physiological, and behavioral adaptations to human sperm competition provides compelling evidence that sperm competition has been a recurrent feature of human evolutionary history.

The Evolution of Sexuality showcases the profound and wide-ranging intellectual value of an interdisciplinary approach to human psychology and behavior. Guided by Darwin's insights, the contributions to this volume provide a compelling case for an evolutionary analysis of sexuality.

Oakland University

Todd K. Shackelford
Ranald D. Hansen

Contents

Contributors

Laura Betzig Ann Arbor, Michigan, USA

Tracey Chapman School of Biological Sciences, University of East Anglia, Norwich, UK

Gregory Cochran Department of Anthropology, University of Utah, Salt Lake City, UT, USA

Steven W. Gangestad Department of Psychology, The University of New Mexico, Albuquerque, NM, USA

Henry Harpending Department of Anthropology, University of Utah, Salt Lake City, UT, USA

Austin John Jeffery Department of Psychology, Oakland University, Oakland, USA

Michael N. Pham Department of Psychology, Oakland University, Rochester, UK

Department of Psychology, Oakland University, Rochester, MI, USA

David A. Puts Department of Anthropology, Pennsylvania State University, State College, PA, USA

James R. Roney Department of Psychological and Brain Sciences, University of California, Santa Barbara, CA, USA

Judith P. Saunders Department of English, Marist College, Poughkeepsie, NY, USA

David P. Schmitt Department of Psychology, Bradley University, Peoria, IL, USA

Yael Sela Department of Psychology, Oakland University, Rochester, UK

Todd K. Shackelford Department of Psychology, Oakland University, Rochester, UK

Department of Psychology, Oakland University, Rochester, MI, USA

Randy Thornhill Department of Biology, The University of New Mexico, Albuquerque, NM, USA

Doug P. VanderLaan Gender Identity Service, Child, Youth and Family Services, Centre for Addiction and Mental Health, Toronto, ON, Canada

Paul L. Vasey Department of Psychology, University of Lethbridge, Lethbridge, Alberta, Canada

John R. Wheatley Department of Anthropology, Pennsylvania State University, State College, PA, USA

Chapter 1
Sexual Conflict and Evolutionary Psychology: Towards a Unified Framework

Tracey Chapman

Sexual Conflict: What Is It, Why Does It Occur, How Do We Measure It, and Why Does It Matter?

It was realized, particularly in the 1970s, that rather than being cooperative ventures, the interactions between the sexes over reproduction are often characterized by conflict rather than cooperation (Parker 1979; Trivers 1972). Therefore, as well as the exquisitely coordinated and spectacular courtship displays of birds (Krebs and Davies 1987), reproductive contests can result in mortal injury to females, the evolution of female-damaging penile spines, and female lifespan-shortening seminal fluids (Chapman et al. 2003a). A gene-centered perspective (Dawkins 1976; Hamilton 1964; Parker 1979; Trivers 1972) explains these seemingly paradoxical phenomena, which are central features of sexual conflict.

The underlying reason for sexual conflict is that males and females often "disagree" about how much energy and resources to invest in each reproductive bout and how often to make that investment. For example, males of many species often gain more fitness from mating frequently than do females (Bateman 1948). This is partly because the costs of each mating are often higher for females than males (Bell and Koufopanou 1986; Chapman et al. 2003a; Partridge and Harvey 1988; Partridge and Hurst 1998). Because mating frequency is an emergent property of the interaction of both sexes, it can only take one value—therefore the different interests of males and females cannot simultaneously be realized (Holland and Rice 1998; Parker 1979, 2006; Partridge and Hurst 1998). In any situation in which this kind of evolutionary tension exists, there is the potential for sexual conflict (Fig. 1.1). Therefore, although reproduction often requires some cooperation to succeed, there are many situations in which one sex can gain fitness (i.e., increased transmission of genes) if they can cause their partner to increase reproductive investment (Krebs and Davies 1987). The strategy by which one sex manipulates the other to increase

T. Chapman (✉)
School of Biological Sciences, University of East Anglia,
Norwich Research Park, Norwich, NR4 7HP, UK
e-mail: tracey.chapman@uea.ac.uk

T. K. Shackelford, R. D. Hansen (eds.), *The Evolution of Sexuality*,
Evolutionary Psychology, DOI 10.1007/978-3-319-09384-0_1

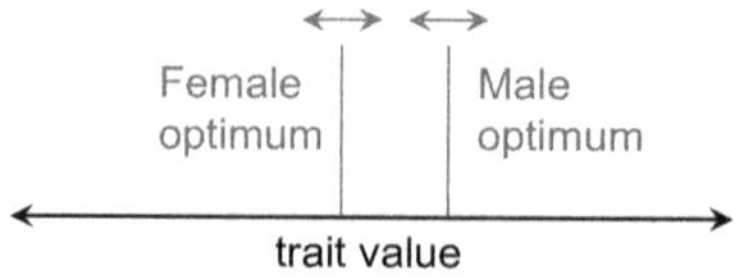

Fig. 1.1 A summary of the basis of sexual conflict. The optimum value of a reproductive trait (trait value) that maximizes male or female lifetime fitness (the number of offspring they produce over their lifetimes) is often different. The sexual conflict that arises from such differences is predicted to be particularly strong when unrelated individuals reproduce and when the mating system is nonmonogamous. These factors reduce the shared interest of each sex in what happens to the other in the future. Therefore, selection to maximize the investment of the other sex in the current reproductive bout at the expense of future costs to the other sex can be selected for. In the example shown here the optimum value of the trait for females is lower than for males. Therefore, a female's lifetime fitness would be higher if the value of the trait were lower. For males, the situation is the opposite. Since the reproductive traits in question are "shared", that is they arise because of an interaction between the sexes (e.g., mating frequency, mating duration, number of eggs laid), the trait can only take one value. Hence, when the trait is expressed at least one sex will not be at its optimum

reproductive investment can be strongly selected even if it results in significant lifetime costs to the manipulated partner. This in turn will select for the manipulated sex to evolve resistance to the manipulation. This creates the evolutionary back and forth that constitutes sexually antagonistic coevolution (Arnqvist and Rowe 2005; Holland and Rice 1998).

Over the last few decades, it has become clear that under any apparent reproductive *status quo*, there is sometimes fast and furious dynamic evolutionary change, driven by adaptation in males followed by counter-adaptation in females (Arnqvist and Rowe 2005). Unless perturbed, this sexually antagonistic coevolution can be difficult to observe at the level of the phenotype, but can productively be investigated via experimentation (Chapman et al. 2003a). Sexually antagonistic coevolution can occur between the same or different loci in males and females (intra- and interlocus sexual conflict, respectively (e.g., Chapman et al. 2003a, and see below) and can generate considerable evolutionary change. Importantly, this coevolution has the potential to lead to reproductive isolation and even speciation (Hayashi et al. 2007; Parker and Partridge 1998; Rice 1998) and is, therefore, of central importance in evolutionary biology.

A huge variety of reproductive traits can be subject to sexually antagonistic selection, from who to mate with, how often to mate, how long to mate, and how much to invest in reproduction (Chapman et al. 2003a). The potential for sexual conflict is universal across all species that engage in sexual reproduction. For instance, it does not even require the existence of separate sexes and can occur over the optimal investment allocated to either male or female sex allocation in hermaphrodites (Charnov 1979). However, whether sexual conflict is realized (Bourke and Franks 1995; Chapman 2006; Ratnieks and Reeve 1992), and the intensity with which it proceeds, depends upon several different factors, as summarized in Table 1.1.

Table 1.1 Factors affecting the expression and intensity of sexual conflict

Factor	Effect on the intensity of sexual conflict
Relatedness	Sexual conflict is exacerbated by low relatedness between reproducing partners (Bourke 2009; Dawkins 1976; Parker 1979, 2006; Rankin 2011). Low relatedness makes it much less likely that currently paired males and females will have a shared evolutionary interest in each other's future
Mating system	Sexual conflict is also exacerbated by nonmonogamous mating systems, as these also promote divergence in the genetic interests of males and females in current versus future offspring. If there is a significant chance of additional reproduction with other partners, then sexual conflict will be increased (Bourke 2009; Dawkins 1976; Parker 1979, 2006; Rankin 2011)
Biologically relevant context	Sexual conflict must be studied in a biologically relevant context. For example, it is difficult to detect patterns of elevated reproduction in populations reproducing at either a low level or at their upper limit (VanderLaan et al. 2012). Nutritional environment can also completely change the direction of sexually antagonistic interactions (Fricke et al. 2009a), so must be considered
Condition	Condition refers to the extent to which individuals can express their genotype, even under poor environmental conditions. It can therefore magnify the differences in quality between individuals (Rowe and Houle 1996) and affect the extent to which they can respond to stimuli. Therefore, condition is likely to alter the intensity of sexual conflict. Variation in condition caused by environmental variation in food availability can alter the extent to which males can express potential sexual conflict adaptations (Amitin and Pitnick 2007; Fricke et al. 2008; McGraw et al. 2007), and also the degree of female responses to them (Fricke et al. 2009a)
Sexual conflict load	Load is the magnitude of the decrease in fitness resulting from each sex not being at its optimum. The larger this difference, the more "distance" required in order to reach the optimum and therefore the stronger the selection arising from sexual conflict
Value of winning	This is the magnitude of the benefit of a sexually antagonistic adaptation in one sex versus the benefits of resisting it in the other, and will determine the strength of sexual conflict
Power	Power refers to the relative costs involved in the value of winning and will thus also determine the strength of sexual conflict
Mechanism	Even if there is the potential for conflict, it may not be manifested if there is no mechanism for each sex to alter the balance of power and winning. This could arise if one sex has effectively "won" the conflict and shut down the opportunities for the other to influence sexual conflict phenotypes (Chapman 2006)
Negative intersexual genetic fitness correlations	Several researchers have found genetic correlations for fitness across generations that vary in direction across sexes. For example, high fitness females may have high fitness sons but average fitness daughters, and vice versa (Chippindale et al. 2001; Foerster et al. 2007). These correlations intensify sexual conflict, because they can constrain the ability of each sex to reach their sex-specific optimum

Concepts and Currency of Sexual Conflict

Costs, Benefits and Fitness

Central to the understanding of sexual conflict is to clarify what is meant by costs and benefits. Though not often explicitly stated, a Hamiltonian (Hamilton 1964) definition is used, such that the outcome of sexual conflict is measured in terms of positive or negative direct fitness consequences for males and females. Direct fitness is defined as that gained through the production of an individual's own offspring. Within a kin-selected context (i.e., within related social groups), sexual conflict is expected to be low (because there are higher shared interests between partners in each other's futures), though it is still possible (Bourke 2009; Parker 1979; Rankin 2011). Some portion of the costs and benefits of sexual conflict can also be measured in terms of indirect fitness (i.e., the portion of fitness gained by helping related individuals to reproduce).

Fitness is a property of a gene, but is often also used with reference to a genotype or individual. This is legitimate in situations where the interests of genes and the individuals in which they reside concur. The currency of sexual conflict is therefore changes in gene frequency, i.e., the basic units of evolutionary change. Conflicts ultimately occur because of differences in how genes maximize their transmission to future generations through males or females. As the interests of genes and the individuals in which they reside are usually (but not inevitably) aligned, it is often convenient to think about sexual conflict in terms of conflicts between individual males and females. We usually use as a proxy for an individual's direct fitness, the number of offspring produced by an individual in their lifetime. This measure may usefully be combined with estimates of survival to give greater weight to offspring produced early in life, when natural selection is strongest (Charlesworth 1980). Fitness is not a property of a group or of a population, as these are seldom units of selection. At a population level, the mean fitness of males and females is equal (Fisher 1930; Arnqvist 2004); instead, it is the relative fitness of individual male and female genotypes that vary.

Measurements of fitness should ideally include a competitive element and a component that can capture the expression of any delayed costs of reproduction (Edward et al. 2011). If the quality of offspring, in addition to the quantity of offspring, is an important component of fitness, then fitness measures necessarily resolve into tests of the effects of the reproductive strategies of parents across generations. In some such studies, the costs experienced by female parents as a result of sexual conflict may be partially compensated for by an increase in offspring quality/fitness (Priest et al. 2008). This has led some researchers to question the sexual conflict paradigm, suggesting that if the costs of sexual conflict are compensated for by genetic benefits to future generations, then there is no conflict. However, it is clear from several studies that the magnitude of direct current costs to parents often exceeds any future indirect genetic benefits to offspring (e.g., Brommer et al. 2012; Lew et al. 2006), hence sexual conflict is the dominant source of selection. The key point is that, through careful measurements of costs and benefits and consideration of the factors summarized in Table 1.1, it is possible to predict the extent and importance of sexual conflict (Fricke et al. 2009b).

Intra- Versus Interlocus Sexual Conflict

The theory and mechanisms of sexual conflict are often cast in terms of whether the conflict is mediated by the same or different loci in males and females (Holland and Rice 1998). This is not purely a semantic categorization—differences in these underlying modes can determine the evolutionary outcome of conflict. Apart from the presence of the Y (or equivalent) sex chromosome in individuals of the heterogametic sex, each cell of an individual carries a full set of all genes in the genome. However, genes can exhibit complex patterns of tissue-, life history stage-, and sex-specific expression. For example, seminal fluid protein-encoding genes are switched on in the reproductive system of sexually mature adult males, but not in females. Genes can also exhibit sex-biased expression, particularly for reproductive genes, i.e., they are expressed at a significantly higher level in one sex in comparison to the other (Civetta and Singh 1999; Ellegren and Parsch 2007; Parisi et al. 2004). Different alleles of the same gene can have different expression levels, or expression levels of the same allele can be regulated differently in males and females. The nature and outcomes of the two major modes of sexual conflict are described below.

Intralocus Sexual Conflict A well-supported example from invertebrates is that of adult locomotory activity in the fruitfly *Drosophila melanogaster*. The positive genetic correlation between movement levels in males and females suggests that the same genes influence adult locomotion in both sexes. However, there is sexual conflict—a high level of activity is beneficial to males as it increases female encounters, leading to increased reproductive success. In contrast, females who are less active have higher fitness, because they have more time available for feeding and egg laying (Long and Rice 2007). Further evidence for intralocus sexual conflict has been gleaned from experiments conducted in fruitflies using ingenious cytogenetic techniques (Chippindale et al. 2001). In these experiments, the same genome can be expressed in males or females and the fitness consequences then measured. These experiments show that there is little fitness difference during larval development (where sexual conflict in minimal), but that at the adult stage, genomes that confer high fitness for males result in low fitness when expressed in females, and vice versa. In addition, experimental evolution experiments have, by various techniques, allowed the genomes of either males or females to evolve free of the constraint of passage through the other sex (Holland and Rice 1999; Rice 1992, 1996). This invariably results in the evolution of a genome that is "better suited" to either male or female interests. The outcome of these intralocus sexual conflicts is predicted to be an evolutionary to and fro depending on whether specific alleles currently reside in males or females (Hayashi et al. 2007; Parker 1979). Resolution of this sexual conflict is also predicted to occur via the evolution of sex limitation, i.e., the silencing of expression of the gene subject to conflict in one sex but not the other. However, this is not necessarily the end of sexual conflict (Chapman et al. 2003a; Chapman 2006).

Interlocus sexual conflict can occur when reproductive traits are influenced by different genes in each sex (e.g., mating frequency; Arnqvist and Rowe 2005).

There is a huge volume of evidence for sexual conflict under this mode (Arnqvist and Rowe 2005), provided by the economic studies of costs and benefits for males and females of manipulating the trait values up and down. A revealing example comes from studies in pondskaters, where the grasping adaptations of males that aid in attachment during mating can be made less effective at achieving matings by experimental manipulation of a specific antigrasping adaptation in females (Arnqvist and Rowe 1995). The predicted evolutionary outcomes of this type of conflict are varied and include evolutionary chases, equilibrium, divergence (Hayashi et al. 2007), or dampening down of the conflict (Arnqvist and Rowe 2005), reflecting a kind of conflict resolution.

Sexual Conflict in Action

The detailed experimental study of sexual conflict has been dominated by studies of just a few species and particularly of invertebrates, notably dungflies (*Scathophaga stercoraria* and *Sepsis cynpisea,* e.g., Martin and Hosken 2003; Parker 1970), fruitflies (*Drosophila melanogaster,* e.g., Chapman et al. 1995; Rice 1996), abalone (*Haliotis* spp, e.g., Clark et al. 2009), and various pondskater species (*Aquarius, Gerris* spp, e.g., Arnqvist and Rowe 2002a, b). Within the vertebrates, important work has been done on Dunnocks (*Prunella modularis,* e.g., Davies 1992) and Red Deer (*Cervus elaphus,* e.g., Foerster et al. 2007). Authors such as Haig (e.g., Haig and Wilczek 2006) have also elegantly outlined the potential for sexual conflict over embryo provisioning in plants, emphasizing the key role of relatedness among the interacting parties in driving conflict not only between the sexes, but also between parents and offspring. These examples of sexual conflict in action have usefully been synthesized in several reviews (e.g., Arnqvist and Rowe 2005; Chapman et al. 2003a). In this section, I use just a few selected examples, some with specific relevance to humans, to illustrate the evolutionary potential of sexual conflict over different reproductive traits in widely different species.

Sexual Conflict over the Effects of Seminal Fluid Proteins in Fruitflies

A valuable model for empirical studies has been the fruitfly *D. melanogaster*, and experimental evolution has been conducted to reveal the covert arms races and evolutionary constraints that accompany sexual conflict (e.g., Rice 1992, 1996). A particular focus of interest has been the actions of the seminal fluid proteins, the nonsperm components of the male ejaculate (Chapman 2001; Ram and Wolfner 2007; Sirot et al. 2014; Wolfner 2002). There are over 100 such proteins and peptides in the *D. melanogaster* male, and they are an extraordinarily diverse group of compounds in terms of structure and function (Mueller et al. 2004, 2005). They can alter female sexual receptivity, ovulation, egg laying, reproductive tract morphology, immunity, feeding behavior, nutrient balancing, sleep patterns, water retention, and lifespan (e.g., Adams and Wolfner 2007; Carvahlo et al 2006; Chapman et al. 2003b; Cognigni et al. 2011; Isaac et al. 2009; Kubli 2003; Liu and Kubli 2003; Peng et al. 2005; Ram and Wolfner 2007; Ribeiro and Dickson 2010; Sirot et al. 2014). Sexual conflict over each of these traits is possible (Sirot et al. 2014).

The genes that encode seminal fluid proteins are among the fastest-evolving (Swanson et al. 2001b). Due to their rapid evolutionary change, there are seminal fluid protein genes that are not found even among close relatives (Wagstaff and Begun 2005a, b; Wong et al. 2008). Though less is known about the seminal fluid proteins of other invertebrates and vertebrates, they appear to be similarly complex (Sirot et al. 2014). Several core seminal fluid functions are maintained across widely different species, even though the genes that encode individual components can be very divergent. These core functional types include: proteases, protease inhibitors, lipases, lectins, cross-linking transaminases, and cysteine rich secretory proteins (CRISPs; e.g., Mueller et al. 2004, 2005; Ram and Wolfner 2007). This finding shows that there can be conservation at the functional, but not genomic, level. The implication is that the lack of genomic conservation might be due, in part, to sexual conflict (Chapman 2001; Clark et al. 2009; Swanson et al. 2001b).

It was discovered some years ago that the transfer of seminal fluid proteins from male to female *D. melanogaster* during mating can decrease female lifespan and reproductive success (Chapman et al. 1995) without any demonstrable benefits for females (e.g., Brommer et al. 2012). Later work identified some of the seminal fluid protein candidates apparently responsible. One example is the 36 amino acid "sex peptide," whose transfer can increase the fitness of males, but decrease that of females that receive it (Fricke et al. 2009c; Wigby and Chapman 2005). The phenotypic effects of sex peptide include decreased female receptivity, increased egg production (Chapman et al. 2003b; Chen et al. 1988; Liu and Kubli 2003), altered female immunity (Domanitskaya et al. 2007; Peng et al. 2005), decreased sleep (Isaac et al. 2009), increased feeding (Carvahlo et al. 2006), and altered nutrient balancing (Ribeiro and Dickson 2010). These findings suggest that the transfer of elevated levels of sex peptide shifts the value of remating and investment into the current batch of eggs towards the male optimum and away from that of the female, with attendant lifespan costs for females. This is consistent with the operation of interlocus sexual conflict. Genome-wide studies of gene expression show that receipt of sex peptide by females causes a change in the expression of many different functional categories of genes, with tissue- and time-specific signatures (Gioti et al. 2012). This suggests that females have many obstacles to overcome to evolve resistance to sex peptide. Other studies indicating toxicity of seminal fluid proteins employed a technique in which seminal fluid proteins were ectopically expressed in females. Four seminal fluid proteins, including sex peptide, were found to decrease female lifespan when expressed in this way (Mueller et al. 2007).

This sexual conflict should select for counter-adaptations in females. However, as yet, we have little understanding of how females respond to the sexually antagonistic effects of male seminal fluid proteins (Wigby and Chapman 2004). Only one receptor to a seminal fluid protein in insects has been identified (Yapici et al. 2008). Hence, the study of the coevolution of interacting molecules in males and females is only just beginning (Clark et al. 2009). That said, there is good evidence for selectable genetic variation in females for the evolution of resistance to the costly effects of male seminal fluid proteins. For instance, females subjected to elevated levels of sexual conflict over evolutionary time evolved to live longer in the presence of males than did controls (Wigby and Chapman 2004). The fact that the longevity differences were not an intrinsic property of the lines and were not observed in the absence of exposure to males suggests that this was a specific response to the manipulation of sexual conflict levels.

Sexual Conflict over Sperm Egg Interactions in Marine Invertebrates

Abalone are a group of marine invertebrates that broadcast their gametes into the water column. In these species, there is good evidence for antagonistic coevolution between male sperm lysin and its female receptor, vitelline envelope receptor for lysin (VERL) (e.g., Clark et al. 2009; Lee et al. 1995; Metz and Palumbi 1996; Swanson et al. 2001a). Lysin is involved in breaking down the vitelline envelope surrounding the egg and permitting further transit of the sperm to effect fertilization. There is evidence for sexual conflict between the male's efforts to have the sperm enter the egg quickly and the female's efforts to avoid costly polyspermy (Frank 2000). The latter refers to the situation in which too many sperm enter the egg, with lethal results. The dynamic pattern of coevolution between VERL and lysin appears to have resulted in divergence in the VERL receptors in females (Clark et al. 2009; Swanson et al. 2001a). This pattern of evolutionary change is a specific prediction of sexual conflict theory (Hayashi et al. 2007) and is therefore strong evidence for sexual conflict in action.

Sexual Conflict over Mating Frequency in Pondskaters and Beetles

A powerful system for demonstrating the existence of sexual conflict is found in the pondskaters (Arnqvist and Rowe 1995, 2002a, b). Experiments show that lengthening of the female's antigrasping spines reduce the length and success of mating attempts, which is consistent with the evolution of antigrasping spines as defenses against male grasping adaptations (Arnqvist and Rowe 1995). A set of species comparisons was also conducted (Arnqvist and Rowe 2002a, b) in which the relative levels of male grasping and female antigrasping armaments were measured. Interestingly, these adaptations were well "matched" across different species in terms of the absolute size of the male graspers versus the female antigraspers.

However, when there were changes in the relative levels of armament and defense between males and females, there were predictable effects on mating rate. For example, when females had relatively lower levels of defense, mating rate was higher, and vice versa. These studies demonstrate the underlying evolutionary tensions in sexual conflict.

Coevolution driven by sexual conflict is also found in a comparative study of *Callosobruchus* seed beetles (Rönn et al. 2007). In these species, there is a predicted sexual conflict over male mating frequency. Penile spines are proposed to anchor the male during mating, but they also damage the female reproductive tract. Counter-adaptation in females is represented by the evolution of a thicker lining in the reproductive tract. The number of spines and amount of damage was found to vary across the different seed beetle species tested, and a combined measure of the degree of harmfulness of male genitalia (based on number, morphology, and distribution of spines) was correlated with the thickness of the female reproductive tract wall. As in pondskaters, the absolute armament level was independent of the degree of damage. It varied instead with the relative armament level, i.e. harm was more evident in species in which the male genitalia were relatively more spiny and where the female tract was relatively less robust. These findings fit the sexual conflict framework well.

Sexual Conflict over Ovulation in Invertebrates and in Humans

Conflicts between the evolutionary interests of males and females can also diverge with respect to different facets of ovulation, some of which have particular relevance to humans. Some evidence consistent with the sexual conflict over ovulation and the efficiency of egg fertilization has been revealed in *D. melanogaster* fruitflies. For example, there can be a break down in the efficiency of egg fertilization in matings between normal females and mutant males that lack a seminal fluid protein (Acp26Aa) that stimulates ovulation (Chapman et al. 2001). In invertebrates, ovulation can occur in response to mating or can be enhanced by signals in the ejaculate transferred during mating (Sirot et al. 2014). However, in mammals, ovulation can often be divorced from mating and the receipt of an ejaculate, and influenced instead by seasonal or cyclical factors. A different aspect of ovulation over which there can be sexual conflict is therefore whether it is concealed or unpredictable. This is of particular interest in humans, in which ovulation is thought to be concealed, it is suggested, in order to reinforce pair bonds, increase male parental care, or to reduce the intensity of male–male competition (e.g., Alexander and Noonan 1979; Benshoof and Thornhill 1979; Hrdy 1979; Strassmann 1981). Whether ovulation is concealed in humans is still a topic of debate (e.g., Kuukasjärvi et al. 2004; Thornhill and Gangestad 1999). Whatever the answer, it is possible that the evolution of even partial concealment may have been selected partly to reduce the intensity of sexual conflict generated as a side effect of competition between males.

Sexual Conflict over Immune Function in Invertebrates and Humans

In invertebrates, there are significant changes to the immune status in females following mating (Lawniczak et al. 2007). Some of these changes are caused by the actions of seminal fluid proteins (e.g., Domanitskaya et al. 2007; Peng et al. 2005; Short et al. 2012). The effect of these changes is to reduce immunity to systemic infection following mating (Fedorka et al. 2011; Short et al. 2012, but see Zhong et al. 2013). Overall, however, the functional significance and fitness effect of these changes is not yet clear. Similar findings are now reported in humans (Sharkey et al. 2012), i.e. changes in the expression of immune genes in the cells of the female cervix following receipt of a male ejaculate. The hypotheses are that post-mating immune changes in females may produce a permissive environment for implantation, may counter sexually transmitted diseases, or may provide a hostile environment to select sperm (e.g., Robertson 2005). There are, therefore, opportunities for sexual conflict over postmating immune function changes in humans and these possibilities would be of great interest for further study.

Sexual Conflict and Homosexuality

The sexual conflict framework has also recently been applied to the understanding of the persistence of male homosexuality in humans. The phenomenon to explain is that the frequency of homosexuality is significant and stable through time and has some genetic basis, yet homosexual individuals have few offspring (Vasey et al. 2007). Recent studies by Vasey and colleagues provide evidence for significant benefits consistent with the operation of kin selection (Vasey et al. 2007; Vasey and VanderLaan 2010). That is, the kin of homosexual individuals have higher than average fecundity and exhibit elevated altruistic behavior towards relatives. However, the elevated fecundity of female relatives has also been interpreted in terms of the sexually antagonistic gene hypothesis (VanderLaan et al. 2012). Because of gene-sharing among kins (e.g., on average, individuals share half of their genes with their full sib brothers), males with female relatives that possess "high female fitness" genotypes will share a portion of that genotype. This might be associated with decreased direct reproduction by such males. One way to distinguish the kin-selected and sexually antagonistic gene hypotheses is to examine their unique predictions. For example, a male with a "high female fitness" genotype might express a variety of traits closer to the female rather than male optimum. These males might, in comparison to controls, show altered patterns of attraction to the opposite sex (PL Vasey, pers. comm.), altered mate searching or mate selection, or altered degrees of investment in parental care.

Sexual Conflict as Social Behavior

It has been noted by Bourke (2009), and also considered by others (e.g., Rankin 2011), that the interactions between males and females that are subject to sexual conflict can be viewed in the wider context of social behaviors. This is because the actions of one sex (i.e., the "actor") can affect the fitness of itself and the other (the "recipient"). Hamilton (1964) classified four types of social behaviors, whereby the social behavior of the actor can:

(i) increase direct fitness of both actor and recipient (mutual benefit)
(ii) increase direct fitness of the actor but decrease that of the recipient (selfishness)
(iii) decrease direct fitness of the actor but increase that of the recipient (altruism)
(iv) decrease direct fitness in both actor and the recipient (spite).

The outcomes of sexual conflict are cast in terms of costs and benefits, which are synonymous with fitness increases or decreases in the sense used in the study of social behaviors (i.e., increases or decreases in individual lifetime offspring). Hence, we can usefully integrate sexual conflict into this wider context (see Table 1.2). Reclassifying the outcome of interactions between males and females in this way, we see that there is the potential for evolutionary disagreements between the sexes, and therefore sexual conflict, in all four types of social behavior classification. However, this is particularly so for selfish behavior, which is generally the outcome assumed to occur most frequently in this field of study.

Sexual Conflict and Selection for "Harm"

An important debate in the study of sexual conflict, and which has relevance for evolutionary psychology, is the nature of the costs inflicted on one sex by the other. In this sense, we can consider whether there is evidence that the costs inflicted by one sex upon the other represent "harm" (Johnstone and Keller 2000; Lessells 2005; Morrow et al. 2003). This is synonymous with the selection for selfishness as an adaptive strategy, unless one is suggesting that harm is different from cost. What is ultimately important whether the recipient suffers a cost in lifetime fitness, or whatever the source of that cost. Therefore, my view is that cost and harm are equivalent in evolutionary terms.

Nevertheless, in the study of sexual conflict, harm is often implied to represent a distinct type of cost. The term harm also has broad usage across different fields of study, but in the context of the social behaviors summarized in Table 1.2, it can be a selfish or spiteful behavior. In sexual conflict, it refers to various kinds of damage, such as lesions, to the female reproductive tract caused by male mating adaptations (e.g., Rönn et al. 2007). Even if, as I suggest, the conceptual basis of harm is unclear, plenty of types of evidence have been advanced as harm. For example, in bedbugs, the costs to females that arise as a side effect of male reproductive strategies can be significant. Some species of male bedbugs inseminate females through their body wall into the free body cavity, and often cause damage to the female as a result (Reinhardt et al. 2003).

Is Sexual Conflict in Humans Qualitatively or Quantitatively Different?

Building on the discussion of harm, here, I consider sexual conflict in reproductive interactions in humans and whether there is anything distinct about the type or intensity of sexual conflict in humans. The potential for sexual conflict is pervasive in humans as in many other plant and animal species. We are diploid outbreeders and individuals coming together for sexual interactions generally exhibit low relatedness. Human mating systems are varied, but importantly there is usually some degree, even if quite low, of mating outside a pair bond. Hence, the stage is set for sexual conflict for all the reasons outlined above, though whether the conflict is realized will depend on the costs and benefits involved, and the opportunity or mechanism to enact selfish behavior (Table 1.2). Costs and benefits of reproduction have been quantified to some extent from the studies of historical human societies for which demographic data records exist. These studies tend to show that women that reproduce at high rates have shorter lifespans (Helle et al. 2002; Westendorp and Kirkwood 1998). The cost of reproduction appears to be particularly high for the production of male children, due to higher gestation costs (Helle et al. 2002). Thus, there is the potential for evolutionary disagreements between male and female parents over the frequency of reproductive bouts and how much to invest in each (Dawkins 1976).

We expect sexual conflict and parent–offspring conflict over nutrient supply to the developing fetus in humans because of relatedness asymmetries. In outbreeding humans, a mother and father will be, on average, 50 % related to all their offspring. A developing offspring is 100 % related to itself, 50 % to full sibs and 25 % to half sibs. However, although there is an equality in the relatedness of male and female parents to their offspring, the residual reproductive value (or propensity to show future discounting) of male and female parents may differ, because of the costs of each reproductive bout. Therefore, a sexual conflict between parents is predicted over nutrient provisioning to the offspring because mothers may gain more than fathers by conserving investment for the future. This conflict is in addition to the parent–offspring conflict derived from relatedness asymmetries between both parents and the developing offspring. The offspring is more related to itself than to either of its parents or any future full or half sibs and will, therefore, be selected to obtain a higher level of nutrients than is optimal for the mother to give.

Sexual conflict and parent–offspring conflict over fetal provisioning is evident when something goes wrong, as in the case of various imprinting diseases. Imprinting in this context refers to a gene that shows a uniparental inheritance of

Table 1.2 Reclassification of interactions between the sexes in terms of social behavior

Effect of actor's social behavior on itself (i.e., effect of one sex on its own fitness)	Effect of actor's social behavior on the recipient (i.e., effect of one sex on the fitness of the other sex)	Consequence of the social behavior	Equivalent consequence in terms of sexual interactions	Potential for the outcome to result in sexual conflict	Examples
Increases fitness	Increases fitness	Mutual benefit (narrow sense cooperation)	Mutual mate choice, obligate monogamy, or lifetime pair bond commitment	No potential for sexual conflict if both sexes gain fitness from the sexual interaction. When there is obligate monogamy or lifetime commitment both sexes share their future interests (Boomsma 2007, 2013)	The outcome of sexual selection can be viewed as mutually beneficial. Here, the outcome of the sexual interactions benefits both sexes through the increased production of numerous, high quality offspring. When there is a shared future interest between males and females reproductive costs may be absent and sexual interactions can have a positive effect on the fitness of both sexes (e.g., Schrempf et al. 2005)
Increases fitness	Decreases fitness	Selfishness	Sexual conflict where the reproductive strategy of one sex benefits themselves but exacts a cost in the other	Strong potential for sexual conflict when the reproductive strategy adopted by one sex benefits themselves but decreases their current mate's fitness (Arnqvist and Rowe 2005; Parker 1979)	Much of the data for sexual conflict can be seen as equivalent to selfish behavior. For example, female fruitflies that mate at high frequency suffer lifetime fitness costs whereas males do not (e.g., Chapman et al. 1995; Cordts and Partridge 1996; Fowler and Partridge 1989; Partridge and Andrews 1985), suggesting selfish behavior on the part of the male

Table 1.2 (continued)

Effect of actor's social behavior on itself (i.e., effect of one sex on its own fitness)	Effect of actor's social behavior on the recipient (i.e., effect of one sex on the fitness of the other sex)	Consequence of the social behavior	Equivalent consequence in terms of sexual interactions	Potential for the outcome to result in sexual conflict	Examples
Decreases fitness	Increases fitness	Altruism	Sexual cannibalism, where, e.g., males suffer the cost of being eaten by their mate, and females gain fitness through the acquisition of resources to boost fecundity	If there is altruism due to sexual cannibalism (e.g., males donate their bodies as a nuptual gift to females), there may nevertheless be sexual conflict over the optimal timing of a male's sacrifice, with perhaps the female benefiting more from earlier sacrifice than the male (e.g., Fromhage and Schneider 2005)	Sexual conflict over the timing of copulation duration is reported in sexually cannibalistic spiders (e.g., Schneider et al. 2006). In addition, males that take opportunistic matings with feeding females significantly reduce their risk of injury and of being eaten (Fromhage and Schneider 2005). Lifetime fitness consequences would be useful, to ascertain whether this represents a case of altruism on the part of the male
Decreases fitness	Decreases fitness	Spite	A reproductive strategy adopted by one sex that is costly to both. Predicted under conditions under which there is negative relatedness, where individuals are less genetically similar at a given locus than are partners on average	Ultimately, to reduce costs of spite then both sexes should become subject to directional selection to minimize the costs involved, under normal patterns of relatedness. The conditions under which spite can evolve are fairly restricted and the potential for sexual conflict is theoretically possible, but likely limited (e.g., West et al. 2007b)	No known examples

expression. In this, it is usually only the maternally inherited or only the paternally inherited copy of an allele that is expressed in the offspring. Many placental genes that influence the flow of nutrients to the developing fetus in humans are imprinted in a way that suggests they are subject to sexual conflict over nutrient provisioning (Haig 1993, 1996, 1997). This perspective predicts that paternally expressed (maternal imprint) alleles will be "greedy" and, conversely, that maternally expressed (paternal imprint) alleles will be less so (e.g., Haig and Graham 1991; Úbeda and Wilkins 2008). Exactly this situation pertains in a condition known as Beckwith Weidemann (BW) syndrome. Here, there is a change to the normal inheritance pattern (i.e., expression of the maternally inherited allele) of IGF-2/H19 genes on chromosome 11. In BW, this gene region is instead expressed solely from the paternally inherited allele. The result is fetal overgrowth disorder due to the misexpression of a "greedier" version of the IGF-2 allele. Similarly, in Prader–Willi (P–W) syndrome, there is a change in the imprinting status of a gene in the q11–13 region of chromosome 15 from the normal expression of the paternal allele to misexpression of the "less greedy" maternally inherited allele. The result is undergrowth disorder during gestation and in the early postnatal period. Such imprinting diseases in humans (Úbeda and Wilkins 2008), and similar examples in mice (Haig and Graham 1991), reveal the underlying conflicts involved (but see Hurst and McVean 1997).

Conflicts over reproductive investment could be especially fierce if there are divergences in residual reproductive value (Stearns 1992; Williams 1966a, b), i.e., the proportion of an individual's reproductive output still to come in the future. I have already invoked this type of conflict, above, when describing a strategy in which females might gain from investment over the longer term (because of their higher residual reproductive value), whereas males might not. If the guarantee of future reproductive success is lower for males than for females, this will result in strong sexual conflict over current investment patterns. The idea is that individuals that live in risky environments should not wait for the future (which they may not live to experience) and so should adopt a strategy that tries to deliver the maximum payoff now, even if the potential future cost is significant, and the payoff carries significant current risk. It seems likely that there will be differences in the optimum discount rate for males and females, with males having higher future discounting.

This topic has been explored by Wilson and Daly (1997), in terms of an especially high disposition for future discounting in males that live in risky environments. The classic work on the Ache by Hill and Hurtado (1996) on predicted age at first reproduction in humans also fits into this framework. Here, age at reproduction is earlier when mortality rates are higher. Another example in humans has been recently been described by Waynforth (2012), in which individuals with chronic health conditions likely to lead to lifespan-shortening reproduce at an earlier age. This phenomenon is also well-described in the animal kingdom. For example, Daphnia water fleas (Stibor 1992) placed in water treated with fish extract (indicating a risk of predation) reproduced earlier with larger and heavier broods than did controls.

Having outlined some evidence that sexual conflict can be predicted using reasoning grounded in evolutionary biology, it is worth considering whether human-specific features will alter the nature of conflict, or make it more or less likely. In terms of behavior within human relationships that we might think of as encompassing sexual conflict, such as punishment, what matters in an ultimate sense is the effect of these behaviors on an individual's lifetime fitness. But does the expression of these behaviors in humans represent something different from that considered in other species?

The obvious and dominant distinct features of human societies are language and cultural evolution. There has been much debate about the extent to which the transmission and evolution of culture can contribute to the evolution of cooperation or, conversely, conflict (e.g., Boyd et al. 2011; Boyd and Richerson 2005; Feldman et al. 1985; Henrich 2004; Lehmann et al. 2008). There is evidence that, in comparison to genetic evolution, cultural evolution may, rather than facilitating cooperation, instead lead to increased selfishness or harm (Lehmann et al. 2008). Future work may clarify whether the dominance of culture leads humans to have an increased frequency of conflicting interactions in comparison to our animal relatives.

There is little evidence that humans are more cooperative than other animals or plants (West et al. 2011), or that humans are atypical in having especially strong enforcement or punishment to reinforce cooperation. Such enforcement mechanisms are widespread across plants and animals (e.g., Frank 2003; Kiers et al. 2003; Mulder and Langmore 1993; Ratneiks et al. 2006; West et al. 2007a; Young et al. 2006). Instead, it is the capacity for information-gathering to assess costs and benefits of cooperation where humans may be distinct, though not qualitatively so, in comparison to our animal relatives. This ability may allow better calibration of benefits and also the likelihood of punishment (e.g., see Fehr and Gächter 2002; West et al. 2011).

Reciprocity, a form of cooperation that can occur between nonrelatives (Trivers 1971), seems to be especially important in humans. Reciprocity occurs when a benefit given by an actor to a recipient is returned in the future by the receiving individual. Unless there is some way to recognize the individual that gave the benefit, reciprocity breaks down. Hence, reciprocity depends on relatively high-level cognition, recognition/enforcement mechanisms, or constructs such as legal agreements (reviewed by West et al. 2011). If an individual receives a benefit but does not then return it, punishment may be inflicted to reinforce the reciprocity. This punishment could represent a type of selfish behavior of relevance to sexual conflict.

Exploitation

Exploitation has been well-researched in humans and may reflect sexual conflict in terms of a strategy to gain more resources by taking advantage of the other sex (Buss and Duntley 2008). The exploiters adopt selfish behavior to gain a benefit for little cost to themselves. Hence, exploitation within human reproductive encounters can be seen in the same cost and benefit terms discussed above. There can be selection for exploitation strategies in men, for example, if there are differences in optimal mating frequency (e.g., a man wanting to have sex with a woman who is not interested) or differences in investment patterns (e.g., a man wanting a brief, sexual relationship and a woman wanting a longer-term, higher-investment relationship; Buss 2003). Which of the nonmutually exclusive types of exploitative behaviors, such as seduction, pressure, deception, or assault, are employed may depend on a woman's exploitability (Goetz et al. 2012) and, thus, the costs and benefits involved. The operation of exploitation strategies therefore depends on the integration of many different and complex signals. It is not necessary that individuals know how and why they might be responding to such cues; it is the ultimate end result that matters in evolutionary terms (Buss and Duntley 2011). The adoption of exploitation should, of course, select for the evolution of mechanisms to avoid exploitation or to reduce the costs of exploitation. Exploitation is therefore of relevance to sexual conflict as it could lead to sexually antagonistic coevolution between exploitation and resistance mechanisms.

Exploitation could be particularly difficult to evolve resistance to, however, if it taps into pathways to which females "need" to respond. This has been explored by West-Eberhard (1979), and others (e.g., Arnqvist 2006), in the context of the "sensory trap" that can be exploited to influence female mate choice. The idea is that males might manipulate a female's investment through a pathway that females necessarily use to gain reproductive fitness. Imagine that the trait in question is the number of fertile eggs laid. A female should start to lay eggs only once a mating has occurred and there are sperm available for fertilization. An efficient way to coordinate this is therefore to respond to a male signal passed during mating that stimulates egg production. If such a male signal evolves to manipulate the investment of the female upwards in terms of the number of eggs to produce then he gains fitness, but the female may suffer costs. However, it may be dangerous for females to evolve complete insensitivity to those signals, because they might easily "overshoot" and reduce their fitness. Hence, the idea of the "trap" from which females may not easily escape. To counter this scenario, females could evolve divergent responses to the same male cues, with the response determined by the context in which the male cues are expressed.

Exploitation is possible in species in which multiple mating occurs, as a side effect of male–male (sperm) competition. If female remating is rapid, then any costs incurred by a later mating male from engaging in sperm competition are lowered. This is because the female reproductive tract is already "primed" for reproduction following earlier matings (e.g., Hodgson and Hosken 2006). The potential for male–male exploitation within females has been supported empirically. For example, male *D. melanogaster* fruitflies are able to tailor their ejaculates to take advantage of the fact that previous mates of a female have already boosted her fecundity. Therefore, subsequent mates can invest less in that specific ejaculate component, while maintaining the levels of others (Sirot et al. 2011). These studies suggest that there is considerable potential for adaptive adjustments to ejaculate composition, and this deserves study across many more species.

I discussed earlier whether sexual conflict can result in "harm" in the animal world, and I stressed the importance of ultimate outcomes, rather than the different mechanisms of inflicting costs, for determining root causes of these phenomena.

The distressingly common examples of violence and aggression within human sexual relationships can also be viewed in this context (Buss and Duntley 2011), that is, as direct or indirect outcomes of adaptive responses to specific scenarios. In terms of the mechanistic underpinning of these ultimate outcomes, our research shows that, even in model systems such as fruitflies, males are capable of astonishingly flexible and precise calibrations in response to the social and sexual environment. These assessments are made to match male reproductive investment in any reproductive bout, to the perceived threat levels from rivals (Bretman et al. 2009) or to the mating status of females (Friberg 2006). I expect the mechanisms employed in human sexual interactions to be especially finely tuned, and hence, it is important to understand those and where possible to measure the fitness consequences of those actions. For example, it would be of interest to better define the costs of aggression (e.g., Bleske and Buss 2001) and the responses of recipients to aggression, which may include a response akin to "convenience polyandry", i.e., matings to avoid or to reduce the costs of aggression (Thornhill and Alcock 1983). This may help to understand better not only the ultimate reasons for aggression, but also how to minimize it. An important point is that here, just as in nonhuman examples, context is crucial to the likelihood and intensity of sexual conflict. For example, conflict may be less prevalent in human relationships when economic resources are abundant (Buss and Duntley 2011) and within specific personality types (Tanha et al. 2010).

Sexual Conflict, Male–Male Competition and Male Mate Choice

Up until now, I have discussed sexual conflict mostly in terms of an interaction between a male and a female. However, across the animal kingdom there is frequent multiple mating by males and multiple mating by females with different males. Females may store the sperm of different males for hours through to months or even years (Neubaum and Wolfner 1999). Wild caught females of many insect species can carry the sperm of more than one male (e.g., Imhof et al. 1998 for *D. melanogaster*). This means that a more realistic biological scenario is for premating contests to occur between multiple males and females, and postmating contests to occur between multiple males within a single female in species that store sperm.

The existence of "lovers triangles" in human social interactions greatly intensifies conflict and competition (Shackelford et al. 2003). This is also true in insects. For example, male *D. melanogaster* fruitflies exposed to rivals prior to encountering females show significantly extended mating durations (Bretman et al. 2009). This gives the focal male a fitness advantage in competition with other males. The longer matings following exposure of males to rivals results in higher fecundity and fertility, decreased female sexual receptivity, and a higher share of paternity. The mechanism by which this is achieved is via the increased transfer of seminal fluid components such as sex peptide (Wigby et al. 2009). Given that the receipt of the sex peptide seminal fluid protein can exact costs on females (Wigby and Chapman

2005), it is evident that the heightened competition between males, as indicated by the increased length of exposure to rivals (which causes more sex peptide to be delivered to females), intensifies the conflict with females (Bretman et al. 2013a).

To detect cues that males use to indicate that they are in the presence of a male of the same species, we tested the effect of removing cues (sight, sound, touch, and smell) by multiple genetic and phenotypic methods (Bretman et al. 2011). For example, we manipulated a male's ability to hear song by exposing focal males to rivals that could not sing (either following wing removal, or by using a *vestigial* wing mutation) or by using focal deaf males. We then investigated the cues that, when removed, rendered males unable to respond to rivals. No single cue on its own was important, but any two cues from the three sensory modes, of (i) smell, (ii) hearing (specifically hearing male courtship song), and (iii) touch, were necessary to allow males to respond adaptively to rivals (Bretman et al. 2011). The finding that there is flexibility but also robustness to the cues used by males to detect rivals suggests that it is important for a male not to make a mistake by responding to a female as if it were a male conspecific, or to a male of a closely related species. This would imply that, as well as significant benefits, there are costs of responding to rivals, which ideally should be avoided unless it is relatively "certain" that rivals are present. The investigations of the lifetime fitness consequences for males of continually responding to rivals confirm the existence of costs in terms of reduced male survival and reduced later-life mating capacity (Bretman et al. 2013b). Whether different strands of information conveyed by each separate cue are weighted differently in this context is not yet known.

We know that contingent responses by males to their rivals are highly individually flexible (Bretman et al. 2012). Individual males can respond to rivals, then not respond, and then respond again across at least three different mating episodes in series. The ways that males can respond to their social and sexual environment, as illustrated by the fruitfly example, are highly sophisticated. Such evolved responses to the social and sexual environment are well known from studies in humans, as for example, in the case of speech accommodation (Giles and Powesland 1975). The outcome of such responses in fitness terms would be interesting to investigate, but presumably the benefits reside in a desire to fit in and/or to become more competitive.

The growing realization that males are sometimes making a substantial investment in mating even in promiscuous mating systems that lack parental care (Dewsbury 1982) has recently been evaluated in the context of a renewed interest in male mate choice (Clutton-Brock 2007; Edward and Chapman 2011). The idea is that if males have substantial investments to make in mating, then they may allocate that investment to specific females. This renewed interest has arisen from the realization that the investment made by males is not necessarily minimal, even in species that appear to contribute little apart from sperm to a reproductive episode.

The key factors that help to determine whether the conditions exist for male mate choice are the degree of reproductive investment a male has available to make, the availability of mates, and his capacity to mate with available mates (Edward and Chapman 2011). If more females are available for mating than a male can mate with, if there is variation in female quality, and if the benefits of choice outweigh the costs of assessment, then male mate choice can evolve. The interesting case is when there is male mate choice but little obvious investment by males other than in ejaculate transfer. However, if that investment is considerable, such as when ejaculate production and courtship costs are high, then male mate choice can still occur. The intensity of sexual conflict is expected to vary according to the magnitude of the investment made by males. This is highly likely in the case of the transfer of ejaculate molecules.

If males exhibit choice and have a substantial investment to make, this may select for intrasexual competition among females for males, or contribute to mutual mate choice. Hence, there is considerably more variation and flexibility in the sex roles adopted by males and females in the animal world than has been assumed to date. The importance of sexual conflict in these scenarios has not yet been fully addressed.

Future Prospects

There is a large literature on behavioral conflict within human and primate relationships and it would be useful to integrate this body of work further into the framework of sexual conflict described above. This might enable the identification of parallels between humans and nonhumans, and also might facilitate investigations into whether there is anything evolutionarily distinct about sexual conflict in humans. There seem to be some candidate features arising from the importance of language and culture in human relationships. For example, traits such as personality, disposition, and humor that contribute to mate choice in humans (Buss 1989) may represent distinct facets of sexual interactions that can intensify competitions or give them greater phenotypic "space" in which to occur. Furthermore, that the intensity of male–male competition can exacerbate the sexual conflict with females is now established from studies on invertebrates (Arnqvist and Rowe 2005; Civetta and Clark 2000). Therefore, the psychological adaptations to sperm competition in humans would also be interesting to investigate in this context. Further work is warranted on whether these features represent qualitative or quantitative differences in sexual conflict. Finally, it is important during any study of sexual conflict to realize that most studies consider only a snapshot in time (Rowe and Day 2006), unless working with a system in which it is possible to follow the development of sexual conflict from initiation, through establishment to eventual outcome (Paterson et al. 2010). Measurements of the current context do not necessarily indicate past selection pressures, and inferences about the ancestral importance and function of traits subject to sexual conflict should be considered with caution. It is therefore important to understand what might be the diagnostic footprint of sexual conflict within the set of observable, currently expressed adaptations (Rice 1998; Rowe and Day 2006).

The involvement of sexual conflict in human health outside the immediacy of sexual relationships would also be worth further study. For example, sexually antagonistic coevolution arising from sexual conflict has been proposed to contribute to maladapted bodily processes that can predispose to increased cancer risk (Summers and Crespi 2008). Imprinted genes, of the type that play such a central role in directing nutrient flow across the placenta to the developing fetus and whose imprinting status is thought to be influenced by sexual conflict, are also often expressed in the brain. The expression patterns of imprinted brain genes are, controversially, proposed to underlie the balance between mental health on an autistic to schizophrenic spectrum (Badcock and Crespi 2008).

In animal systems, we are just beginning to develop a predictive framework for the expected signature of sexual conflict at the genomic level (Sirot et al. 2014). For example, across different species of mammals there are correlations between the pattern of seminal fluid evolution and mating systems (Wong 2010, 2011; Wyckoff and Wu 1997). Given the wealth of genomic data from humans, there are opportunities to make significant advances in this area, once the traits and molecular mechanisms underlying sexual conflict in human sexual interactions are better known.

Acknowledgements I thank Todd Shackelford and Ranald Hansen for the invitation to the *2013 Evolution of Sexuality Conference* and for the invitation to write this chapter. I thank Andrew Bourke, David Buss and Todd Shackelford for their generous and insightful comments, and Andrew Bourke for troubleshooting EndNote. Finally, I thank the University of East Anglia, the Natural Environment Resources Council, and the Biotechnology and Biological Sciences Research Council for financial support.

References

Adams, E. M., & Wolfner, M. F. (2007). Seminal proteins but not sperm induce morphological changes in the *Drosophila melanogaster* female reproductive tract during sperm storage. *Journal of Insect Physiology, 53,* 319–331.

Alexander, R. D., & Noonan, K. M. (1979). Concealment of ovulation, parental care, and human social evolution. In Chagnon, N. & Trions, W.E. (Eds.), *Evolutionary biology and human social behavior: An anthropological perspective* (pp. 436–453). North Scituate: Duxbury.

Amitin, E.G., & Pitnick, S. (2007). Influence of developmental environment on male- and female-mediated sperm precedence in *Drosophila melanogaster*. *Journal of Evolutionary Biology, 20,* 381–392.

Arnqvist, G. (2004). Sexual conflict and sexual selection: Lost in the chase. *Evolution, 58,* 1383–1388.

Arnqvist, G. (2006). Sensory exploitation and sexual conflict. *Philosophical Transactions of the Royal Society B, 361,* 227–386.

Arnqvist, G., & Rowe, L. (1995). Sexual conflict and arms races between the sexes: A morphological adaptation for control of mating in a female insect. *Proceedings of the Royal Society B, 261,* 123–127.

Arnqvist, G., & Rowe, L. (2002a). Antagonistic coevolution between the sexes in a group of insects. *Nature, 415,* 787–789.

Arnqvist, G., & Rowe, L. (2002b). Correlated evolution of male and female morphologies in water striders. *Evolution, 56,* 936–947.

Arnqvist, G., & Rowe, L. (2005). *Sexual conflict*. Princeton: Princeton University Press.
Badcock, C., & Crespi, B. (2008). Battle of the sexes may set the brain. *Nature, 454,* 1054–1055.
Bateman, A. J. (1948). Intrasexual selection in *Drosophila. Heredity, 2,* 349–368.
Bell, P. D., & Koufopanou, V. (1986). The cost of reproduction. In Dawkins, R. & Ridley, M. (Eds.), *Oxford surveys in evolutionary biology*. (Vol. 3, pp. 83–131). Oxford: Oxford University Press.
Benshoof, L., & Thornhill, R. (1979). The evolution of monogamy and loss of estrus in humans. *Journal of Society and Biological Structures, 2,* 95–106.
Bleske, A. L., & Buss, D. M. (2001). Opposite sex friendship: Sex differences and similarities in initiation, selection and dissolution. *Personality and Social Psychology Bulletin, 27,* 1310–1323.
Boomsma, J. J. (2007). Kin selection versus sexual selection: Why the ends do not meet. *Current Biology, 17,* R673–R683.
Boomsma, J. J. (2013). Beyond promiscuity: Mate-choice commitments in social breeding. *Philosophilcal Transactions of the Royal Society B, 368,* 20120050.
Bourke, A. F. G. (2009). The kin structure of sexual interactions. *Biology Letters, 5,* 689–692.
Bourke, A. F. G., & Franks, N. F. (1995). *Social evolution in ants*. Princeton: Princeton University Press.
Boyd, R., & Richerson, P. J. (2005). *The origin and evolution of cultures*. Oxford: Oxford University Press.
Boyd, R., Richerson, P. J., & Henrich, J. (2011). Rapid cultural adaptation can facilitate the evolution of large-scale cooperation. *Behavioural Ecology and Sociobiology, 65,* 431–444.
Bretman, A., Fricke, C., & Chapman, T. (2009). Plastic responses of male *D. melanogaster* to the level of sperm competition increase male reproductive fitness. *Proceedings of the Royal Society B, 276,* 1705–1711.
Bretman, A., Westmancoat, J. D., Gage, M. J. G., & Chapman, T. (2011). Multiple, redundant cues used by males to detect mating rivals. *Current Biology, 21,* 617–622.
Bretman, A., Westmancoat, J. D., Gage, M. J. G., & Chapman, T. (2012). Individual plastic responses by males to rivals reveal mismatches between behaviour and fitness outcomes. *Proceedings of the Royal Society B, 279,* 2868–2876.
Bretman, A., Westmancoat, J. D., & Chapman, T. (2013a). Male control of mating duration following exposure to rivals in fruitflies. *Journal of Insect Physiology, 59,* 824–827.
Bretman, A., Westmancoat, J. D., Gage, M. J. G., & Chapman, T. (2013b). Costs and benefits of lifetime exposure to mating rivals in male *Drosophila melanogaster*. *Evolution, 67,* 2413–2422.
Brommer, J. E., Fricke, C., Edward, D. A., & Chapman, T. (2012). Interactions between genotype and sexual conflict environment influence transgenerational fitness in *Drosophila melanogaster*. *Evolution, 66,* 517–531.
Buss, D. M. (1989). Sex differences in human mate preferences: Evolutionary hypothesis testing in 37 cultures. *Behavioral and Brain Sciences, 12,* 1–49.
Buss, D. M. (2003). *The evolution of desire: Strategies of human mating*. New York: Basic Books.
Buss, D. M., & Duntley, J.D. (2008). Adaptations for exploitation. *Group Dynamics: Theory, Research and Practice, 12,* 105–126.
Buss, D. M., & Duntley, J.D. (2011). The evolution of intimate partner violence. *Agression and Violent Behavior, 16,* 411–419.
Carvahlo, G. B., Kapahi, P., Anderson, D. J, & Benzer, S. (2006). Allocrine modulation of feeding behavior by the sex peptide of *Drosophila*. *Current Biology, 16,* 692–696.
Chapman, T. (2001). Seminal fluid-mediated fitness traits in *Drosophila*. *Heredity, 87,* 511–521.
Chapman, T. (2006). Evolutionary conflicts of interest between males and females. *Current Biology, 16,* 744–754.
Chapman, T., Liddle, L.F., Kalb, J.M., Wolfner, M.F., Partridge, L. (1995). Cost of mating in *Drosophila melanogaster* females is mediated by male accessory gland products. *Nature,* 373, 241–244.
Chapman, T., Herndon, L. A., Heifetz, Y., Partridge, L., & Wolfner, M. F. (2001). The Acp26Aa seminal fluid protein is a modulator of early egg-hatchability in *Drosophila melanogaster*. *Proceedings of the Royal Society B, 268,* 1647–1654.

Chapman, T., Arnqvist, G., Bangham, J., & Rowe, L. (2003a). Sexual conflict. *Trends in Ecology and Evolution, 18,* 41–47.
Chapman, T., Bangham, J., Vinti, G., Seifried, B., Lung, O., Wolfner, M. F., Smith, H. K., & Partridge, L. (2003b). The sex peptide of *Drosophila melanogaster*: Female post-mating responses analyzed by using RNA interference. *Proceedings of the National Academy of Sciences USA, 100,* 9923–9928.
Charlesworth, B. (1980). *Evolution in age-structured populations*. Cambridge: Cambridge University Press.
Charnov, E. L. (1979). Simultaneous hermaphroditism and sexual selection. *Proceedings of the National Academy of Sciences USA, 76,* 2480–2484.
Chen, P. S., Stumm-Zollinger, E., Aigaki, T., Balmer, J., Bienz, M., & Bohlen, P. (1988). A male accessory gland peptide that regulates reproductive behaviour of female D. melanogaster. *Cell, 54,* 291–298.
Chippindale, A. K., Gibson, J. R., & Rice, W. R. (2001). Negative genetic correlation for adult fitness between sexes reveals ontogenetic conflict in Drosophila. *Proceedings of the National Academy of Sciences, USA, 98,* 1671–1675.
Civetta, A., & Clark, A. (2000). Correlated effects of sperm competition and postmating female mortality. *Proceedings of the National Academy of Sciences USA, 97,* 13162–13165.
Civetta, A., & Singh, R.S. (1999). Broad-sense sexual selection, sex gene pool evolution, and speciation. *Genome, 42,* 1033–1041.
Clark, N. L., Gasper, J., Sekino, M., Springer, S. A., Aquadro, C. F., & Swanson, W. J. (2009). Coevolution of interacting fertilization proteins. *PLoS Genetics, 5,* e1000570.
Clutton-Brock, T. H. (2007). Sexual selection in males and females. *Science, 318,* 1882–1885
Cognigni, P., Bailey, A. P., & Miguel-Aliaga, I. (2011). Enteric neurons and systemic signals couple nutritional and reproductive status with intestinal homeostasis. *Cell Metabolism, 13,* 92–104.
Cordts, R., & Partridge, L. (1996). Courtship reduces longevity of male *Drosophila melanogaster*. *Animal Behaviour, 52,* 269–278.
Davies, N. B. (1992). *Dunnock behaviour and social evolution*. Oxford: Oxford University Press.
Dawkins, R. (1976). *The selfish gene*. Oxford: Oxford University Press.
Dewsbury, D. A. (1982). Ejaculate cost and male choice. *American Naturalist, 119,* 601–610.
Domanitskaya, E. V., Liu, H., Chen, S., & Kubli, E. (2007). The hydroxyproline motif of male sex peptide elicits the innate immune response in Drosophila females. *FEBS Journal, 274,* 5659–5668.
Edward, D. A., & Chapman, T. (2011). Evolutionary significance of male mate choice. *Trends in Ecology and Evolution, 12,* 647–654.
Edward, D. A., Fricke, C., Gerrard, D. T., & Chapman, T. (2011). Quantifying the life history response to increased male exposure in female. *Drosophila melanogaster. Evolution, 65,* 564–573.
Ellegren, H., & Parsch, J. (2007). The evolution of sex-biased genes and sex-biased gene expression. *Nature Reviews Genetics, 8,* 689–698.
Fedorka, K. M., Winterhalter, W. E., & Ware, B. (2011). Perceived sperm competition intensity influences seminal fluid protein production prior to courtship and mating. *Evolution, 65,* 584–590.
Fehr, E., & Gächter, S. (2002). The nature of human altruism. *Nature, 415,* 137–140.
Feldman, M. W., Cavalli-Sforza, L. L., & Peck, J. R. (1985). Gene-culture coevolution: Models for the evolution of altruism with cultural transmission. *Proceedings of the National Academy of Sciences USA, 82,* 5814–5818.
Fisher, R. A. (1930). *The genetical theory of natural selection*. New York: Dover.
Foerster, K., Coulson, T., Sheldon, B. C., Pemberton, J. M., Clutton-Brock, T. H., & Kruuk, L. E. B. (2007). Sexually antagonistic genetic variation for fitnes in red deer. *Nature, 447,* 1107–1111.
Fowler, K., Partridge, L. (1989). A cost of mating in female fruitflies. *Nature, 338,* 760–761.
Frank, S. A. (2000). Sperm competition and female avoidance of polyspermy mediated by sperm-egg biochemistry. *Evolutionary Ecology Research, 2,* 613–625.
Frank, S. A. (2003). Repression of competition and the evolution of cooperation. *Evolution, 57,* 693–705.
Friberg, U. (2006). Male perception of female mating status: Its effect on copulation duration, sperm defence and female fitness. *Animal Behaviour, 72,* 1259–1268.

Fricke, C., Bretman, A., & Chapman, T. (2008). Adult male nutrition and reproductive success in *Drosophila melanogaster Evolution, 62,* 3170–3177.
Fricke, C., Bretman, A. & Chapman, T. (2009a). Female nutritional status determines the magnitude and sign of responses to a male ejaculate signal in *Drosophila melanogaster. Journal of Evolutionary Biology, 23,* 157–165.
Fricke, C., Perry, J., Chapman, T., & Rowe, L. (2009b). Conditional economics of sexual conflict. *Biology Letters, 5,* 671–674.
Fricke, C., Wigby, S., Hobbs, R., & Chapman, T. (2009c). The benefits of male ejaculate sex peptide transfer in *Drosophila melanogaster. Journal of Evolutionary Biology, 22,* 275–286.
Fromhage, L., & Schneider, J. M. (2005). Safer sex with feeding females: Sexual conflict in a cannibalistic spider. *Behavioural Ecology, 16,* 377–382.
Giles, H., & Powesland, P. F. (1975). *Speech style and social evaluation.* Oxford: Academic.
Gioti, A., Wigby, S., Wertheim, B., Schuster, E., Martinez, P., Pennington, C.J., et al. (2012). Sex peptide of *D. melanogaster* males is a global regulator of reproductive processes in females. *Proceedings of the Royal Society B, 279,* 4423–4432.
Goetz, C. D., Easton, J. A., Lewis, D. M. G., & Buss, D. M. (2012). Sexual exploitability: Observable cues and their link to sexual attraction. *Evolution and Human Behavior, 33,* 417–426.
Haig, D. (1993). Genetic conflicts in human pregnancy. *Quarterly Review of Biology, 68,* 495–531.
Haig, D. (1996). Gestational drive and the green-bearded placenta. *Proceedings of the National Academy of Sciences USA, 93,* 6547–6551.
Haig, D. (1997). Parental antagonism, relatedness asymmetries, and genomic imprinting. *Proceedings of the Royal Society B, 264,* 1657–1662.
Haig, D., & Graham, C. (1991). Genomic imprinting and the strange case of the insulin-like growth factor II receptor. *Cell, 64,* 1045–1046.
Haig, D., & Wilczek, A. (2006). Sexual conflict and the alternation of haploid and diploid generations. *Philosophical Transactions of the Royal Society B, 361,* 227–386.
Hamilton, W. D. (1964). The genetical evolution of social behaviour I, II. *Journal of Theoretical Biology, 7,* 1–52.
Hayashi, T. I., Vose, M., Gavrilets, S. (2007). Genetic differentiation by sexual conflict. *Evolution, 61,* 516–529.
Helle, S., Lummaa, V., & Jokela, J. (2002). Sons reduced maternal longevity in preindustrial humans. *Science, 296,* 1085.
Henrich, J. (2004). Cultural groups selection, coevolutioary processes and large-scale cooperation. *Journal of Economic Behaviour and Organization, 53,* 3–35.
Hill, K. R., & Hurtado, A. M. (1996). *Ache life history: The ecology and demography of a foraging people.* Hawthorne: Walter de Gruyter Inc.
Hodgson, D. J., & Hosken, D. J. (2006). Sperm competition promotes the exploitation of rival ejaculates. *Journal of Theoretical Biology, 243,* 230–234.
Holland, B., & Rice, W. (1998). Chase-away sexual selection: Antagonistic seduction versus resistance. *Evolution, 52,* 1–7.
Holland, B., & Rice, W. R. (1999). Experimental removal of sexual selection reverses intersexual antagonistic coevolution and removes a reproductive load. *Proceedings of the National Academy of Sciences USA, 96,* 5083–5088.
Hrdy, S. B. (1979). Infanticide among animals—review, classification, and examination of the implications for the reproductive strategies of females. *Ethology and Sociobiology, 1,* 13–40.
Hurst, L. D., & McVean, G. T. (1997). Growth effects of uniparental disomies and the conflict theory of genomic imprinting. *Trends in Genetics, 13,* 436–443.
Imhof, M., Harr, B., Brem, G., & Schlotterer, C. (1998). Multiple mating in wild *Drosophila melanogaster* revisited by microsatellite analysis. *Molecular Ecology, 7,* 915–917.
Isaac, R. E., Li. C., Leedale, A. E., & Shirras, A. D. (2009). Drosophila male sex peptide inhibits siesta sleep and promotes locomotor activity in the post-mated female. *Proceedings of the Royal Society B, 277,* 65–70.
Johnstone, R. A., & Keller, L. (2000). How males can gain by harming their males: Sexual conflict, seminal toxins and the cost of mating. *American Naturalist, 156,* 368–377.

Kiers, E. T., Rousseau, R. A., West, S. A., & Denison, R. F. (2003). Host sanctions and the legume-rhizobium mutualism. *Nature, 425,* 78–81.
Krebs, J.R., Davies, N.B. (1987). Sexual conflict and sexual selection. In *An Introduction to behavioural ecology* (2nd ed, pp. 161–190). Sunderland: Sinhauer Asociates.
Kubli, E. (2003). Sex peptides: Seminal peptides of the Drosophila male. *Cellular and Molecular Life Sciences, 60,* 1689–1704.
Kuukasjärvi, S., Eriksson, C. J. P., Koskela, E., Mappes, T., Nissinen, K., & Rantala, M. J. (2004). Attractiveness of women's body odors over the menstrual cycle: The role of oral contraceptives and receiver sex. *Behavioural Ecology, 15,* 579–584.
Lawniczak, M. K. N., Barnes, A. I., Linklater, J. R., Boone, J. M., Wigby, S., & Chapman, T. (2007). Mating and immunity in invertebrates. *Trends in Ecology and Evolution, 22,* 48–55.
Lee, Y.-H., Tatsuya, O., & Vacquier, V. D. (1995). Positive selection is a general phenomena in the evolution of abalone sperm lysis. *Molecular Biology and Evolution, 12,* 231–238.
Lehmann, L., Feldman, M. W., & Foster, K. R. (2008). Cultural transmission can inhibit the evolution of altruistic helping. *American Natualist, 172,* 12–24.
Lessells, C. M. (2005). Why are males bad for females? Models for the evolution of damaging male mating behaviour. *American Naturalist, 165,* S46–S63.
Lew, T. A., Morrow, E. H., & Rice, W. R. (2006). Standing genetic variance for female resistance to harm from males and its relationship to intralocus sexual conflict. *Evolution, 60,* 97–105.
Liu, H., & Kubli, E. (2003). Sex peptide is the molecular basis of the sperm effect in *Drosophila melanogaster*. *Proceedings of the National Academy of Sciences USA, 100,* 9929–9933.
Long, T. A. F., & Rice, W. R. (2007). Adult locomotory activity mediates intralocus sexual conflict in a laboratory-adapted population of *Drosophila melanogaster*. *Proceedings of the Royal Society B, 274,* 3105–3112.
Martin, O. Y., & Hosken, D. J. (2003). The evolution of reproductive isolation through sexual conflict. *Nature, 423,* 979–982.
McGraw, L. A., Fiumera, A. C., Ramakrishnan, M., Madhavarapu, S., Clark, A. G., & Wolfner, M. F. (2007). Larval rearing environment affects several post-copulatory traits in *Drosophila melanogaster*. *Biology Letters, 3,* 607–610.
Metz, E. C., & Palumbi, S. R. (1996). Positive selection for sequence rearrangements generates extensive sequence polymorphism in the gamete recognition protein bindin. *Molecular Biology and Evolution, 13,* 391–406.
Morrow, E. H., Arnqvist, G., & Pitnick, S. (2003). Adaptation versus pleiotropy: Why do males harm their mates? *Behavioural Ecology, 14,* 802–806.
Mueller, J. L., Ripoll, D. R., Aquadro, C. F., Wolfner, M. F. (2004). Comparative structural modeling and inference of conserved protein classes in Drosophila seminal fluid *Proceedings of the National Academy of Sciences USA, 101,* 13542–13547.
Mueller, J. L., Ram, K. R., McGraw, L. A., Qazi, M. C. B., Siggia, E. D., Clark, A. G., Aquadro, C. F., & Wolfner, M. F. (2005). Cross-species comparison of Drosophila male accessory gland protein genes. *Genetics, 171,* 131–143.
Mueller, J. L., Page, J. L., & Wolfner, M. F. (2007). An ectopic expression screen reveals the protective and toxic effects of Drosophila seminal fluid proteins *Genetics, 175,* 777–783.
Mulder, R. A., Langmore, N. E. (1993). Dominant males punish helpers for temporary defection in superb fairly wrens. *Animal Behaviour, 45,* 830–833.
Neubaum, D. M., Wolfner, M. F. (1999). Wise, winsome, or weird? Mechanisms of sperm storage in female animals. *Current Topics in Developmental Biology, 41,* 67–97.
Parisi, M., Nuttall, R., Edwards, P., Minor, J., Naiman, D., Lu, J. N., Lu, J. N., Doctolero, M., Vainer, M., Chan, C., Malley, J., Eastman, S., & Oliver, B. (2004). A survey of ovary-, testis-, and soma-biased gene expression in *Drosophila melanogaster* adults. *Genome Biology, 5,* R40.
Parker, G. A. (1970). Sperm competition and its evolutionary consequences in the insects. *Biological Reviews, 45,* 525–567.
Parker, G. A. (1979). *Sexual selection and sexual conflict.* New York: Academic.
Parker, G. A. (2006). Sexual conflict over mating and fertilisation: An overview. *Philosophical Transactions of the Royal Society B, 361,* 235–260.

Parker, G. A., & Partridge, L. (1998). Sexual conflict and speciation. *Philosophical Transactions of the Royal Society B, 353,* 261–274.
Partridge, L., & Andrews, R. (1985). The effect of reproductive activity on the longevity of male *Drosophila melanogaster* is not caused by an acceleration of ageing. *Journal of Insect Physiology, 31,* 393–395.
Partridge, L., & Harvey, P. H. (1988). The ecological context of life history evolution. *Science, 241,* 1449–1454.
Partridge, L., & Hurst, L. D. (1998). Sex and conflict. *Science, 281,* 2003–2008.
Paterson, S., Vogwill, T., Buckling, A., Benmayor, R., Spiers, A. J., Thomson, N.R., Quail, M., Smith, F., Walker, D., Libberton, B., Fenton, A., Hall, N., & Brockhurst, M. A. (2010). Antagonistic coevolution accelerates molecular evolution. *Nature, 464,* 275–278.
Peng, J., Zipperlen, P., & Kubli, E. (2005). Drosophila sex peptide stimulates female innate immune system after mating via the Toll and Imd pathways. *Current Biology, 15,* 1690–1694.
Priest, N. K., Galloway, L. F., Roach, D. A. (2008). Mating frequency and inclusive fitness in *Drosophila melanogaster. American Naturalist, 171,* 10–21.
Ram, K. R., Wolfner, M. F. (2007). Seminal influences: Drosophila Acps and the molecular interplay between males and females during reproduction. *Integrative and Comparative Biology, 47,* 427–445.
Rankin, D. J. (2011). Kin selection and the evolution of sexual conflict. *Journal of Evolutionary Biology, 24,* 71–81.
Ratnieks, F. L. W., & Reeve, H. K. (1992). Conflict in single-queen Hymenopteran societies: The structure of conflict and processes that reduce conflict in advanced eusocial societies. *Journal of Theoretical Biology, 158,* 33–65.
Ratnieks, F. L. W., Foster, K. R., Wenseleers, T. (2006). Conflict resolution in insect societies. *Annual Review of Entomology, 51,* 581–608.
Reinhardt, K., Naylor, R., Siva-Jothy, M. T. (2003). Reducing a cost of traumatic insemination: Female bedbugs evolve a unique organ. *Proceedings of the Royal Society B, 270,* 2371–2375.
Ribeiro, C., Dickson, B. J. (2010). Sex peptide receptor and neuronal TOR/S6K signaling modulate nutrient balancing in Drosophila. *Current Biology, 20,* 1000–1005.
Rice, W. R. (1992). Sexually antagonistic genes—experimental-evidence. *Science, 256,* 1436–1439.
Rice, W. R. (1996). Sexually antagonistic male adaptation triggered by experimental arrest of female evolution. *Nature, 381,* 232–234.
Rice, W. R. (1998). Intergenomic conflict, interlocus antagonistic coevolution and the evolution of reproductive isolation. In Howard, D.J. & Berlocher, S.H. (Eds.), *Endless forms—species and speciation* (pp. 261–270). Oxford: Oxford University Press.
Robertson, S. A. (2005). Seminal plasma and male factor signalling in the female reproductive tract. *Cell and Tissue Research, 322,* 43–52.
Rönn, J., Katvala, M., & Arnqvist, G. (2007). Coevolution between harmful male genitalia and female resistance in seed beetles. *Proceedings of the National Academy of Sciences USA, 104,* 10921–10925.
Rowe, L., & Day, T. (2006). Detecting sexual conflict and sexually antagonistic coevolution. *Philosophical Transactions of The Royal Society B, 361,* 277–285.
Rowe, L., & Houle, D. (1996). The lek paradox and the capture of genetic variance by condition dependent traits. *Proceedings of the Royal Society B, 263,* 1415–1421.
Schneider, J. M., Gilberg, S., Fromhage, L., & Uhl, G. (2006). Sexual conflict over copulation duration in a cannibalistic spider. *Animal Behaviour, 71,* 781–788.
Schrempf, A., Heinze, J., & Cremer, S. (2005). Sexual cooperation: Mating increases longevity in ant queens. *Current Biology, 15,* 267–270.
Shackelford, T. K., Buss, D. M., Weeks-Shackelford, V. (2003). Wife-killings committed in the context of a lovers triangle. *Journal of Basic and Applied Social Psychology, 25,* 137–143.
Sharkey, D. J., Tremellen, K. P., Jasper, M. J., Gemzell-Danielsson, K., Robertson, S. A. (2012). Seminal fluid induces leukocyte recruitment and cytokine and chemokine mRNA expression in the human cervix after coitus. *Journal of Immunology, 188,* 2445–2454.

Short, S. M., Wolfner, M. F., Lazzaro, B. P. (2012). Female *Drosophila melanogaster* suffer reduced defense against infection due to seminal fluid components. *Journal of Insect Physiology, 58,* 1192–1201.
Sirot, L. K., Wolfner, M. F., Wigby, S. (2011). Protein-specific manipulation of ejaculate composition in response to female mating status in *Drosophila melanogaster*. *Proceedings of the National Academy of Sciences USA, 24,* 9922–9926.
Sirot, L. K., Wong, A., Chapman, T., Wolfner, M. F. (2014). Sexual conflict and seminal fluid proteins: A dynamic landscape of sexual interactions. In Rice, W.R. & Gavrilets, S. (Eds.), *Sexual conflict*. Cold Spring Harbor: CSHL (in press).
Stearns, S. C. (1992). *The evolution of life histories*. Oxford: Oxford University Press.
Stibor, H. (1992). Predator induced life-history shifts in a freshwater cladoceran. *Oecologia, 92,* 162–165.
Strassmann, B. I. (1981). Sexual selection, paternal care, and concealed ovulation in humans. *Ethology and Sociobiology, 2,* 31–40.
Summers, K., & Crespi, B. (2008). The androgen receptor and prostate cancer: A role for sexual selection and sexual conflict? *Medical Hypotheses, 70,* 435–443.
Swanson, W. J., Aquadro, C. F., & Vacquier, V. D. (2001a). Polymorphism in abalone fertilisation proteins is consistent with neutral evolution of the egg's receptor for lysin (VERL) and positive Darwinian selection of sperm lysin. *Molecular Biology and Evolution, 18,* 376–383.
Swanson, W. J., Yang, Z., Wolfner, M. F., Aquadro, C. F. (2001b). Positive Darwinian selection drives the evolution of several reproductive proteins in mammals. *Proceedings of the National Academy of Sciences USA, 98,* 2509–2514.
Tanha, M., Beck, C. J. A., Figueredo, A. J., & Raghavan, C. (2010). Sex differences in intimate partners violence and the use of coercive control as a motivational factor for intimate partner violence. *Journal of Interpersonal Violence, 25,* 1836–1854.
Thornhill, R., & Alcock, J. (1983). *The evolution of insect mating systems*. Cambridge: Harvard University Press.
Thornhill, R., & Gangestad, S.W. (1999). The scent of symmetry: A human sex pheromone that signals fitness? *Evolution and Human Behaviour, 20,* 175–201.
Trivers, R. L. (1971). The evolution of reciprocal altruism. *Quarterly Review of Biology, 46,* 35–57.
Trivers, R.L. (1972). Parental in investment and sexual selection. In Campbell, B. (Ed) *Sexual selection and the descent of man*. (pp. 136–179). London: Heinemann.
Úbeda, F., & Wilkins, J. F. (2008). Imprinted genes and human disease: An evolutionary perspective. *Advances in Experimental Medicine and Biology, 626,* 101–115.
VanderLaan, D. P., Forrester, D. L., Petterson, L. J., Vasey, P. L. (2012). Offspring production among the extended relatives of samoan men and Fa'afafine. *PLoS One, 7,* e36088.
Vasey, P. L., Pocock, D. S., VanderLaan, D. P. (2007). Kin selection and male androphilia in Samoan fa'afaine. *Evolution and Human Behaviour, 28,* 159–167.
Vasey, P. L., & VanderLaan, D. P. (2010). Avuncular tendencies and the evolution of male androphilia in Samoan fa'afaine. *Archives of Sexual Behavior, 39,* 821–830.
Wagstaff, B. J., & Begun, D. J. (2005a). Comparative genomics of accessory gland protein genes in *Drosophila melanogaster* and *D. pseudoobscura*. *Molecular Biology and Evolution, 22,* 818–832.
Wagstaff, B. J., & Begun, D. J. (2005b). Molecular population genetics of accessory gland protein genes and testis-expressed genes in *Drosophila mojavensis* and *D. arizonae*. *Genetics, 171,* 1083–1101.
Waynforth, D. (2012). Life-history theory, chronic childhood illness and the timing of first reproduction in a British birth cohort. *Proceedings of the Royal Society B, 279,* 2998–3002.
West, S. A., Griffin, A. S., & Gardner, A. (2007a). Evolutionary explanations for cooperation. *Current Biology, 17,* R661–R672.
West, S. A., Griffin, A. S., & Gardner, A. (2007b). Social semantics: Altruism, cooperation, mutualism, strong reciprocity and group selection. *Journal of Evolutionary Biology, 20,* 415–432.
West, S. A., El Mouden, C., & Gardner, A. (2011). Sixteen misconceptions about the evolution of cooperation in humans. *Evolution and Human Behaviour, 32,* 231–262.

West-Eberhard, M. J. (1979). Sexual selection, social competition, and evolution. *Proceedings of the American Philosophy Society, 123,* 222–234.

Westendorp, R., & Kirkwood, T. (1998). Human longevity at the cost of reproductive success. *Nature, 396,* 743–746.

Wigby, S., & Chapman, T. (2004). Female resistance to male harm evolves in response to manipulation of sexual conflict. *Evolution, 58,* 1028–1037.

Wigby, S., & Chapman, T. (2005). Sex peptide causes mating costs in female *Drosophila melanogaster. Current Biology, 15,* 316–321.

Wigby, S., Sirot, L. K., Linklater, J. R., Buehner, N., Calboli, F. C. F., Bretman, A., Wolfner, M. F., & Chapman, T. (2009). Seminal fluid protein allocation and male reproductive success. *Current Biology, 19,* 1–7.

Williams, G. C. (1966a). *Adaptation and natural selection.* Princeton: Princeton University Press.

Williams, G. C. (1966b). Natural selection, the costs of reproduction and a refinement of Lack's principle. *American Naturalist, 100,* 687–690.

Wilson, M., & Daly, M. (1997). Life expectancy, economic inequality, homicide, and reproductive timing in Chicago neighbourhoods. *British Medical Journal, 314,* 1271.

Wolfner, M. F. (2002). The gifts that keep on giving: Physiological functions and evolutionary dynamics of male seminal proteins in *Drosophila. Heredity, 88,* 85–93.

Wong, A. (2010). Testing the effects of mating system variation on rates of molecular evolution in primates. *Evolution, 64,* 2779–2785.

Wong, A. (2011). The molecular evolution of animal reproductive tract proteins: What have we learned from mating-system comparisons? *International Journal of Evolutionary Biology,* 2011, 2011.

Wong, A., Turchin, M. C., Wolfner, M. F., & Aquadro, C. F. (2008). Evidence for positive selection on *Drosophila melanogaster* seminal fluid protease homologs. *Moleulcar Biology and Evolution, 25,* 497–506.

Wyckoff, G. J., Wu, C. I. (1997). Sexual selection at the molecular level:evolutionary analysis of two sperm associated proteins in primates. *American Journal of Human Genetics, 61,* A215.

Yapici, N., Kim, Y.-J., Ribeiro, C., & Dickson, B. J. (2008). A receptor that mediates the post-mating switch in Drosophila reproductive behaviour. *Nature, 451,* 33–37.

Young, A. J., Carlson, A. A., Monfort, S. L., Russell, A. F., Bennett, N. C., & Clutton-Brock, T. H. (2006). Stress and the suppression of subordinate reproduction in cooperatively breeding meerkats. *Proceedings of the National Academy of Sciences USA, 103,* 12005–12010.

Zhong, W., McClure, C. D., Evans, C. R., Mlynski, D. T., Immonen, E., Ritchie, M. G., & Priest, N. K. (2013). Immune anticipation of mating in Drosophila: Turandot M promotes immunity against sexually transmitted fungal infections. *Proceedings of the Royal Society B, 280,* 2013–2018.

Chapter 2
Darwinian Literary Analysis of Sexuality

Judith P. Saunders

Darwinian Literary Analysis of Sexuality

Literary Darwinists draw upon research and theory from evolutionary psychology to analyze fictional, dramatic, and poetic representations of human behavior. A product of imaginative and aesthetic energies, literature offers special insight into universals of human nature. In the arena of literary make-believe, characters confront choices and difficulties mimicking those in real life, enabling readers to rehearse behavioral options, ponder social complexities, and study hypothetical life histories. From problem-solving to wish-fulfillment, art consistently engages deep-seated human concerns. Prominent among these is a preoccupation with the human condition itself. Literature serves as a forum in which writers and readers can examine, celebrate, question, deplore, and defy the forces constraining their existence. Individual texts do not merely illustrate the operations of evolved adaptations; they scrutinize and evaluate these in specific environmental contexts. They offer fascinating glimpses into the psyche of an animal intelligent enough to discern and assess the workings of its own mental and emotional processes.

Literary plots and themes focus with unsurprising persistence on activities with direct or indirect impact on fitness. Given the evolutionarily central role of reproduction in all plant and animal life, a concentration of interest on behavior that facilitates the passing on of genes is only to be expected. Historically and cross-culturally, human narratives return with unfailing regularity to topics emphasizing erotic desire, courtship tactics, marital strife, and parental commitment. Relations between the sexes, including the imperfect overlap of male and female reproductive interests, are by turns exalted, lamented, and ridiculed in countless poems, plays, and tales. Sexual passion, together with cultural norms regulating its expression, serves as the stuff of genres ranging from tragedy and romance to comedy, satire, and farce. Because human sexual behaviors entail complex assessments and inter-

J. P. Saunders (✉)
Department of English, Marist College, Poughkeepsie, NY 12601, USA
e-mail: Judith.Saunders@Marist.edu

T. K. Shackelford, R. D. Hansen (eds.), *The Evolution of Sexuality*,
Evolutionary Psychology, DOI 10.1007/978-3-319-09384-0_2

actions, including strategic choices influenced by environmental and phenotypic considerations, literary depictions of erotically driven behaviors and choices are correspondingly complex and varied. Deft handling of rhetorical devices such as point of view, metaphor, allusion, and irony enables writers to explore the multifaceted psychosocial impact of sexual impulses and choices. Darwinian literary criticism examines richly detailed, artfully indirect representations of eros in literary works, remaining alert to the evolutionary implications of the feelings and actions described.

Organized around different types of sexually motivated behavior, from mate selection and courtship to infidelity and mate-poaching, the discussion that follows is intended to be representative rather than exhaustive. Its purpose is twofold: first, to confirm the pervasive presence of sexual concerns in literature, and second, to show Darwinian literary criticism in action, clarifying its methods and purposes. Literary Darwinism, also known as evolutionary criticism or biopoetics, has established itself in recent decades as a fast growing sub-discipline, attracting notice and stimulating debate. Theoretical questions concerning the origin and adaptive value of art have drawn substantial scholarly attention; interrelationships between imagination, aesthetics, and other aspects of human cognition likewise have undergone exploration. Because contemporary evolutionary studies is an essentially interdisciplinary endeavor, it has stimulated fruitful collaboration across traditional disciplinary boundaries, bringing literary scholars together with cohorts from psychology, anthropology, linguistics, economics, philosophy, and numerous others fields. Appendix A lists important foundational works in Literary Darwinism, particularly those with strong interdisciplinary or theoretical emphasis; Appendix B lists works of practical Darwinian criticism, materials featuring illustrative analyses of individual texts. These are selective surveys, intended to provide useful points of access to an internationally active, rapidly growing field of study.

Section I: Bateman's Principle: Male Ardor and Female Reserve

When choosing and pursuing mates, men and women invoke criteria and employ strategies that are only partially congruent. A difference in lifetime reproductive potential accounts in large part for the asymmetry: men are limited in the number of children they can sire chiefly by the number of women to whom they gain sexual access, whereas, women are limited by the physiological demands of ovulation, pregnancy, and lactation. Unlike men, women cannot increase their reproductive success simply by increasing the number of their sexual partners. Given the finite limits to the number of offspring potential mothers can conceive and bear, each reproductive effort is critical: the costs of poor mating decisions are much higher for women than for men (Buss 2003, pp. 19–20, 45). It proves adaptive, typically, for women to proceed circumspectly when choosing a mate, taking time to assess genetic quality, resource access, and character traits. Men, in contrast, can compensate for less selective mating choices by employing mating strategies that emphasize frequency, quantity, and variety. Bateman's principle sums up the implications of these fundamental sex differences: men tend to be sexually eager, often recklessly so, while women tend to be sexually cautious (Buss 2003, p. 77). These divergent, sex-based tendencies exercise influence on all aspects of mating, from courtship to fidelity. Literary texts frequently focus on these differences, as well as on the intersexual conflict they precipitate.

Well-known poems by Andrew Marvell and Robert Herrick present the conflict between male ardor and female reserve from the male point of view. The speaker in Marvell's "To His Coy Mistress" (1681) reproaches his beloved for her "coyness," albeit mockingly (line 2). Her reserve is slowing down the progress of his courtship, and the poem expresses his consequent frustration. Evidently she has been enforcing a careful and deliberate pace in their courtship. This is adaptive for her, as she takes time to assess his qualities as a partner, but it interferes with the quick access to a mating opportunity that would better serve his evolutionary interest. He attempts to increase the pace she finds comfortable by emphasizing the brevity of mortal existence: "Time's winged chariot" is "hurrying near," he warns her (line 31). If death is close at hand, it follows that there is no "time" for protracted courtship (line 1). Because of the woman's delaying tactics, he furthermore insists, they risk losing all opportunity to consummate their love. As he puts it, "the grave's a fine and private place/ But none, I think, do there embrace" (lines 30–31). With the humorously grotesque image of her "long preserved virginity" yielding only to the "worms" that will penetrate her decaying corpse, the speaker underlines the futility of prolonged hesitation (lines 27–28).

It's worth noting that he does not offer as inducement any promises of long-term commitment; he emphasizes instead the gratification of proximal impulses, i.e., "all [his] lust" (line 30). Like "amorous birds of prey," he tells his beloved, they should "devour" the pleasures of their intimacy all at once (line 38):

> Let us roll all our strength and all
> Our sweetness up into one ball
> And tear our pleasures with rough strife
> Through the iron gates of life. (lines 41–44).

This final set of images emphasizes immediacy, invoking a *let's have it all right now and never mind the future* spirit. Instead of appealing to female interests by depicting a future of mutual fidelity, the speaker concentrates on the fulfillment of distinctly masculine desires. The goal of deriving maximum enjoyment from one brief but gloriously satisfying expenditure of sexual energy is not calculated to appeal to a woman's point of view. She is not likely to embark upon a sexual encounter, however potentially pleasurable, if afterwards she might find herself pregnant with no committed partner, no dependable access to resources. Marvell's speaker presents an unapologetically masculine perspective on courtship and mating, and readers sense that he is, at least to some extent, inviting mockery of male eagerness. His failure to present arguments against "coyness" that a female addressee might find seriously persuasive, together with his hyperbolic evocations of tombs and worms, suggests that his plea is tongue-in-cheek. The poem is a vehicle for a fantasy of male wish-fulfillment, allowing him to imagine the pleasures of immediate gratification if only he could infect his beloved with masculine haste and eliminate the strategic interference her reticence represents.

Like Marvell's poem, Herrick's "To the Virgins, to Make Much of Time" (1646) seeks to undermine female reserve by emphasizing the rapid passage of time. The speaker presents himself as an avuncular advisor rather than a suitor. He speaks to all young women, with ostensibly disinterested purpose. He counsels them to "use [their] time" properly and choose mates without undue hesitation, since those who "tarry" past "their prime" likely will be doomed to spinsterhood (lines 13, 15). Thus he reminds young girls that markers of female fertility—youth and beauty—peak early in adulthood, declining thereafter: "that age is best which is the first" (line 9). He attempts to foil the female inclination to court slowly—that is, to compare the potential advantages of various mating options before making a selection—by instilling a sense of urgency. Like the rosebuds to which he compares them, the girls are losing the first freshness of youth, and he exaggerates the rapidity of that aging process: time is not merely passing, it is "a-flying" (line 2). Taunting the girls with the transience of their physical loveliness, he indicates that remaining "forever" unmated is the just fate of those exercising selective caution (line 16).

That his audience might need his warning seems unlikely: every young woman is aware (although perhaps not consciously) that she is balancing a limited period of fertility against the need to choose wisely before committing her reproductive resources. In counseling more speed and less care in the female selection process, Herrick's speaker is promoting his own interests. He stands to benefit if his advice creates a less reserved, less choosy population of young girls. Reduced to its essential message, the poem attempts to frighten girls into behavior that would serve male evolutionary interests more than their own. Almost certainly, as the speaker seems aware, his efforts to modify female mating strategies are unlikely to succeed: the poem is playful in tone and intention. The poet-speaker appears to be offering his counsel as a tactical move in the eternal mating game, a competition in which members of each sex seek to foster their own advantage. Lyric poetry provides him with a culturally acceptable, appealingly aesthetic framework for presenting his only half-serious, transparently selfish propositions.

Section II: Mate Selection and Courtship

The mechanisms motivating male ardency and female reserve also guide preferences for mate selection and courtship styles more generally. Men who secure a relative abundance of female reproductive resources leave the largest genetic legacy; a potential partner's probable fertility is a *sine qua non*, therefore, in male choice. In seeking mates, men are alert to signs of fertility—most obviously, to health and youth. Research suggests that female beauty is largely defined by these qualities: a waist-hip ratio consistent with fecundity, childbearing, and lactation, together with the hair color, skin tone, and physical vitality associated with young adulthood, comprise crucial components, cross-culturally, of female attractiveness (Buss 2003, pp. 50–58). It is easy to see why Edgar Allan Poe declares in "The Philosophy of Composition" that "the death of a beautiful woman" is the "most poetical topic in the world" (1846, p. 1621). The physiological and temporal costs of female reproduction, from large eggs to gestation and lactation, render women the limiting resource in human reproduction: female fertility is precious. The death of a "beautiful" woman, that is, a young and fertile woman whose residual reproductive potential still encompasses many years, is poignant because it represents the loss of this resource precisely at its moment of greatest value. Such loss is a topic calculated to evoke aesthetic and emotional intensity.

In literature, as in life, we observe a strong male tendency to focus on the physical attributes of potential mates. Other criteria also come into play, particularly in the context of long-term commitment—intelligence, empathy, humor, compatibility, for example—but these play a secondary role in comparison with fertility. When the scholarly Roger Chillngworth from *The Scarlet Letter* seeks to marry "in the autumn of [his] days" and "already in decay," he does not select a middle-aged lady who shares his intellectual interests; despite his "misshapen" torso and small stature, he chooses the "budding" Hester, whose "youth and beauty" constitute significant appeal for him (Hawthorne 1850, p. 1392, 1445). In George Eliot's *Middlemarch*, the "ugly and learned" Revered Casaubon, a "dried bookworm towards fifty," whose face is disfigured by "two white moles with hairs on them," similarly selects the youthful, lovely Dorothea Brooke as his wife (1871–1872, p. 35, 17, 15). Gilbert and Sullivan stage a farcical version of this mate selection pattern in *The Mikado* (1885). This comic opera opens with preparations for a marriage between the nubile Yum-Yum, only just out of school, and Ko-Ko, her middle-aged and physically unprepossessing guardian. In all these cases, the markers of fertility evident in the much younger woman's figure and face override considerations of compatibility.

Sonnets by Petrarch, Spenser, and countless other male poets' lavish praise upon female bodily beauty: bright eyes, rosy cheeks, full lips, glossy hair, full bosom, narrow waist, and rounded buttocks all are celebrated, along with grace and ease of movement. Literary narratives of mate quests likewise reflect these male preferences. Sickly female characters typically are not successful in attracting suitors, even when they display other valued qualities such as status, wealth, intelligence, or kindness. Young Laura in *The Glass Menagerie*, for example, slightly "crippled" physically and "a little peculiar" in personality, is unable to attract the eligible "gentleman callers" her mother so desires for her (Williams 1945, scene 1). Miss Anne De Bourgh, Darcy's sickly cousin in Jane Austen's *Pride and Prejudice*, likewise fails to win suitors. Her mate selection standards (and her mother's on her behalf) arguably are too high, but she is a conspicuously wealthy, upper-class young lady: one might anticipate that she would at least have the pleasure of refusing impecunious and ambitious men yearning to profit from her social status and material resources. In a community full of men seeking rich brides with substantial dowries, Anne nevertheless remains unsought: her defective health suggests all too strongly that she is reproductively unviable. Preferring Elizabeth Bennett to his invalid cousin, despite family pressures to wed Anne, Darcy finds himself drawn at the outset to Elizabeth's physical energy as well as to her vivacious personality. Her ability to walk several miles across uneven, muddy terrain wins masculine approval: he notes that her "fine eyes" are "brightened by the exercise" (Austen 1813, p. 26). The "loveliness" attributed to Emma Woodhouse similarly is grounded in excellent health: she presents "the complete picture of grown-up health" (Austen 1816, p. 68).

In her 1925 novel, *The Mother's Recompense*, Edith Wharton presents an intriguing plot to underscore the robustness of male preferences. She introduces readers to a mother and daughter, Kate and Anne Clephane, who resemble each other to a remarkable degree. In the course of the narrative, the two love and are loved by the same man, Chris Fenno, whose age puts him approximately mid-way between them: he is 13 years younger than Kate, 10 years older than Anne. Engaging in a romance with the divorced Kate that lasted three or four years, when he is aged approximately 26 to 28, she approximately 39 to 41, Chris ends the affair—to Kate's intense disappointment. Not long thereafter he meets her 21 year-old, look-alike daughter Anne and instigates a serious courtship ending in engagement and marriage. This long-term commitment contrasts with his earlier decision to discard the middle-aged Kate, whose residual reproductive value rapidly is approaching zero. Indeed, given the absence of reliable contraception at the time, Kate's failure to become pregnant during the 3 years of their intense involvement provides a powerful signal of diminished fertility. The physical resemblance between mother and daughter, reaffirmed by at least one character who mistakes Anne for Kate, enables Wharton to come as close as possible to suggesting that the two women are almost indistinguishable. What would happen, she asks, if a man had to choose between *two versions of the same woman*, each representing a different stage of life? Age, she concludes, would be the deciding factor. A man will follow evolutionary self-interest by choosing the woman whose reproductive value is higher; this is a cruel inevitability. Chris is able to "love" Kate as much as Anne, evidently, and to value the personal qualities of both perhaps equally, but he makes his long-term mating commitment to the younger version of the pair. Wharton's narrative promotes empathy with the plight of aging women who must learn that potential partners will reject them in favor of the young and the fertile—even when the attractions offered by those younger rivals do not in other respects exceed their own.

Female courtship behavior is influenced in numerous ways by these critical male preferences (Buss 2003, p. 133). Women enhance bodily shape with corsets, bustles, and other undergarments or draping effects designed to suggest ideal waist-hip ratio (i.e., WHR). They employ cosmetics to improve the appearance of skin, disguise wrinkles, minimize complexion flaws, or enlarge the appearance of eyes. Literature repeatedly illustrates women's anxiety about their ability to attract men based on their physical traits: characters agonize over flaws, real or perceived, in their own appearance; they worry about comparisons between other women's looks and their own; they expend time and energy acquiring clothing and accessories; they lament the appearance of grey hair or wrinkles. Such preoccupations are often stereotyped and ridiculed: "a woman's dress, at least, is never done," Thoreau derisively comments (1855, p. 1883). Just as often, however, women's obsessive interest in their outward appearance is presented sympathetically, as a realistic feminine concern. In *The House of Mirth*, Wharton shows the beautiful Lily Bart's aghast reaction to the first indications of aging: "she was frightened by two little lines near her mouth, faint flaws in the smooth curve of the cheek" (1905, p. 43). Lenina Crowne from *Brave New World*, attractive though she is, frets that a lover may find her "too plump" (Huxley 1932, p. 93). Elizabeth Bowen's story "The Needle Case" sketches the plight of upper-class, impecunious girls whose chances to marry suitably are imperiled by deficiencies in wardrobe. The family estate no longer generates sufficient income to maintain itself ("this well of a house drank money"), and consequently, "its daughters were likely to wither" for want of 'advantages' (1934, p. 456). "Balked" by the inability to dress with sufficient allure, one sister voices her desperation to the seamstress hired to refurbish her shabby old gowns for the summer season: "I've *got* to look nice" (p. 456).

Women's efforts to maximize their physical attractions are so extreme, at times, that they cross the boundary between enhancement and deception. Blanche Dubois in *A Streetcar Named Desire* worries about her advancing age and takes obsessive interest in her looks: "you've got to be … *attractive*," she laments, "and I—I'm fading now! I don't know how much longer I can turn the trick" (Williams 1947, scene 5). She takes steps to conceals her age, particularly in the presence of Mitch, whom she has singled out as a potential husband: "I want to *deceive* him enough to make him—want me," she explains (scene 5). She refuses to go out with him by daylight, and she covers the light bulb in her sister's apartment with a "colored paper lantern" to ensure that he will see her only dimly illuminated (scene 3). Louisa May Alcott's gothically inspired "Behind the Mask" speaks to this same theme. Alcott introduces a "haggard, worn woman" whose misleadingly youthful appearance is created by an astonishing array of artificial aids. Her "scanty locks" have been replaced by a wig of "long abundant braids"; her smile is augmented by "several pearly false teeth"; her cheeks are rouged "pink"; her scarred breast is concealed by an attractive gown (1866, p. 12). Such duplicitous self-presentation targets male fears that women may disguise their physical imperfections, particularly signs of aging, so successfully that they foil male reproductive agendas.

Duessa from *The Fairie Queen* is a still more horrifying agent of male fears, a female figure who misrepresents her bodily decay in order to captivate, exploit, and destroy men. With "forged beauty," Duessa seeks to win lovers by means of "guile" and "hellish science" (Spenser 1596, 36, line 1; 28, lines 3, 4). Seen in her true form, she is "a filthy foule old woman," repulsive to men: "euer to haue toucht her, I did deadly rew" (40, line 9). Her reproductive organs in particular ("neather partes"), are described as "hideous," "misshapen, monstruous" (41, lines 1, 3). Here, an older woman who succeeds in creating a false impression of youth is associated with moral and spiritual evil; she is a sorceress wielding wicked powers. Such a character triggers evolutionarily based male anxieties. Since female fertility is the key to men's genetic legacy, any tactics that trick men into choosing post-reproductive women as long-term mates represent deadly danger, threatening a duped man's hope of biological continuity.

Just as men focus for evolutionarily sound reasons on fertility when selecting a mate, women focus on resources (Buss 2003, p. 22). Given men's typically life-long production of sperm, women do not need to worry much about a potential mate's ability to impregnate. They are concerned instead with an access to critical material assets, such as food and shelter, which will enable them to survive pregnancy and raise offspring through infancy and early childhood. Women seek partners who have resources on hand and who possess, in addition, character traits associated with the ability to manage existing resources wisely and to obtain more over time. Various kinds of competencies, depending on physical and social environmental context, may prove relevant to the mate quest; typically women judge that men with flexible intelligence, persevering ambition, and income-generating education or skills will make suitable mates. Social dominance also is associated with the ability to command resources and thus enhances a man's attractiveness to women (Buss 2003, pp. 23–35).

As frequently as literature documents and sometimes satirizes male emphasis on women's physical attributes, it illustrates and sometimes mocks the equally conspicuous female demand for plentiful resources. Dorothy Parker, for instance, pokes fun at the avaricious inclinations of her own sex in a poem. The speaker laments men's tendency to woo women with romantic offerings—"one perfect rose"—in place of more substantial courtship gifts such as "one perfect limousine" ("One Perfect Rose," 1944, lines 4, 10). As Jane Austen wryly observes, "it is a truth universally acknowledged" that unmarried women strive mightily to attract the attention of "a single man in possession of a good fortune" (1813, p. 1). Even before she has met the two Bertram boys, Mary Crawford from *Mansfield Park* feels a "presentiment that she should like the eldest best," since it is he who will inherit a title and large estate (Austen 1814, p. 80). Trying desperately to find husbands for her five daughters, Mrs. Allaby from *The Way of All Flesh* is happy to snatch at the "second son" of a man so rich that even a younger son "should have something very comfortable" (Butler 1903, p. 70). The appearance of wealthy young Percy Gryce in upper-crust New York social circles is greeted with great enthusiasm by marriageable girls and their mothers in *The House of Mirth*. Even though he is socially insipid, personally dull, and a bit of a Mama's boy, altogether lacking in traits suggestive of dominance, Percy is regarded as a highly desirable catch; his material resources are so enormous that they offset his personal deficiencies. Friends assist Lily Bart to win Percy's favor, even as they lament her probable fate with a husband so lacking in companionable appeal: "What an awful life you'll lead [with him]" (Wharton 1905, p. 72).

Male courtship tactics respond to women's preferences: when seeking to impress potential mates, men flaunt wealth or reputation—or both (Buss 2003, p. 99). Othello woos Desdemona with a classic display of male dominance, for example, describing confrontations with human enemies and natural disasters; his account includes fierce battles, "hairbreadth 'scapes," and adventures with "cannibals" (Shakespeare 1602, I, iii, lines 136, 143). His ability to triumph over "insolent foe" and "disastrous … accidents" renders him desirable to Desdemona (lines 135–136, 137). Bold, brave, and persevering, he has demonstrated the physical prowess and mental toughness necessary to achieve high male ranking in this social environment. She loves him, Othello explains, "for the dangers" he has so gallantly surmounted and so effectively reported to her (line 167).

Taking an equally adaptive approach, Jay Gatsby targets the female concern with resources: in his courtship of Daisy Buchanan he commits himself to the accumulation and display of wealth. He spends years building a vast personal fortune, showcasing his success with conspicuous examples of material luxury: a European-style mansion, fancy cars, expensive clothing, and elaborate parties featuring "celebrated people" (Fitzgerald 1925, p. 90). He plans his reunion with Daisy carefully, ensuring that she will see his fantastically "huge" dwelling—and thus grasp the extent of his wealth—immediately. Taking her on a tour of his "Marie Antoinette music-rooms and Restoration salons," the "period bedrooms" and "sunken baths," he shows off his exquisite possessions in loving detail (pp. 90, 91). Moving deliberately from the exterior to the interior of his home, he ends the tour in his bedroom, tacitly inviting Daisy to associate his riches with sexuality and mating opportunities. He shows her his "toilet set of pure dull gold" and is delighted when she immediately begins to smooth her hair with his brush, a subtle sign that she is responding positively to this spectacular display of resources. The scene reaches its well-known climax when he piles his high-priced British shirts before her in "many-colored disarray": the intimate extravagance of this "soft rich heap" of "beautiful shirts" moves Daisy to stormy tears (p. 92). Exhibiting magnificent apparel that has clothed his own body, Gatsby triggers an emotionally intense reaction that promises to lead, as it in fact does, to a romantically charged sexual affair.

Section III: Mate Value and Competition for Mates

Although fertility and resources—together with related attributes such as health, vigor, competence, status, or dominance—emerge as centrally important mate selection criteria for men and women, respectively, members of both sexes seek other qualities in long-term mates as well. Worldwide, as Buss has demonstrated, men and women value traits such as dependability, kindness, generosity, intelligence, honesty, humor, and social skills in their partners. They also value compatibility, as measured by shared interests and values and by similarity in social background and status (2003, pp. 35–38, 179–181, 210). The quest for a long-term mate requires individuals to assess their own qualities as well as those of potential mates. Since very few individuals possess every desirable feature in equal measure, it is necessary to weigh the importance of one quality against another, to decide whether evident deficits are sufficiently offset by other advantages. Plentiful resources may compensate for social ineptness; youthful beauty may compensate for a moody temperament; dependability may compensate for mediocre social status. When Ben Franklin discovers, for instance, that his perceived value is lower than he had imagined, "the Business of a Printer being generally thought a poor one," he is forced to downgrade his demands for a dowry: "I was not to expect Money with a Wife, unless with such a one as I should not otherwise think agreeable" (*Autobiography* 1794, p. 56). A match is judged to be suitable, as Franklin implies, when both parties are satisfied that they have obtained a partner whose value, overall, equals or exceeds their own (Buss 2003, pp. 37–38).

In *The House of Mirth* (1905) Edith Wharton portrays two characters engaging in explicit discussions of their relative mate values. Simon Rosedale makes a proposal of marriage to Lily Bart that sounds very much like "a plain business" arrangement: he considers her genteel background and social skills equivalent in worth to his self-acquired fortune (p. 285). Combining his money and business know-how with her respectability and elegant social presence will give both of them, he suggests, exactly what they want in life. Unwilling to assess his mate value as high as her own, Lily refuses the *nouveau riche* outsider. Many months later, however, her reputation and finances have suffered severe blows; she has been dropped by most members of her social circle and disinherited by her aunt. She decides, in consequence, that marriage to Rosedale will solve her problems, and she informs him that she now is prepared to accept him as a husband. Due to her damaged reputation, however, she no longer can help him gain entrée into elite social circles, and her value to him as a potential wife has dropped decisively:

> Last year I was wild to marry you, and you wouldn't look at me; this year—well, you appear to be wiling. Now, what has changed in the interval? Your situation, that's all. Then you thought you could do better; now—"You think you can?" broke from her ironically."Why, yes, I do." (p. 412).

Both participants in this conversation recognize marriage as an exchange of benefits, a partnership to which each person brings a variety of assets and seeks equivalent value.

Mistakes in assessing potential mates typically prove costly in terms of both individual and inclusive fitness; thus mating decisions are among the most important any individual makes in the course of a lifetime. At the same time, however, it is difficult to assess the complex constellation of traits presented by others, and even more difficult to assess one's own with accuracy. Theory of Mind plays a critical role in mate selection, both in penetrating the deliberate deceptions of others and in assessing the probable reactions of others to one's personality, reputation, and appearance. Hence, the conflicts central to literary plots frequently highlight the causes and consequences of mating mistakes. Dorothea Brooke and David Copperfield illustrate the painful results of inaccurate assessment, indicating that youthful inexperience can lead even intelligent people toward poor choices. Edward Casaubon lacks the noble qualities of mind and character Dorothea naively attributes to him: his scholarship proves to be mostly sham; he fears and repels emotional warmth; he is capable of mean-spirited jealousy (Eliot, *Middlemarch*, 1871–1872). Like Dorothea, David Copperfield discovers only after marriage the unsuitability of the partner he chose with unquestioning fervency: not only does Dora lack domestic skills and financial prudence, she is wholly unable to share his intellectual and creative interests (Dickens 1850). Both protagonists are released from their unsuitable marriages by the death of their spouses; relatively early in life, they are allowed by their creators to profit from their experience and select new mates from a more mature vantage point. These novels present the mating errors of these central characters with a mixture of exasperation and sympathy.

When mating decisions involve significant socioeconomic disparities, one partner experiences a distinct reduction in status and resources. Often such matches threaten the social standing—and indirectly the fitness—of a whole extended family. Hence, relatives often intervene to prevent such romances from flourishing: another common source of conflict in literature. In *Mansfield Park*, the heroine's mother had fallen in love with a sailor (a common seaman rather than an officer), and she married, accordingly, "to disoblige her family" (Austen 1814, p. 41). Living in straightened, even sordid, circumstances with a family of nine children, she is an object lesson in the perils of hypogamy. Without assistance from wealthier relatives, none of Mrs. Price's children would have access to vocational training or social advantages. Predictably, self-interest influences families to discourage marriages that appear to have an unstable financial basis. Elizabeth Bennet's maternal aunt cautions her against a serious courtship with Mr. Wickham, for instance, not because she intuits the all-too-real deficiencies in his character but because she deplores his "want of income" (1813, p. 109).

In the Elizabeth–Wickham courtship, Austen also introduces the issue of deception, underlining the importance of making efforts to unmask suitors who misrepresent themselves. Wickham lies to Elizabeth about his past, concealing his misdeeds and blackening the reputation of the Darcy family. Until she gets accurate information from a different source, the acute but insufficiently alert Elizabeth sees no reason to distrust this pleasing, apparently candid young man. Here, as in many fictional stories, the dangers of credulity—together with the importance of fact-checking—contribute to building suspense and resolving the plot. Isabel Archer in *Portrait of Lady* is a victim of conspicuously deceptive courtship tactics: Gilbert Osmond conceals important aspects of his past from Isabel, particularly his intimate relationship with Madame Merle, who turns out to be the unacknowledged mother of his daughter. Osmond's chief object in marrying Isabel is to gain access to her newly inherited fortune, and his style of wooing is calculated to blind her to that exploitative purpose. His outward charm and old World gentility effectively conceal his coldness and cupidity (James 1881).

Even in the absence of deliberate deception, faulty evaluative processes frequently thwart discerning selection. Perceived mate value is a central issue in Austen's *Emma*: the narrative explores Emma's erroneous assessment of Harriet Smith in the social environment of a small English village. Charmed by Harriet's physical beauty and agreeable personality, Emma overlooks her distinctly average intelligence and nondescript education. She also interprets Harriet's illegitimacy unrealistically, investing it with upper-class mystery: "There can be no doubt that of your being a gentleman's daughter, and you must support your claim to that station," she tells her (1816, p. 60). Not only does she conclude that Harriet would make a suitable mate for the socially superior Mr. Elton, she discourages her from accepting Robert Brown, a respectable young farmer. Emma's plans for Harriet bring her into conflict with Mr. Knightly, who judges Harriet's value much more dispassionately. He points out that Harriet has no "claims, either of birth, nature or education, to any connection higher than Robert Martin" (p. 87). She is "the natural daughter of nobody knows whom, with probably no settled provision" (p. 87). This lack of resources, coupled with an uncertain social background and mediocre education, all suggest, Knightly insists, that Harriet is meant to remain in the social circle from which Emma has plucked her: "She is pretty and she is good tempered, and that is all" (p. 87). Meanwhile, in yet another instance of conflicting perceptions, Mr. Elton is angered by Emma's supposition that Harriet Smith could be worthy of serious attention on his part, and he earns Emma's ire by aspiring to *her* hand: an over-evaluation of his mate value, in her view, and a decided under-evaluation of hers.

Events correct Emma's willful misperception of Harriet Smith: when she realizes that she has inflated her friend's sense of her own value so preposterously that Harriet actually aspires to marry Mr. Knightly—a wealthy man whose high status and dominant personality render him the most important man in the community—Emma is shaken. At a stroke she recognizes her own "blindness," her "irrational" analysis of Harriet's prospects (p. 398). Indeed, the strength of her revulsion triggers important realizations about her own inclinations, namely, that "Mr. Knightley must marry no one but herself!" (p. 398). Her rapid re-assessment of Harriet's claims to an upper-class marriage coincides with the happy resolution of questions concerning her own domestic future. The lasting harm Emma might have visited upon her friend by infecting her with an exaggerated sense of self-worth (and concomitantly exaggerated expectations regarding her future husband) remains in the background of reader awareness, a dark undercurrent in a comedic plot: had Robert Brown not renewed his proposals, Harriet might have missed her best chance to be "safe, respectable, and happy for ever" (Austen 1816, p. 90).

In *Bleak House* (1853), Charles Dickens explores changes in real as well as perceived mate value, focusing on Esther Summerson. Esther is endowed with many excellent qualities of temperament and character, including, for example, industry, reliability, intelligence, generosity, and warmth; she is, in addition, young and strikingly beautiful. The principal obstacle to her marrying well is her illegitimate birth. As the ward of Mr. Jarndyce of Bleak House, however, she retains a socially respectable position. She emphatically refuses a proposal of marriage from Guppy, a law clerk from a lower station in life who is smitten with Esther's looks: clearly she regards his offer as unworthy of consideration. When Alan Woodcourt, a young physician with talent and social standing but not much money, begins a tentative courtship, both Esther and her guardian appear to think this a suitable match. Before the courtship can culminate in a proposal, however, her facial beauty is destroyed by a serious, smallpox-like illness.

Because of her "scarred face" Esther assumes that she now possesses "no attractions," that her mate value has been irretrievably lowered (pp. 463, 630). Guppy takes the trouble to explain that his earlier, rejected proposal has been definitively withdrawn: "circumstances" … put it out of my power ever to … renew it in any shape or form whatever," further evidence that she is correct in her pessimistic re-evaluation of her mating options (p. 412). Eventually, however, Guppy renews his offer of marriage, a gesture he characterizes as "magnanimous" but which clearly indicates that Esther's essential value remains high—or higher, at any rate, than Guppy's (p. 653). She is certain enough of this to refuse him again, as decisively as before. Mr. Jarndyce, Esther's guardian, also offers to marry her, and she interprets this proposal, too, as recognition of her decline in mate value. An admirer who refrained from proposing when her beauty was intact, Jarndyce now steps forward because he assumes that Esther's ability to attract more age-appropriate suitors is diminished: his value as a husband is now more nearly equivalent to her newly reduced value as a wife. Against her expectations, however, Woodcourt reappears to renew his courtship.

In defiance of credibility, readers learn that during the course of the first few years of her married life with Alan Woodcourt Esther's looks mysteriously return. This twist in the plot of her personal history shines a spotlight on the relationship between facial beauty and mate value. Dickens puts forward the cheering proposition that so long as a woman's reproductive capacity remains undamaged, perceptive men can dispense with smooth skin and luminous complexion, particularly when a woman possesses superior personal and social attributes. Once she regains her customary robust health, there is no reason to suppose that Esther's fertility has been impaired by her illness; presumably, too, she retains the genetic capacity to transmit her once great beauty to offspring (as she later, in fact, does). The sub-plot concerning her fluctuating male value—a tale of lost beauty mysteriously regained—effectively rebukes the narrow focus of typical male preference. Esther's regenerated loveliness might be interpreted as Woodcourt's reward for overlooking an accidental and superficial disfigurement; he manifests a rare ability to focus holistically on attributes of character, temperament, and vigor.

Individuals whose high mate value is widely recognized inevitably attract many suitors. Since there are not enough of these extremely desirable individuals to go around, competition plays a complicating role in courtship (Buss 2003, pp. 8–9). Sometimes suitors resort to physical altercation, as Lysander and Demetrius do in *A Midsummer Night's Dream* at the point when both are in love with Helena. Only supernatural interference from the fairy world prevents these two young men from injuring or killing one another. In *The Sun Also Rises*, Robert Cohen's jealous response to Brett Ashley's promiscuity likewise expresses itself in physical violence: he throws punches at all three of his male rivals, reserving his most punitive attack for her current paramour, the young bullfighter Pedro Romero, whom he hurts "most badly" (Hemingway 1926, p. 210). The two women in *A Midsummer Night's Dream*, who have been mutually friendly and supportive until Lysander suddenly rejects Hermia for Helena, also grow hostile in the face of increased competitive pressures. They engage in physical combat, which takes the form of a nonlethal but recklessly excited tussle. Additionally, they trade verbal abuse: Helena draws negative notice to Hermia's short stature, for example, and accuses her of having a quarrelsome disposition (Shakespeare 1595).

Such derogation of rivals is a tactic favored by both sexes (Buss 2003, pp. 97–98). Tom Buchanan prevents his wife from leaving him by characterizing her lover, Jay Gatsby, as a "common swindler" and "bootlegger," a "Mr. Nobody from Nowhere" (Fitzgerald 1925, pp. 132, 133). Competing for Darcy's attention in *Pride and Prejudice*, Miss Bingley makes snide comments to him concerning Elizabeth Bennet's personal appearance, social presence, and family connections, repeatedly noting that some of Elizabeth's relatives are *in trade*. Mrs. Bennet is similarly guilty of disparaging her daughters' rivals, attempting to excite admiration for her girls by pointing out that Charlotte Lucas is not nearly as good-looking (Austen 1813). Such acts of verbal aggression are intended to secure advantage, at the expense of others, in the competition for desirable mates.

Varying with auctorial purpose and tone, competition for mates may be limited to entertaining put-downs or escalate to life-threatening violence. Numerous Western writers have turned to the Trojan War to illustrate the large-scale destruction sexual rivalry can inspire (Gottschall 2008). When two powerful, high-status men both wish to possess an unusually desirable individual, *the most beautiful woman in the world*, nations are drawn into the fray; countless heroes fall, and a culture lies in ruins. With bitter brevity, Yeats summarizes the after-effects of this mating competition gone global: "the burning roof and tower/ And Agamemnon dead" ("Leda and the Swan" 1924, lines 10–11). He points out in his poem that this fierce battle for a mating opportunity is itself the result of a heedless short-term sexual encounter, when an alpha male (king of the Greco-Roman gods) forces himself upon a futilely resisting female. Despite his divine foreknowledge of the bloodshed to follow (i.e., the battle at Troy to possess Helen, the child thus conceived), the disguised god acts on masculine ardor. With "a shudder in the loins" he "engenders," simultaneously, a lovely child and a brutal war (line 9). Yeats thus strips the Trojan War to its essence, destruction on a vast scale originating in the aggressive urgency of proximal motives—in this instance, masculine ardor.

Literature illumines problems and probes complexities inherent in human mate selection and courtship behaviors. Often people have trouble judging the intentions, circumstances, and characters of potential partners. Frequently, too, they do not accurately perceive their own value, let alone the biases and preconceptions they bring to the selection process. People often choose long-term mates when they are young, so that inexperience accounts for many poor outcomes. Having located suitable mates, they may be required to fend off hostile competitors. Another complicating factor is that mate value can change over time due to fluctuations in wealth, reputation, or health—or the waning fertility of aging females (Buss 2003, pp. 184, 185). Such changes provide incentives to reconsider previous choices, leading to union-disrupting behaviors such as mate-switching and mate-poaching.

Section IV: Short-Term and Mixed Mating Strategies

Women and men tend to value similar personal qualities in long-term partners: intelligent, good-humored, honest, generous, agreeable, capable individuals likely command more long-term mating options than those who lack some or all of those qualities. Not all mating efforts foresee long-term involvement, however; individuals of both sexes sometimes take advantage of casual mating opportunities (Buss 2003, pp. 74–75). Short-term strategies and preferences may differ considerably from those prevailing in quests for long-term mates. For obvious reasons, male standards for short-term partners typically are lower than those by which they assess long-term partners. If a man does not plan to invest substantial, long-term resources in a woman or her offspring, he can afford to settle for less in the way of youth, beauty, personality, status, and reputation (Buss 2003, pp. 78–79). He may confine his efforts to a brief courtship and a few token gifts, deserting a pregnant woman to raise the child on her own. A man who engages in repeated short-term involvements may leave a considerable genetic legacy, having invested very little beyond his sperm.

Men sometimes pursue a mixed mating strategy: a man commits himself to a long-term partnership, investing in the offspring it yields, meanwhile exploiting short-term opportunities. He may abandon short-term partners and their children altogether or, alternatively, offer them limited support in terms of material resources, time, and attention. Over the course of a lifetime a man may change his mating strategy, devoting his youthful energies to short-term affairs before committing himself to marriage. Once married, he may remain faithful or he may at some point combine long-term commitment with short-term involvements.

Unlike men, women cannot increase the number of their offspring by increasing the numbers of their partners and matings: the biological limitations posed by gestation and lactation set upper limits to a woman's reproductive capacity. Women nevertheless employ short-term or mixed mating strategies for identifiable reasons, chief among them these: to gain access to otherwise unobtainable resources; to attach themselves to men who may become long-term partners; to achieve conception or, alternatively, to conceive genetically high-quality children with men unavailable to them as long-term mates (Buss 2003, pp. 86–88, 90–91, 121). When extraction of resources provides the chief motivation for a liaison, women may be relatively unconcerned with the personal qualities of short-term partners. In general, however, they assess short-term and long-term partners by similar standards. Since the possibility of converting a short-term involvement to long-term commitment contributes to the motivation for many affairs, and because pregnancy (intended or not) is a possible outcome, equivalent selection standards tend to prevail (Buss 2003, p. 88).

Literature reflects and explores these facts indirectly, often emphasizing the frequency with which men choose to pursue short-term strategies: the womanizing male, irresistibly charming, handsome, and persuasive, is a staple of drama and fiction. The Don Juan character memorably depicted by Mozart and Byron pursues women with no intention of making an enduring commitment, often luring naïve maidens to his bed with deceptive promises. Less exaggerated forms of this behavior abound. One commonly observed pattern is exemplified by Goethe's Faust: an upper-class man woos a lovely, unsophisticated girl from a much lower social stratum, sweeping her off her feet with the aura of refinement and prosperity he projects and cementing the seduction with dazzling gifts of jewelry (1808). In George Eliot's *Adam Bede*, this pattern repeats itself: Hetty Sorrell, the niece of a dairy farmer, is seduced by Arthur Donnithorne, the local squire's son, who lures Hetty to his bed with a combination of admiration, affection, and "pretty ear-rings" (1858, p. 242). Like Goethe's Gretchen, Hetty resorts to infanticide when she finds herself pregnant and abandoned. Thomas Hood strips this tragic outline to its essence in his popular poem, "The Bridge of Sighs," a schoolroom staple from the Victorian era through the mid-twentieth century. The lilting momentum of its insistently dactylic metre lends a slightly ironic, *here we go again* tone to Hood's narrative of "one more Unfortunate" girl ("fashion'd so slenderly/ Young and so fair!") exploited by a "dissolute Man" (1844, lines 1, 7–8, 77). Hood's unnamed "Unfortunate" puts an end to her miseries with a melodramatic "plunge[]" into the icy waters of a "black flowing river" (line 77).

For every story inviting sympathy on behalf of hoodwinked women whose reproductive resources have been co-opted by deceitful men, there are others examining short-term strategies from a perspective placing less emphasis on women as victims. Such stories indicate, directly or indirectly, the fitness benefits women may hope to obtain from short-term matings. Even girls like Gretchen or Hetty, whose inexperience makes them easy dupes, presumably believe they have something to gain from the premarital intimacies they permit. Typically, they hope that such intimacies will foster an enduring attachment or an advantageous rise in status; thus they employ a short-term strategy chiefly in the hope of obtaining more long-lasting benefits (Buss 2003, pp. 121, 216). The example of Julius Beaufort in *The Age of Innocence*, who marries his long-term mistress after the death of his wife, provides evidence that such conversion can and does occur (Wharton 1920). Optimistically, Hetty fantasizes that Arthur's love for her will grow and that she will be elevated to upper-crust prosperity: "she was to be a lady in silks and satins" (Eliot, p. 301). For a girl whose prospects in life are severely limited, even the small possibility of a gloriously hypergamous union may prove inducement enough to accept the risks of a short-term liaison.

Less naïve women, who recognize the limited nature of a short-term lover's involvement, still may have reasons to value the gifts and attentions of a rich, high-status man. Lawrence Lefferts in *The Age of Innocence* is presented by Wharton (1920) as a mixed strategist who carries on affairs with women of distinctly lower socioeconomic stations: a typist, a postmistress. He does not mean for these involvements to disturb his marital commitment, and he goes to considerable trouble to conceal his affairs from his wife. Readers do not observe his interactions with any of his temporary mistresses, but evidently his extra-pair partners are content with whatever Lefferts lavishes on them in the way of entertainment and gifts. There is no suggestion that he forces women into his bed or lures them with elaborate deceptions; rather, he seeks out those who will be sufficiently impressed by his wealth and social position to accept the distinctly limited investment he offers (1920). In *The Great Gatsby*, Myrtle Wilson demonstrates mixed motivation when she consents to become Tom Buchanan's mistress. She would like to convert their secret affair into marriage, and he encourages her, misleadingly, in the belief that this may one day occur. Meanwhile, however, she obtains enough benefits from the involvement to continue it as is: she enjoys spending Tom's money, furnishing the flat he has taken for their rendezvous, and playing the part of the rich, high-status woman she dreams of becoming (Fitzgerald 1925).

The childless Myrtle also may be motivated, consciously or unconsciously, by the desire to test her fertility with extramarital matings. A desire for successful conception constitutes yet another reason why women sometimes might find short-term affairs beneficial, particularly if they can locate a man whose genetic quality appears high or, in any event, better than other alternatives, including that of a husband (Buss 2003, p. 76). In "Roman Fever" (1936) and "The Old Maid" (1924), Edith Wharton explores the life histories of women who risk short-term involvement with the hope, certainly, of winning long-term commitment from their lovers, but who are satisfied to have obtained high-quality offspring from the affairs. They find ways to support the children born of these furtive encounters, which are conducted against the backdrop of prosperous families and communities (Saunders 2009).

The far less financially secure Louise Trunnion in Sherwood Anderson's story "Nobody Knows" makes overtures to George Willard, an attractive, up-and-coming young man in town. "Whispered tales" concerning Louise suggest that she is already committed to short-term sexual strategies, and she writes a note offering herself to George: "I'm yours if you want me" (1919, pp. 60, 61). For George, this is an unexpected mating opportunity; since the girl has initiated the encounter he does not even have to expend any effort in courtship. The omniscient narrator focuses selectively on George's motives and reactions during the encounter. Readers observe a male lowering his standards in the context of short-term mating, finding himself unaccountably attracted to qualities he normally would find off-putting: "just to touch the folds of the soiled gingham dress would … be an exquisite pleasure" (p. 60). George's concern with secrecy, his obsessive need to reassure himself that "nobody knows" he has slept with Louise, presumably reflects his determination to expend no resources on her and also, perhaps, to deny paternity if conception should occur.

Louise's motives require (and receive) no explicit discussion in Anderson's narrative. By signaling sexual availability she has succeeded in capturing the attention of a young man whose status puts him far out of her reach as a long-term mate; she has little to lose from employing a short-term strategy. It is possible, she may reason, that George will like her enough to continue seeking intimacies with her, perhaps providing her with temporary resources, and unlikely though this seems, gradually be drawn into more substantial commitment. If she should become pregnant, she can attempt to extract support from George; as a decent young man from a well-known Winesburg family, he probably could be coerced, however reluctantly, into assuming some paternal responsibility. Even in the absence of material help from George, Louise would have secured from him a genetic contribution of higher quality than any she otherwise might hope to obtain in the small local population. Her history demonstrates that women with few social and economic options frequently have the most to gain—and the least to risk—in short-term matings.

Section V: Fidelity and Infidelity

Consideration of short-term sexual strategies leads inevitably to the topic of fidelity: in many temporary liaisons, at least one partner already has made a long-term commitment to a third party. Given the large number of casual affairs littering the social landscape, instances of adultery occur regularly: literary texts represent and interpret such activity, assessing community context along with personal and familial reactions to extra-pair liaisons. Members of both sexes have powerful reasons to enforce fidelity in long-partners, although their reasons differ. Cuckolded husbands risk investing paternal care in children fathered by other men; betrayed wives risk losing important material resources, which may be redirected toward mistresses and children outside the primary union (Buss 2003, pp. 10–11). Mutual demands for fidelity function, furthermore, as a common source of conflict between the sexes, serving as a locus of strategic interference. Men's desire to reap the fitness benefits of a mixed mating strategy interferes with women's desire to lay exclusive claim to paternal resources. Men's desire to ensure paternal confidence and, concomitantly, to monopolize a wife's reproductive potential, interferes with women's desire to garner the benefits, genetic or material, available from short-term affairs.

Men whose wives bear children fathered by casual lovers suffer a double loss, first in terms of fitness and second in terms of misdirected resource investment. Since female infidelity poses a potentially catastrophic threat to a man's reproductive success, men employ a variety of tactics to prevent, detect, and punish sexual straying on the part of wives. Such tactics are motivated by jealousy, a psychological mechanism "our ancestors evolved … for solving the paternity problem" (Buss 2003, p. 126). The first line of defense against sexual defection is mate-guarding. Literature highlights intriguingly varied manifestations of this behavior, which historically has taken notoriously extreme forms, from veiling and claustration to foot-binding and gynecological surgery. In *The Egoist*, for example, Sir Willoughby Patterne's attempts to assert intellectual, spiritual, and moral ownership of his fiancée illumine the claustrophobic effects of mate-guarding tactics, even the subtle forms observed in highly refined social environments (Meredith 1879). In *Middlemarch*, Edward Casaubon's efforts to control his wife's sexual behavior from beyond the grave (by means of a will disinheriting her if she should wed the man Casaubon suspects she prefers) bear eerie witness to the tenacity of sexual possessiveness (Eliot 1871–72).

Often coupled with rage and aggression, male jealousy can lead to physical violence, directed against a cheating wife or her extra-pair partner, and to complicated schemes of revenge. Even the most dispassionate and mild-mannered man is apt to react with unaccustomed intensity to a wife's infidelity. The plot of *The Scarlet Letter* is driven in large part by the jealous machinations of Roger Chillingworth, a slight, stooped, scholarly man who devotes himself with unappeasable resentment to discovering the identity of his wife's secret lover. He then exacts a devious, psychologically torturous revenge, "which he analyzed and gloated over" (Hawthorne 1850, p. 1443). Chillingworth is sufficiently introspective to marvel at his own transformation: once "studious, thoughtful," "peaceful and innocent," he discovers that in yielding so completely to jealousy he has, in effect, become a different person.

Another highly civilized man who finds himself unexpectedly overcome by punitive anger—here directed toward the wife rather than her paramour—is Hewson Blair from Eliabeth Bowen's 1924 story "Making Arrangements." Like

Chillingworth, Hewson believes he is a wholly rational man, immune to emotional outbursts and vengeful thoughts. When his beautiful young wife not only leaves him but has the temerity to write and ask him to send on her clothing, he goes to her room with the intention of methodically packing the finery exactly as requested. As he handles the expensive gowns, all of which he has purchased for her, he begins to see them as different manifestations of his wife's kaleidoscopic personality: the dresses enable her to assume a dazzling variety of social selves. In leaving the husband whose Oxford education and prosperous business provide the resources necessary for her expensive social life, Margery has "committed suicide," he decides (1924, p. 177). The scene ends in a curious spectacle of sublimated violence. Spreading on the bed the dresses that have come to stand for his wife ("stretched out and provocative"), the betrayed husband tears them to pieces with his bare hands (p. 178). He then packs the costly shreds of velvet, satin, fur, and silk into his wife's trunk, addresses it, and arranges to have it sent to her. Readers are left to imagine her reaction as she is confronted with the loss of her wardrobe and, simultaneously, the evidence of her husband's murderous rage at her defection. Hewson Blair's behavior illustrates the universal susceptibility of men, no matter how genteel, logical, or even-tempered they may be, to jealousy and the behaviors it inspires. Faced with the loss of reproductive resources to extra-pair rivals, men remain capable of responses that proved effective in Paleolithic times. Hawthorne and Bowen remind readers, by means of compelling fictions, that evolved adaptations continue to operate in cultural conditions very different, technologically and socially, from those of the ancestral past.

Although they run no risk of misplaced maternal investment, women risk losing resources important to offspring survival if their long-term mates are unfaithful (Buss 2003, p. 10). The more tenuous a couple's economic situation, the more serious the consequences if a man's investment in his wife and her children is diluted by the needs of a mistress and out-of-wedlock offspring. Just as a cuckolded man often becomes an object of derision, an abandoned wife is apt to incur social costs. Women take measures, therefore, to prevent masculine infidelity, mounting spirited defenses against sexual intruders (Buss 2003, p. 9). Women resort less frequently to the physically violent interventions so often implemented by men, but there are notable exceptions. Medea, for example, acts decisively when her husband announces that he is "supplanting" her with a new wife. Unwilling to suffer "the loss of married love/ Which Jason has betrayed," she reviles him as a "breaker of oaths" and "deceiver" and arranges to have her rival murdered (Euripides 431 BC, lines 678, 970–971, 1367).

More typically, women favor less obviously aggressive tactics such as spying, snooping, and interrogation (Buss 2003, pp. 33–36). Alert to signals of wavering spousal commitment, they erect psychological barriers to cheating. Once May Welland from *The Age of Innocence* suspects that her husband has fallen in love with her cousin Ellen Olenska, she employs a variety of strategic maneuvers to rid herself of a dangerous, mate-poaching rival. She rightly intuits that this extramarital romance has begun to trigger mate-switching impulses in her husband, and she accordingly enlists help from her family of origin to stave off this threat. Her relatives exercise vigilance on May's behalf, keeping her informed about her husband's activities. She engineers a devious victory over her rival by telling her—and her husband—that she is pregnant: this premature announcement of an as yet unconfirmed reproductive event induces guilt and excites paternal commitment. Self-interest (i.e., the expectation of fitness benefits that are immediate and certain) supplies an unstated but powerful motive for May's husband to renounce his plans for deserting her (Wharton 1920).

In addition to detecting and discouraging unfaithfulness in their long-term partners, women must contend with uncompromising male demands for female fidelity. As the reproductive stakes are so high, men typically insist on exclusive sexual access to long-term mates. In addition to the preventative restriction of female activity and vengeful punishment of transgressions already described, men worldwide seek mates whose behavior signals sexual loyalty. Such signals vary from culture to culture but generally include premarital chastity and exaggerated sexual reserve (Buss 2003, p. 67). Demure or even prudish behavior around men, for example, usually proves reassuring and thus attractive to suitors; a willingness to interact easily or boldly with men socially, in contrast, may be interpreted as a sign of future sexual license. Literature focuses with great frequency on the male demand for chaste and sexually inhibited women as long-term mates, examining the costs women may incur in both fitness and reputation if their behavior suggests that they cannot be relied on to safeguard male reproductive interests. Even the victims of male seduction like Goethe's Gretchen, Eliot's Hetty, or Hardy's Tess, are punished by their communities for their "shameful" conduct (Eliot 1871–72, p. 324). "Unfortunate" though she is, the suicidal girl in Hood's poem is said to be guilty of "dishonor" and "impurity"; she, rather than her "dissolute" seducer, is charged with "evil behaviour" (1844, lines 24, 91, 103). Women engaging deliberately in strategic short-term matings, if publicly detected, are equally liable to be reviled and ostracized. Such reputational damage frequently extends to relatives, imperiling the fitness of entire families. When Lydia Bennet runs away with Wickham in the expectation that short-term intimacies will lead to marital commitment, onlookers predict that "this false step … will be injurious to the fortunes of all [her sisters]" (Austen 1813, p. 221). Stephen Crane's protagonist accepts her brother's friend as her lover in the hopes that marriage will follow, only to have her closest relatives cast her off and join the community outcry against her ("Maggie's gone the deh devil") because they fear reputational contamination: "dis t'ing queers us …. See? We're queered!" (1893, *Maggie and A Girl of the Streets*, pp. 32, 40).

The seduction of Little Em'ly by Steerforth, an important sub-plot in *David Copperfield*, demonstrates that greater blame nearly always attaches to the woman in short-term alliances. The exquisitely beautiful Emily yields to the blandishments of a handsome, gentlemanly young man who is her socioeconomic superior, not only because his courtship offers romantic excitement that her engagement to her fisherman cousin Ham cannot match, but also because she hopes to benefit her family by an elevation in status: "we would all be gentlefolks together" (p. 41). Even as a child she dreamed of being "a lady," able to give the uncle she adores a set of fancy clothes "with diamond buttons," a "large gold watch," and "a box of money" (p. 41). Denouncing herself as "wicked," she accepts short-term conditions because the possibility of an eventual marriage, however slim, would convert her lifelong fantasy of wealth and status into reality. She accepts in advance the cost of a negative outcome, vowing "never to come back, unless he brings me back a lady" (Dickens 1850, pp. 382, 383).

Aghast at Emily's defection, her betrayed fiancé immediately asks God to strike her dead: "I pray my good and gracious God to kill her … sooner than let her come to ruin and disgrace" (p. 382). In a moment of foreshadowing earlier in the novel, David himself has said something similar. Remembering a risky maneuver Emily executed as a child on a "jagged timber" above the sea, he asks himself, "would it have been better for little Em'ly to have had the waters close above her head that morning in my sight;" and finds that he sometimes has "answered Yes" (p. 42). Two different men, both loving and valuing Emily, nonetheless judge that she would be better off dead than as a woman guilty of sexual impropriety. Her fiancé's reaction may be explained in part as jealousy, but David has no personal claims on Emily's reproductive potential and he, too, finds her behavior irredeemably culpable.

Women who guard their chastity, and their reputations for chastity, are winners in the long-term mate-selection process. Richardson's Mr. B. weds the virtuous Pamela (a marriage that from his point of view is conspicuously hypogamous) because she successfully resists his seductions (Richardson 1740). Her story is an extreme version of an often observed phenomenon: men test women by pressing for intimacies, then penalize those who yield, making serious courtship only to the staunch few who pass the test by continuing their resistance in the face of powerful inducements (Wright 1994, pp. 72–74); such inducements typically take the form of provisions, gifts, assistance to family members, or deceptive assurances of commitment. Men frequently choose women exhibiting less reserved behavior as short-term partners, preferring for brief and uncommitted affairs behavior they would deplore (i.e., quick and easy sexual access) in potential long-term partners (Buss 2003, p. 79). Cultural norms and legal systems generally favor male reproductive interests in this respect, levying heavy penalties against sexually assertive women and disloyal wives. The painful histories of women whose infidelities are detected dominate numerous stories, plays, and poems, eliciting varying degrees of empathy.

John Milton's *Comus* celebrates female chastity, investing it with otherworldly powers. The premise of the plot is that a truly virtuous woman can ward off threats of seduction, or even rape, by sheer purity of intention. Thus a "Lady" wandering alone in a dark forest, eventually made captive by a lecherous male, is not as "defenseless" as she appears. The "sacred rays of Chastity" provide her with "a hidden strength," effectively armoring her against assault even from a male commanding a necromancer's wiles and weapons (1634, lines 414, 415, 425). Because she is "sincerely" dedicated to "the sublime notion and high mystery … of Virginity," the Lady's powers of resistance protect her like an invisible shield: she "is clad in complete steel" (lines 421, 453, 785, 787). It follows, such claims imply, that women who fall victim to the "base forgery" or "brew'd enchantments" of a "foul deceiver" are themselves at fault: their dedication to female purity proves insufficiently steadfast and "sincere" (lines 696, 698). Valorizing "chaste" and "virtuous" behavior (and validating its signaling efficacy), Milton's masque encourages women to internalize the sexual standards to which men wish them to adhere, to assume full responsibility for preserving those standards even against men's aggressive efforts to violate them. Compliance with guidelines thus internalized need not be interpreted as coercive (even by the women themselves), since no palpable external force is employed: the force of cultural norms is sufficient to ensure that male preferences in female behavior will prevail.

In the *Age of Innocence*, readers encounter a very different reaction to conventions shaping female purity. As her title suggest, Wharton examines the "innocence" required of marriageable girls in cultural–historical context. She laments the restrictions on women's education: young girls are carefully guarded from sexual experience and sexual knowledge of any kind, so that much of the world's great art is off limits to them. The narrowness of their horizons, intellectual as well as experiential, can serve only to stunt their development permanently: they remain "childish" and "incapable of growth" (1920, pp. 78, 274). Reared to project the abstract virtues Milton lauds ("whiteness, radiance, goodness"), they lack well-defined personal values and opinions (p. 31). They are less fully adult, less fully human, than they might have been without such exaggerated demands for sexual ignorance, and this wasted potential represents an unmitigated loss—not only to women themselves but to the men whose long-term partners they become. Men seeking "passionate and tender comradeship" discover that the girls in their society have been "carefully trained not to possess" the very qualities that would make such relationships possible, namely, "experience … versatility … freedom of judgment" (p. 46). Wharton presents a powerful argument against demands for "innocence," noting the hypocrisy of the prevailing double standard: many of the male characters engage in adulterous behavior, yet they suffer no significant social penalties. Her novel protests an unjust and harmful regulation of female sexual strategies to suit male preferences.

For literary purposes, as foregoing examples illustrate, the selection, acquisition, and retention of sexual partners provide unfailing sources of drama and suspense. People remain intrigued by the many forms desire, courtship, competition, jealousy, infidelity, and mate-poaching can assume. They cherish opportunities to project themselves into fictively conceived dilemmas, to rehearse behavioral options, to ponder culturally imposed limitations on strategic choices. Creating hypothetical situations remarkable for their diversity as well as their evocative intensity, literature exploits the "versatile repertoire" humans command for "solving the adaptive problems of mating" (Buss 2003, p. 209). Auctorial intention positions sexual behaviors in varied social environments, promoting consideration of adaptive choices from divergent angles and encouraging reader reactions ranging from sympathy to outrage. One author may encourage praise for the very motivations or tactics that another author sets up for ridicule or condemnation; behavior triggering cautionary concern in one work may present itself as noble rebelliousness in another. Such differences can be seen even in works stemming from similar historical periods and cultural contexts or even, at times, in works by the same author. Literature would not speak to readers as powerfully as it does if it were not grounded in a shared sense of a universal human nature, but it does not reflect evolutionary theory in a simplistic fashion. Enriched but never subsumed by interdisciplinary outreach, literary Darwinism draws upon theory and research originating in the biological and social sciences while remaining a humanistic endeavor.

References

Alcott, L. M. (1866). Behind a mask, or, a woman's power. In M. Stern (Ed.) *Behind a mask: The unknown thrillers of Louisa May Alcott* (pp. 1–104). New York: William Morrow (1975).
Anderson, S. (1919). Nobody knows. In J. H. Ferres (Ed.), *Winesburg, Ohio: Text and criticism* (pp. 58–62). New York: Viking (1966).
Austen, J. (1813). *Pride and prejudice*. Boston: Houghton Mifflin (1956).
Austen, J. (1814). *Mansfield park*. Baltimore: Penguin (1966).
Austen, J. (1816). *Emma*. London: Penguin (1987).
Bowen, E. (1924). Making arrangements. In *The Collected stories of Elizabeth Bowen* (pp. 170–199). New York: Knopf (1981).
Bowen, E. (1934). The needle case. In *The collected stories of Elizabeth Bowen* (pp. 453–460). New York: Knopf (1981).
Buss, D. M. (2003). *The evolution of desire: Strategies of human mating* (Revised ed.). New York: Basic Books.
Butler, S. (1903). *The way of all flesh*. London: Grant Richards.
Crane, S. (1893). Maggie: A girl of the streets. T. Gullason (Ed.). New York: W. W. Norton (1979).
Dickens, C. (1850). *David Copperfield*. New York: Airmont (1965).
Dickens, C. (1853). *Bleak house*. Boston: Houghton Mifflin (1956).
Eliot, G. (1858). *Adam Bede*. London/New York: Dent/Dutton (1960).
Eliot, G. (1871–72). *Middlemarch*. Cambridge: Houghton Mifflin (1956).
Euripides. (431 BC). Medea. In M. Mack et al. (Eds.), *The Norton anthology of world masterpieces* (Vol. 1, 5th ed., pp. 741–777). New York: Norton (1985).
Fitzgerald, F. S. (1925). *The great Gatsby*. New York: Scribner (1953).
Franklin, B. (1794). *The autobiography*. In J. A. Leo Lemay & P. M. Zall (Eds.), *Benjamin Franklin's autobiography: An authoritative text, backgrounds, criticism* (pp. 1–146). New York: Norton (1986).
Gilbert. W. S. (1885). *The Mikado*. New York: Dover (1992).
Gottschall, J. (2008). *The rape of Troy: Evolution, violence, and the world of Homer*. Cambridge: Cambridge University Press.
Goethe, J. W. (1808). *Faust* (Trans. W. Kaufmann). New York: Anchor Press (1962).
Hawthorne, N. (1850). *The scarlet letter*. In N. Baym et al. (Eds.), *The Norton anthology of American literature* (vol. B, pp. 1352–1493). New York: Norton (2007).
Hemingway, E. (1926). *The sun also rises*. New York: Scribner (2003).
Herrick, R. (1646). To the virgins, to make much of time. In J. Ferguson, M. J. Salter, & J. Stallworthy (Eds.), *The Norton anthology of poetry* (pp. 211–212).
Hood, T. (1844). The Bridge of Sighs. In O. Williams (Ed.) *F. T. Palgrave's The golden treasury of the best songs and lyrical poems* (pp. 191–192) New York: New American Library (Revised Edition), (1953).
Huxley, A. (1932). *Brave new world*. New York: HarperCollins (1989).
James, H. (1881). *The portrait of a lady*. Boston: Houghton Mifflin (1963).
Marvell, A. (1681). To his coy mistress. In M. Ferguson, M. J. Salter, & J. Stallworthy (Eds.), *The Norton anthology of poetry* (pp. 271–272). New York: Norton (1970).
Meredith, G. (1879). *The egoist: A comedy in narrative*. Boston: Houghton Mifflin (1958)
Milton, J. (1634). *A mask* (Comus). In M. Y. Hughes (Ed.), *John Milton: The complete poems and major prose* (pp. 86–114). New York: Odyssey (1957).
Parker, D. (1944). One perfect rose. In D. Parker (Ed.), *The portable Dorothy Parker* (p. 104). New York: Viking (1973).

Poe, E. A. (1846). The philosophy of composition. In Baym, N. et al. (Eds.), *The Norton anthology of American literature* (Vol. B, pp. 1617–1625). New York: Norton (2007).
Richardson, S. (1740). *Pamela, or virtue rewarded.* New York: Norton (1958).
Saunders, J. P. (2009). Reading Edith Wharton through a Darwinian lens: Evolutionary biological issues in her fiction. Jefferson: McFarland.
Shakespeare, W. (1595). A midsummer night's dream. In G. B. Harrison (Ed.), *Shakespeare: The complete works* (pp. 516–540). New York: Harcourt Brace & World (1952).
Shakespeare, W. (1602). The tragedy of Othello: The moor of Venice. In G. B. Harrison (Ed.), *Shakespeare: The complete works* (pp. 1059–1099). New York: Harcourt Brace & World (1952).
Spenser, E. The fairie queen. (1596). In L. Kirschbaum (Ed.), *Selected poetry* (pp. 7–580). New York: Holt, Rinehart and Winston (1961).
Thoreau, H. D. *Walden* (1855). In N. Baym et al. (Eds.), *The Norton anthology of American literature* (Vol. B, pp. 1872–2046). New York: Norton (2007).
Wharton, E. (1905). *The house of mirth.* New York: Charles Scribner's Sons.
Wharton, E. (1920). *The age of innocence.* In C. J. Singley (Ed.), *Complete text with introduction, historical contexts, critical essays* (pp. 13–285). Boston: Houghton Mifflin (2000).
Wharton, E. (1924). *The old maid (the fifties).* In *Old New York.* New York: Appleton.
Wharton, E. (1925). *The mother's recompense.* New York: D. Appleton.
Wharton, E. (1936). Roman fever. In *The World over* (pp. 215–239). New York: D. Appleton-Century.
Williams, T. (1945). *The glass menagerie.* New York: New Directions (1970).
Williams, T. (1947). *A streetcar named desire.* New York: New Directions (1980).
Wright, R. (1994). *The moral animal: Evolutionary psychology and everyday life.* New York: Vintage.
Yeats, W. B. (1924). Leda and the Swan. In M. Ferguson, M. J. Salter, & J. Stallworthy (Eds.), *The Norton anthology of poetry* (pp. 682). New York: Norton (2007).

Appendix A

Literary Darwinism: Selected Bibliography of Foundational and Theoretical Works

Boyd, B. (2005a). Evolutionary theories of art. In J. Gottschall & D. S. Wilson (Eds.), *The literary animal* (pp. 147–176).
Boyd, B. (2005b). Literature and evolution: A bio-cultural approach. *Philosophy and Literature, 29,* 1–23.
Boyd, B. (2008). The art of literature and the science of literature. *The American Scholar, 77*(2), 31–57.
Boyd, B. (2009). *The origin of stories: Evolution, cognition, and fiction.* Cambridge: Harvard University Press.
Boyd, B. (2012). Evolution and literary response. In C. Gansel & D. Vanderbeke (Eds.), *Telling stories* (pp. 64–76).
Boyd, B., Carroll, J., & Gottschall, J. (Eds.). (2010). *Evolution, literature, and film: A reader.* New York: Columbia University Press.
Carroll, J. (1995a). *Evolution and literary theory.* Columbia: University of Missouri Press.
Carroll, J. (1995b). Literature and evolutionary psychology. In D. M. Buss (Ed.), *The handbook of evolutionary psychology* (pp. 931–952). Hoboken: Wiley.
Carroll, J. (2004). *Literary Darwinism: Evolution, human nature, and literature.* New York: Routledge.

Carroll, J. (2012). Adaptive function of the arts: Alternative evolutionary hypotheses. In C. Gansel & D. Vanderbeke (Eds.), *Telling stories*. Geschichten erzählen: Literature and evolution/ Literatur und evolution (pp. 50–76).
Cooke, B. (1999). Biopoetics: The new synthesis. In B. Cooke & F. Turner (Eds.), *Biopoetics: Evolutionary explorations in the arts* (pp. 3–25). Lexington: ICUS.
Cooke, B. & Turner, F. (Eds.) (1999). *Biopoetics: Evolutionary explorations in the arts*. Lexington: ICUS.
Dissanayake, E. (1988). *What is art for*? Seattle: University of Washington Press.
Dissanayake, E. (1995). *Homo aestheticus: Where art comes from and why*. Seattle: University of Washington Press.
Dissanayake, E. (2000). *Art and intimacy: How the arts began.* Seattle: University of Washington Press.
Dutton, D. (2009). *The art instinct: Beauty, pleasure, and human evolution.* New York: Bloomsbury.
Eible, K. (2009). The induction instinct: The evolution and poetic application of a cognitive tool. *Studies in the Literary Imagination, 42*(2), 43–60.
Eible, K. (2012). How can evolutionary biology enrich the study of literature? In C. Gansel & D. Vanderbeke (Eds.), *Telling stories* (pp. 11–29).
Easterlin, N. (1999). Do cognitive predispositions predict or determine literary value judgements? In B. Cooke & F. Turner (Eds.), *Biopoetics: Evolutionary explorations in the arts* (pp. 241–282). Lexington: ICUS.
Easterlin, N. (2012). *A biocultural approach to literary theory and interpretation.* Baltimore: Johns Hopkins University Press.
Gansel, C. (2012) Storytelling from the perspective of evolutionary theory. In C. Gansel & D. Vanderbeke (Eds.), *Telling stories* (pp. 77–109).
Gansel, C. & Vanderbeke, D. (Eds.) (2012). *Telling stories/Geschichten erzählen: Literature and evolution/ Literatur und evolution.* Berlin and Boston: De Gruyter.
Gottschall, J. (2008). *Literature, science, and a new Humanities*. New York: Palgrave Macmillan.
Gottschall, J. (2011). *Reading human nature: Literary darwinism in theory and practice*. Albany: State University of New York Press.
Gottschall, J. (2012). *The storytelling animal: How stories make us human.* Boston: Houghton Mifflin Harcourt.
Gottschall, J., & Wilson, D. S. (Eds.). (2005). *The literary animal: Evolution and the nature of narrative*. Evanston: Northwestern University Press.
Mellman, K. (2012). Is storytelling a biological adaptation? Preliminary thoughts on how to pose that question. In C. Gansel & D. Vanderbeke (Eds.), *Telling stories* (pp. 30–49).
Miller, Geoffrey. (2000). Arts of seduction. In B. Boyd, J. Carroll, & J. Gottschall (Eds.), *Evolution, literature, and film* (pp. 116–173).
Nettle, D. (2005). The wheel of fire and the mating game: Explaining the origins of tragedy and comedy. *Journal of Cultural and Evolutionary Psychology, 3*(1), 39–56.
Nordlund, M. (2002). Consilient literary interpretation. *Philosophy and Literature*, 26, 312–333.
Pinker, S. (2010). Art and adaptation. In B. Boyd, J. Carroll, & J. Gottschall (Eds.), *Evolution, literature, and film* (pp. 125–134).
Storey, R. (1996). *Memesis and the human animal: On the biogenetic foundations of literary representation*. Evanston: Northwestern University Press.
Sugiyama, M. S. (1996). On the origins of narrative: Storyteller bias as a fitness-enhancing strategy. *Human Nature, 7,* 403–425.
Sugiyama, M. S. (2001a). Food, foragers, and folklore: The role of narrative in human subsistence. *Evolution and Human Behavior, 22,* 221–240.
Sugiyama, M. S. (2001b). Narrative theory and function: Why evolution matters. *Philosophy and Literature, 25,* 233–250.
Tooby, J. & Cosmides, L. (2001). Does beauty build adapted minds? Toward an evolutionary theory of aesthetics, fiction, and the arts. *SubStance, 30,* 6–27.

Vanderbeke, D. (2012). The unlovely little sister: Storytelling as an exaptive phenomenon. In C. Gansel & D. Vanderbeke (Eds.), *Telling stories* (pp. 110–120).
Vermeule, B. (2010). *Why do we care about literary characters*? Baltimore: Johns Hopkins University Press.
Wilson, E. O. (1998). *Consilience: The unity of knowledge*. New York: Knopf.
Zunshine, Lisa (2006). *Why we read fiction: Theory of mind and the novel*. Columbus: Ohio State University Press.

Appendix B

Literary Darwinism: Selected Bibliography of Works of Practical Criticism

Boyd, B. (2009). *On the origin of stories: Evolution, cognition, and fiction*. Cambridge: Harvard University Press.
Boyd, B., Carroll, J., & Gottschall, J. (Eds.). (2010). *Evolution, literature, and film: A reader*. New York: Columbia University Press.
Carroll, J. (2004a). Adaptationist criteria of literary value: Assessing Kurtén's *Dance of the tiger, Auel's The clan of the cave bear, and Golding's The inheritors*. In J. Carroll (Ed.), *Literary Darwinism* (pp. 163–185).
Carroll, J. (2004b). Human ature and literary meaning: A theoretical model illustrated with a critique of *Pride and prejudice*. In J. Carroll (Ed.), *Literary Darwinism* (pp. 187–216).
Carroll, J. (2004c). Human universals and literary meaning: Sociobiological critique of *Pride and prejudice, Vilette, O pioneers! Anna of the five towns, and Tess of the D'urbervilles*. In J. Carroll (Ed.), *Literary Darwinism* (pp. 129–145).
Carroll, J. (2004d). *Literary Darwinism: Evolution, human nature, and literature*. New York: Routledge.
Carroll, J. (2011a). Aestheticism, homoeroticism, and Christian guilt in *The picture of Dorian Gray*. In J. Carroll (Ed.), *Reading human nature* (pp. 91–108).
Carroll, J. (2011b). The Cuckoo's history: Human nature in *Wuthering heights*. In J. Carroll (Ed.), *Reading human nature* (pp. 109–122).
Carroll, J. (2011c). Intentional meaning in *Hamlet*: An evolutionary perspective. In J. Carroll (Ed.), *Reading human nature* (pp. 123–147).
Clasen, M. (2012a). Can't sleep: Clowns will eat me. In C. Gansel & D. Vanderbeke (Eds.), *Telling stories* (pp. 338–360).
Clasen, M. (2012b). Attention, predation, counterintuition: Why Dracula won't die. *Style, 46*(3–4), 378–398.
Cooke, B. (1999). Sexual property in Pushkin's 'The snowstorm': A Darwinian perspective. In B. Cooke & F. Turner (Eds.) *Biopoetics* (pp. 175–204).
Cooke, B. (2002). *Human nature in Utopia: Zamyatin's* We. Evanston: Northwestern University Press.
Cooke, B. (2012). Wasted reproductivity in Wagner's operas. In C. Gansel & D. Vanderbeke (Eds.), *Telling stories* (pp. 327–337).
Cooke, B. & Machann, C. (2012). Applied evolutionary criticism. *Style, 46*(3–4), 277–296.
Cooke, B. & Turner, F. (Eds.). (1999). *Biopoetics: Evolutionary explorations in the arts*. Lexington: ICUS.
Dolak, T. (2012). Homo oneginesis: Pushkin and evo-cognitive approaches to literature. *Style, 46*(3–4), 338–354.
Easterlin, N. (2000). Wordsworth, psychoanalysis, and the 'discipline of love.' *Philosophy and Literature, 24*(2), 261–279.

Easterlin, N. (2009). Who was it if it wasn't me? The problem of orientation in Alice Munro's 'Trespasses': A cognitive ecological analysis. *Studies in the Literary Imagination, 42*(2), 79–102.
Easterlin, N. (2012). Aesthetics and ideology in Felicia Hemans's The Forest Sanctuary. *Style, 46*(3–4), 461–478.
Flesh, W. (2010). Vindication and vindictiveness: *Oliver twist.* In B. Boyd, J. Carroll, & J. Gottschall (Eds.), *Evolution, literature, and film* (pp. 360–366).
Gansel, C., & Vanderbeke, D. (Eds.). (2012). *Telling stories/Geschichten erzählen: Literature and evolution/ Literatur und evolution.* Berlin/Boston: De Gruyter.
Gottschall, J. (2008). *The rape of Troy: Evolution, violence, and the world of Homer.* Cambridge: Cambridge University Press.
Hart, K. (2012). Strangers to ourselves: Animality and theory of mind in Honoré de Balzac's A passion in the desert (1830). *Style, 46*(3–4), 399–419.
Keener, Joe. (2010). Romeo and Juliet are animals. *Soundings: An Interdisciplinary Journal, XCIII* 93 (3–4), 341–357.
Keener, Joe. (2012). Evolving Hamlet: Brains, behavior, and the Bard. *Interdisciplinary Literary Studies*, 14 (2), 150–163.
Keller W. (2012). Evolutionary art: Charles Dickens's *Bleak house* and the Ovidian metamorphosis. In C. Gansel & D. Vanderbeke (Eds.), *Telling stories* (pp. 310–326).
Kruger, D., Fischer, M., & Jobling, I. (2003). Proper hero dads and dark hero cads: Alternative mating strategies in British and Romantic literature. *Human Nature, 14,* 305–317.
Müller-Wood, A. (2012a). Shakespeare and the (evolved) pleasures of revenge. In C. Gansel & D. Vanderbeke (Eds.), *Telling stories* (pp. 279–295).
Müller-Wood, A. (2012b). No ideology without psychology: The emotional effects of Shakespeare's *Henry V. Style, 46*(3–4), 355–377.
Nordlund, M. (2007). *Shakespeare and the nature of love: Literature, culture, evolution.* Evanston: Northwestern University Press.
Salmon, C., & Symons, D. (2004). Slash fiction and human mating psychology. *The Journal of Sex Research, 41*(1), 94–200.
Saunders, J. P. (2007). Male reproductive strategies in Sherwood Anderson's 'The untold lie.' *Philosophy and Literature, 31,* 311–322.
Saunders, J. P. (2009). *Reading Edith Wharton through a Darwinian lens: Evolutionary biological issues in her fiction.* Jefferson: McFarland.
Saunders, J. P. (2010a). Paternal confidence in Hurston's 'The gilded six-bits.' In B. Boyd., J. Carroll, & J. Gottschall (Eds.), *Evolution, literature, and film* (pp. 392–408).
Saunders, J. P. (2010b). Franklin's Autobiography: The story of a successful social animal. *Politics and Culture*, 1–16. Web.
Saunders, J. P. (2012a). Hawthorne's theory of mind: An evolutionary psychological approach to 'The minister's black veil.' *Style, 46*(3–4), 420–438.
Saunders, J. P. (2012b). Female mate-guarding in D. H. Lawrence's 'wintry peacock': An evolutionary perspective. *College Literature, 39*(4), 69–83.
Saunders, J. P. (2012c). Nepotism in Hawthorne's 'My Kinsman, Major Molineux.' In C. Gansel & D. Vanderbeke (Eds.), *Telling stories* (pp. 296–309).
Sugiyama, M. S. (2001). New science, old myth: An evolutionary critique of the Oedipal paradigm. *Mosaic, 34,* 121–136.
Sugiyama, M. S. (2003). Cultural variation is part of human nature: Literary universals, Context-sensitivity, and 'Shakespeare in the bush.' *Human Nature* 14, 383–396.
Turpin, J. & R. W. Fuhreman (2012). Adaptive and maladaptive poetry in Plath, Roethke, Kunitz, and Moraga. *Style, 46*(3–4), 479–499.
Vanderbeke, D. (2012). *Fanny Hill*, or: What's so exciting about mossy mounts? *Style, 46*(3–4), 439–460.
Zunshine, L. (2007). Why Jane Austen was different, and why we may need cognitive science to see it. *Style, 41,* 277–299.

Chapter 3
Assortative Mating, Class, and Caste

Henry Harpending and Gregory Cochran

Assortative Mating, Class, and Caste

Assortative mating in humans has interested human geneticists for decades. It is widely found and it is difficult to make sense out of it. For example, tall people are likely to have tall mates, short people are likely to have short mates, people with more years of education have more highly educated spouses, and so on. These phenomena are usually not treated as selective processes but they can and likely do cause the equivalent of very strong natural and diversifying selection, leading to persistent inherited group differences.

Selection in the Old Order Amish

The Amish are an Anabaptist sect whose members immigrated to Pennsylvania in colonial times, mostly between 1727 and 1770 (Kraybill 2001; Hostetler 1980). Their way of life is unusual; they refrain from using many forms of modern technology, and they generally try to separate themselves from modern society. Some of their customs have biological significance, that we will discuss further. We think that analogous customs—endogamous marriage, in particular—may have occurred often in human history.

The Amish marry within their faith. Although they accept converts, there are a very few, so there is almost no inward gene flow. They descend almost entirely from about 200 eighteenth century founders. On the other hand, there is considerable outward gene flow, since a significant fraction of Amish youth do not choose to adopt the Amish way of life. In recent years, approximately 10–15 % of young Amish left the community. In the past, the defection rate seems to have been higher,

H. Harpending (✉) · G. Cochran
Department of Anthropology, University of Utah, Salt Lake City, UT 84112, USA
e-mail: harpend@gmail.com

T. K. Shackelford, R. D. Hansen (eds.), *The Evolution of Sexuality*,
Evolutionary Psychology, DOI 10.1007/978-3-319-09384-0_3

around 18–24%. Defection is up to the individual—there are no barriers against Amish who want to participate in modern society.

Since the Amish have very high birth rates (>six children per family), their numbers have increased very rapidly, even though there is a substantial defection rate. There were about 5000 descendants of the original 200 by 1920, and today (2013) there are about 280,000 Amish.

Every way of life selects for *something*, but the Amish way of life is so different that natural selection in that population should be noticeably different from that in the general US population. It seems likely that the Amish have undergone selection for two specific traits due to their unusual social and reproductive pattern. First, they were almost certainly selected for higher fertility. A recent study (Milot et al. 2011) found evidence of this kind of selection in preindustrial French Canadians, who, like the Amish, went through a very rapid population expansion. Second, and more interestingly, the Amish have probably experienced selection for increased "Amishness"—an increase in the degree to which Amish find their lifestyle congenial, since those who like it least, leave. We term this kind of differential emigration "boiling-off." Obviously, if some of the soup boils off, what is left is more concentrated.

This boiling off is essentially truncation selection. If we assume a normal distribution, the loss of the least plain 10% corresponds to the loss of everyone more than 1.25 standard deviations (SD) below the Amish mean. If we assume a narrow-sense heritability of 0.3 and use a scale similar to that for IQ, the Amish gain about one point of plainness per generation. Not counting possible selection for this kind of personality in Europe, before they settled in North America, the Amish have spent about 10 generations under this kind of selection. Therefore their "plainness," their "Amish Quotient" (AQ), might have increased by about 0.6 SD. During most of the period for which we have sufficient information, the defection rate was significantly higher than 10%, so this may be a conservative estimate. Although there are certainly other factors that might influence the defection rate, such as increasing differences between the Amish way of life and that of their neighbors, increasing plainness would tend to reduce the defection rate over time.

The Amish have some genetic problems because of genetic drift in a small population, and these problems have received a fair amount of attention from medical geneticists. However, their social pattern probably drives strong selection for a particular constellation of personality traits, which is fascinating and worthy of further investigation. One could, with difficulty and significant investment, identify personality traits associated with a hypothetical AQ. These traits would likely include affinity for work, perseverance, low status competition, respect for authority, conscientiousness, and community orientation. We proposed (Cochran et al. 2006) a similar mechanism to account for Ashkenazi Jewish evolution in Medieval times selecting for ability and success in white collar occupations.

Recall that there is no selection at all in this model; AQ is selectively neutral. Seen from within the group, this is very strong selection. Mean population fitness does not increase, but mean AQ of Amish increases rapidly. The increase in retention, that is the decline in rate of boiling off, is very likely due to this mechanism of selection as the Amish population's average affinity for being Amish increases.

Some Background

For much of the twentieth century, the study of heredity has had two faces reflecting the old conflict between the biometricians and the Mendelians early in the century. The biometricians, associated especially with the Victorian genius and polymath Francis Galton, sought to understand heredity and its workings by collecting a lot of data and measuring everything that could be measured, or so it seems. The Mendelians based their approaches on the newly-discovered laws of inheritance of Mendel. The root of the conflict was whether the path to understanding heredity was to be through careful analysis of quantitative traits, regarding Mendelian mechanisms as side issues, or through elaborated Mendelian analyses, regarding quantitative traits as side issues.

Darwin was the grand inspiration behind this outbreak of biology, but Darwin's own theory of evolution by natural selection was not faring well at the end of the nineteenth and the early twentieth century. Evidence from selection experiments often did not look good for Darwin's theory. For example, Pearl and Surface (1909) conducted selection experiments on egg production by chickens. They found that there was no cumulative effect of selection. They did not realize at the time that chickens had been selected for many generations for egg production and there was little or no genetic variation in the trait left in the chickens.

The divergence in biology was resolved with a single paper by R. A. Fisher (1918) about "the correlation between relatives under the supposition of Mendelian inheritance." The paper is somewhat dense so it does not enjoy the acclaim that it deserves. What Fisher did was to reconcile approaches of the biometricians and the Mendelians, ending at once the disagreements between them. This paper was, in the view of many, the real beginning of evolutionary biology as a coherent discipline.

Fisher proposed a model of an inherited trait whose phenotype was determined by the additive effects of a very large number of Mendelian loci, as well as an effect of the environment. Because of the additivity, the phenotype should have a normal distribution with variance determined by the variance of genic effects across all the loci plus the variance of environmental effects. What Fisher was able to predict was the extent of similarities between relatives. The model worked, it described the world that we observe, and it is widely applicable.

In the years since, the old divide between Mendelian genetics, on the one hand, and Galton and biometrics and quantitative genetics, on the other, has persisted in many ways. Much of academic genetics remains firmly Mendelian in the sense that it focuses on molecular biology, molecular genetics, and biochemistry. The current triumph of this approach is the Human Genome Project and our ability to sequence thousands of complete human genomes at reasonable cost. Human genetics has been concerned, obsessed perhaps, with finding Mendelian roots of medical disorders and diseases. In the trade, this is called the search for "genesfor,"—that is, genes for diabetes or schizophrenia or Obsessive Compulsive Disorder or bad breath. It has occasionally worked out: for example, human skin color is mostly determined by six or so Mendelian loci. On the other hand, it has completely failed in other domains, including fingerprint ridge counts or intelligence or stature, which otherwise appear to follow closely the Fisher (1918) model. An excellent introduction to and survey of this divide and its history can be found in Razib Khan's (2013) recent blog post.

In this chapter, we explore some simple implications of quantitative genetic transmission of important traits or characteristics within a population with assortative mating. The model is quite general and the implications are startling, at least to us.

A Model of Assortative Mating

We begin by describing a simple model of assortative mating for a quantitative trait. The model is quite general. It may be useful for understanding social phenomena such as change in income inequality over time, social and economic class stratification, assortative mating by education, and the persistence of ethnicity.

We assume that there are two classes in a population called the right class and the left class. These are not necessarily of equal size, but the size of each class, and the size of the total population, is constant over time. There is no selection in this model: everything is selectively neutral.

An important parameter in quantitative genetics is the heritability of a trait, the fraction of the population variance of a trait that is attributable to genetic differences. This idea does not fit well into our model since we are not studying the change in a trait in the population, but the growth of differences in the population. Heritability proceeds from a model in which an individual's genes determine his underlying or genotypic value while the actual value, or phenotype, is the sum of the genotypic value and a random perturbation that is, unfortunately, called "environment." The perturbation reflects many effects, including developmental errors, dominance, gene interactions and interactions among loci, as well as effects conventionally associated with environment, such as nutrition, culturally-transmitted traits, or disease. Low heritability is often taken, among journalists for example, to imply sensitivity to environmental manipulation, but it implies no such thing.

Unfortunately for our model of assortative mating, the accounting is very difficult, so we study instead an analogue of heritability that we call h, which is the fraction of matings that are assortative. A fraction $1\text{-}h$ are random matings within classes. If h is 0, then there is no assortative mating, while h of 1 implies maximum assortative mating by group. Our results are the outcome of computation rather than of analysis (Mathematica 2012). We have made no progress toward deriving explicit algebraic descriptions of the demographic process that we study.

We begin with the simplest case of two classes that mate assortatively for a heritable trait. Initially, there is a single random mating population in which mating suddenly becomes assortative. We will show our results graphically with a scale of mean of 0 and standard deviation of 1, describing the original distribution of the trait under global random mating. If we were studying stature, the mean would be 70 in. and the standard deviation 2.5 in. Another familiar scale is IQ, with a mean of

100 and standard deviation of 15. We find it helpful to think of these scales when we consider how large the differences that evolve are in familiar and plausible terms.

Figure. 3.1 illustrates the process we are modeling. The top panel in both columns shows the initial distribution of the trait, normal with mean of 0 and standard deviation of 1. Speaking of stature for illustration, we think of that mean as at 70 in. rather than at 0 standard units and the standard deviation scaled to 2.5 in. rather than 1 standard unit. Strict assorative mating then occurs. The rule is that everyone shorter than the average moves to the left class and everyone taller than the average moves to the right class. This separates the population into short people in the left distribution and tall people in the right distribution, as shown in the middle panel. Mating occurs within class, and the shuffling that occurs leads to normal distributions *within classes* in the offspring generation, as shown in the bottom panel of Fig. 3.1.

Much of quantitative genetics depends on random mating, but it is not clear what random mating is in this context. It might mean random within class or random over the whole population. The latter will attenuate the effects of assortment: this is shown in the right column of Fig. 3.1. In this column, 5 % of the population mates at random from both classes.

The new mating rule has split our single population into two classes. After only a single generation, the classes differ notably. The average stature in the short class is now 68 in. and in the tall class it is 72 in.. The difference is 4 in., 1.6 SD on the original scale. With 5 % random mating imposed on each generation, the difference is reduced to 3.75 in.—that is, the effect of the random mating is very slight. Over time it becomes important. Without the random gene exchange, the two classes continue to separate over time, albeit very slowly, on their way to becoming genetic castes. With some random mating between classes the process of separation stops so that over time there is not much change beyond what has occurred in one generation.

We could not quite instantly look at someone and identify their class, since there is overlap as shown in the double-shaded area in the bottom panel of Fig. 3.1. After a few more generations of the process, we will be close to being able to glance at someone and identify their class, absent a large amount of random mating between classes. How do the class differences in genes here compare to differences among major human continental groups? Lewontin (1972) investigated diversity in a collection of genetic markers from human populations around the world and estimated that the fraction of variance that was between-populations was about 15 % of overall human diversity, such that 85 % of the variance was within populations. His original estimate has held up well for 40 years as the standard data set has gone from a small handful of loci to over a million. However, at the end of his otherwise fine and pioneering paper, Lewontin asserted that this 15/85 ratio implies that human differences must be small and of no significance. Although widely cited, this claim is unsupported by Lewontin's results, and unfortunately gives the impression of being politically motivated.

We can compare this global estimate of group difference to group difference in genes influencing stature between the short people and the tall people in our model. The separation between the group means is about 1.6 graph units, corresponding to

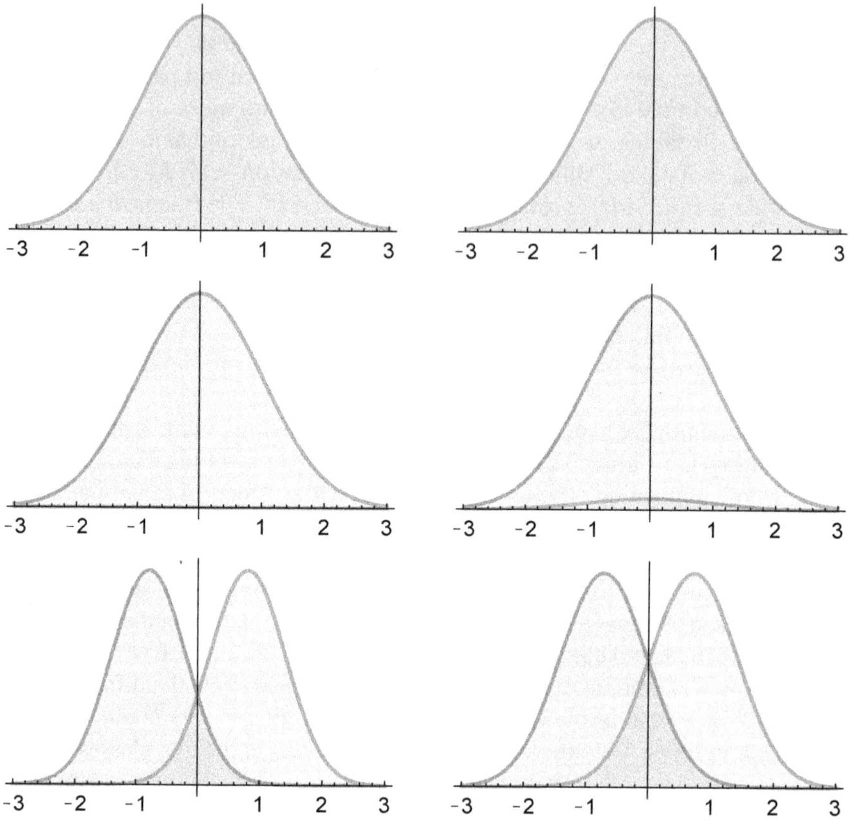

Fig. 3.1 The onset of assortative mating according to class. In the top panels, both classes share the same unit normal distribution of the trait. In the *second row*, everyone with a value greater than 0 has moved to the right class and everyone with a value less than 0 has moved to the left class. Random mating follows and the distribution within each class becomes normal in the *third row*. The *left column* shows the process with no other mating between classes. The *right column* shows the effects of 5 % random mating between classes. See text for elaboration

4 in. The standard deviation within each group is about 0.6 on the original scale, or 1.5 in.. These numbers correspond to a ratio of 0.64 for between-group variance relative to total variance. One generation of assortative mating caused the difference between short people and tall people at loci influencing stature to be four times greater than the differences between human continental groups.

The assumption of harsh and immediate imposition of the new mating rule is not realistic. On the other hand, something not so far from that apparently happened after World War II, with the sudden availability of college education and massive college enrollment booms for both men and women. Suddenly both young men and women from Dubuque went off to the state university and met and married each other there. The boys and girls next door who stayed behind ended up marrying each other and, consequently, there was a very rapid assortment for education.

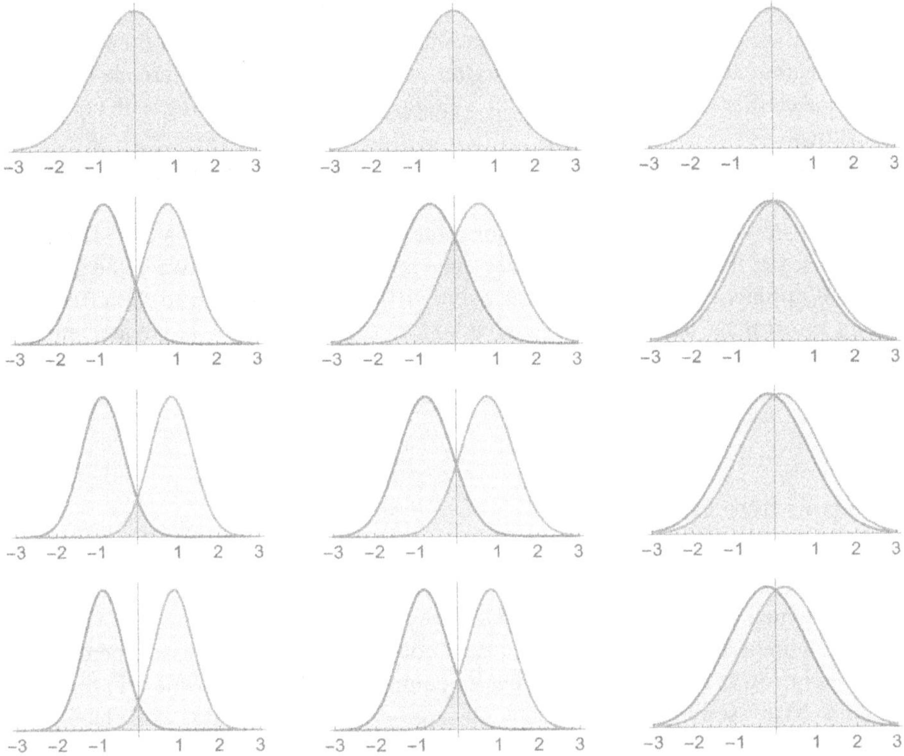

Fig. 3.2 These columns show the accumulating effects of assortative mating over three generations. The *left column* shows changes over time with maximum possible assortment ($h=1$), the *middle column* when half of matings are assortative and the other half random within class ($h=0.5$), and the *right column* when only 10% of matings are assortative and the rest random ($h=0.1$)

Varying Strength of Assortment

The example in Fig. 3.1 illustrated the initial changes when all matings were positively assortative by group in a single generation. The left column of Fig. 3.2 tracks ongoing changes over three generations (corresponding to 75 human years) with the same mating rule. After the first generation, the difference between group means continues to increase at a slower rate. The "raw material" for ongoing change is the area of overlap between the left and right distributions. This area decreases over time as the separation of the left and right distributions grows. After three generations, the difference between group means has only grown to 1.75 SD on the original scale.

The middle column of Fig. 3.2 models the same process, except that only half the matings are assortative rather than all of them. This provides an analogue of a trait that is moderately heritable. Studies of the heritability of traits like religiosity or aggressiveness usually estimate heritabilities in the neighborhood of 0.50. Differences accumulate more slowly so that after two generations the difference between groups is similar to the difference after a single generation of mating with complete assortment.

The third column shows the process when only 10 % of matings are assortative by group and the rest random within groups. The group difference continues to grow, but slowly. After three generations, the difference in means is 0.4 SD on the original scale; think of a one-inch difference in average stature between the groups, or six IQ points. These are not impressive differences compared to the effects of strong assortment, but they are not trivial. How much of income or socioeconomic group differences in stature are consequences of assortative mating by stature? A mere one-inch is not great, but it is surely noticeable.

Unequal Size Groups

Many situations in which we might expect the assortment mechanism to operate are those in which group sizes differ considerably. Figure 3.3 shows the consequences of three generations of assortment, in each column, for complete assortment in the left column, 50 % assortment in the center column, and 10 % assortment in the right column. Many social concerns focus on large differences in class size: labor and management, poor and rich, etc.

With complete assortment, the right class quickly and prominently separates from the left class, but without much apparent consequence for the left class. Here the mean of the left class decreases by 0.2 SD while the smaller right class's mean has increased by 2 SD. On the stature scale, the left class is one-half inch shorter on average than it was before the assortment while the right class is fully 5 in. taller, on average. On the IQ scale, there is a three-point decrease in the left class, on average, and a 30-point increase in the right class, on average.

The recipe for instant elite creation requires strong assortment. If assortment is 10 % rather than 100 %, shown in the right column of Fig. 3.3, differences accumulate very slowly. After three generations of low assortment, shown in the bottom right panel of Fig. 3.3, the left class has lost 0.05 SD, on average, or between an eighth and a sixteenth of an inch on the stature scale and less than one IQ point on the IQ scale. The right class mean has gained one-half standard deviation, on average, or three quarters of an inch on the stature scale and seven points on the IQ scale.

Applications?

Patterns like these generated by assortative mating would warm the hearts of Victorian eugenicists and might inspire governments not in the mold of modern liberal democracies. An implication is that a totalitarian regime could impose mating rules

Fig. 3.3 In this model, the left group in each column is 10 times the size of the right group. The volumes of the left and right distributions reflect the difference in group size. The *left*, *middle*, and *right columns* show results for h equal to 1.0, 0.5, and 0.1, respectively

and create an instant (i.e., in a quarter to a half century) elite class to populate its universities and professions, with little social disruption and damage to anyone. When assortment is strong, most of the effects happen within a generation or two, at the beginning of the process, so this mechanism is an attractive target for social improvers, just as it was a century ago before eugenics went out of fashion in favor of the compelling new fad of social science.

Richard Herrnstein (1973), late professor of psychology at Harvard University, was concerned for much of his career with the process we present in this chapter, wary that society was fragmenting with the creation of a meritocracy that would be a hereditary meritocracy. With Charles Murray, he published *The Bell Curve* (Herrnstein and Murray 1996) in 1995, a data-intensive exploration of the social and personal correlates of intelligence. A central focus of the book was the creation of a meritocracy in America. Unfortunately, they included a chapter on race differences that touched the American nerve and the message of the book was obscured by the cacophony from the politically correct. More recently Murray (2013) published a kind of ethnographic exploration of the lives of the new right and left classes in America, titled *Coming Apart.* He does not include much explicit discussion of the quantitative genetics of what he describes, but he would certainly find such data congenial to his arguments.

The President of the USA recently declared that income inequality is the "defining challenge of our time," referring to the large and growing income and wealth gap between the wealthy and the rest in the USA. There are also many references in the press to the vanishing middle class, a phenomenon which is expected with strong assortment, as we have discussed in this chapter.

Economists have a lively interest in income comparisons between parents and their children. There are serious issues with data quality and availability for empirical study of these issues, however. For example, one can find data on father's and son's incomes, but the son may be aged 30 and the father aged 60, so researchers must rely on cumbersome corrections and adjustments. A classic paper (Zimmerman 1992) from 1992 investigated transition rates between income quartiles, a convenient trick since quartiles are by definition the same size. The general pattern showed that retention rates in the bottom and top quartiles—the probability that a son is in the same quartile as the father—were about 40 % for the top and bottom quartiles and about 30 % for the second and third quartiles. Several years later, a comparison of mobility between the USA and Sweden (Björklund and Jäntti 1997) gave compatible results for the USA and showed that retention rates of the wealthy and poor groups were somewhat lower in Sweden, implying greater income mobility in Sweden. This is to be expected since the level of income variation in Nordic countries is much less than that in the USA.

An important study of trends over time in educational assortative mating in the USA was published by Schwartz and Mare (Schwartz and Mare 2005) in 2005 using census data. The researchers generated tables of matings between couples by complete education of the partners. Overall attainment changed over time; in 1940, most marriages were between people with 10 or fewer years of schooling, whereas in 2000 the most common marriage was between two college graduates. In a blog post (Harpending and Cochran 2013), we used a simple technique to estimate underlying "attraction" to remove the effect of changing average overall population levels, estimating in effect which marriages would occur if all educational levels were equally frequent in the overall population. The results were remarkable: the tables were essentially identical for all periods from 1944 to 2000 in spite of the great increase in population average educational attainment.

Data on educational attainment offer an excellent platform for investigating our proposed mechanisms in detail. Unfortunately, the data provided by Schwartz and Mare are too coarse for our purposes; the categories of educational attainment established in the 1940s are less useful today when a huge fraction of the population graduates from college. A check on our model would require the ability to drill into educational attainment to investigate differences between Liberal Arts and STEM graduates, for example, adding another category or two at the top.

References

Björklund, A., & Jäntti, M. (1997). Intergenerational income mobility in Sweden compared to the United States. *The American Economic Review, 87*(5), 1009–1018.

Cochran, G., Hardy, J., & Harpending, H. (2006). Natural history of Ashkenazi intelligence. *Journal of Biosocial Science, 38*(5), 659–693.

Fisher, R. A. (1918). The correlation between relatives on the supposition of Mendelian inheritance. *Proceedings of the Royal Society of Edinburgh, 52*(2), 399–433.

Harpending, H., & Cochran, G. (2013). Raking a table. http://westhunt.wordpress.com/2013/08/04/raking-a-table/. Acessed 15 Sep 2013.

Herrnstein, R. J. (1973). *I.Q. in the meritocracy* (1st ed.). New York: Little, Brown.

Herrnstein, R. J., & Murray, C. (1996). *The bell curve: Intelligence and class structure in American life (A Free Press Paperbacks book)* (1st Free Press pbk.). New York: Free Press.

Hostetler, J. A. (1980). *Amish society* (3rd ed.). Baltimore: Johns Hopkins University Press.

Khan, R. (2013). Genetics is one: Mendelism and quantitative traits—The Unz review. http://www.unz.com/gnxp/genetics-is-one-mendelism-and-quantitative-traits/. Acessed 15 Sep 2013.

Kraybill, D. B. (2001). *The riddle of Amish culture*. Baltimore: Johns Hopkins University Press. (Rev. ed.).

Lewontin, R. C. (1972). The apportionment of human diversity. *Evolutionary Biology, 6*(38), 381–398.

Mathematica. (2012). Mathematica. Version 9, Wolfram Research.

Milot, E., Mayer, F. M., Nussey, D. H., Boisvert, M., Pelletier, F., & Réale, D. (2011). Evidence for evolution in response to natural selection in a contemporary human population. *Proceedings of the National Academy of Sciences, 108*(41), 17040–17045.

Murray, C. (2013). *Coming apart: The state of white America, 1960–2010* (Reprint ed.). Crown Forum.

Pearl, R., & Surface, F. M. (1909). Is there a cumulative effect of selection? *Molecular Genetics and Genomics, 2*(1), 257–275.

Schwartz, C. R., & Mare, R. D. (2005). Trends in educational assortative marriage from 1940 to 2003. *Journal of the American Statistical Association, 42*(4), 621–646.

Zimmerman, D. J. (1992). Regression toward mediocrity in economic stature. *The American Economic Review, 82,* 409–429.

Chapter 4
Do Men and Women Perform Oral Sex as Mate Retention Behavior?

Yael Sela, Michael N. Pham and Todd K. Shackelford

Do Men and Women Perform Oral Sex as Mate Retention Behavior?

Over human evolution, individuals whose partner was unfaithful or who defected from the relationship experienced reproductive costs (Buss 1988; Buss and Shackelford 1997). Both men and women have had to solve the adaptive problem of retaining a mate. However, men and women face different costs as a consequence of their partner's infidelity. Men whose partner was sexually unfaithful risked investing in offspring sired by another male and diversion of their partner's maternal investment away from his offspring and towards another male's offspring. Men whose partner defected from the relationship altogether suffered multiple costs, including the loss of reproductive resources, a potential rival's gain of those resources, and replacement costs incurred to attract an alternative mate (Buss 1988; Buss and Shackelford 1997).

Women whose partner was unfaithful risked losing the paternal investment of their partner by the way of diverting those resources away from her offspring and towards a potential rival female's offspring. Women whose partner defected from the relationship altogether suffered multiple costs, including loss of a valuable provider and protector, a potential rival's gain of those benefits, a reduction of their "mate value" (e.g., not being virgins, having children from other men) and, therefore, high replacement costs incurred to attract an alternative mate (Buss 1988; Buss and Shackelford 1997)

Because of these selection pressures, men and women may have evolved psychological mechanisms that motivated the performance of "mate retention" behaviors (Buss 1988; Buss and Shackelford 1997). These behaviors decrease the likelihood of a partner's temporary or permanent relationship defection by influencing them to perceive alternative options as unattractive or unattainable. Further, these mate retention behaviors can change the perceptions that potential rivals have of one's

Y. Sela (✉) · M. N. Pham · T. K. Shackelford
Department of Psychology, Oakland University, 130 Pryale Hall, Rochester 48309, UK
e-mail: ysela@oakland.edu

T. K. Shackelford, R. D. Hansen (eds.), *The Evolution of Sexuality*,
Evolutionary Psychology, DOI 10.1007/978-3-319-09384-0_4

own mate, dissuade potential rivals through displays of resources and ornamentation, and can involve direct threats to potential rivals (Buss 1988; Buss and Shackelford 1997).

Buss (1988) identified 104 acts clustered into 19 mate retention "tactics" ranging from subtle to overt (see Table 4.1). Buss organized these tactics into five "categories": direct guarding, intersexual negative inducements, intrasexual negative inducements, positive inducements, and public signals of possession. Miner, Starratt, and Shackelford (2009) grouped the five categories into two superordinate "domains": cost-inflicting mate retention behaviors and benefit-provisioning mate retention behaviors (see Table 4.1). Behaviors in the cost-inflicting domain reduce the risk of partner infidelity by lowering a partner's self-esteem, thereby causing that partner to feel unworthy of the current relationship, but especially of any other potential relationship (Miner et al. 2009). In contrast, behaviors in the benefit-provisioning domain reduce the risk of partner infidelity by increasing a partner's relationship satisfaction (Miner et al. 2009).

Sex differences in mate retention behaviors have been documented for some tactics but not others. For example, men more frequently display their resources and threaten potential rivals, whereas women more frequently enhance their appearance and punish a mate's infidelity threat. Men and women are equally likely to be vigilant about a partner's whereabouts and to attempt to monopolize their partner's time (Buss 1988; Buss and Shackelford 1997).

Oral Sex Behaviors

Oral sex is a common sexual activity (e.g., Chambers 2007; Santilla et al. 2008; Vannier and O'Sullivan 2012) that is linked with sexual (Brody and Costa 2009) and relationship satisfaction (Ashdown et al. 2011; cf. Brody and Costa 2009). Men are equally likely to report their female partner performing oral sex on them (i.e., *fellatio*) and performing oral sex on their female partner (i.e., *cunnilingus*) at their most recent sexual encounters as women are to report performing fellatio and receiving cunnilingus (Vannier and O'Sullivan 2012; cf. de Visser et al. 2003). Most men and women report experiencing oral sex at least once in their life (de Visser et al. 2003), and both desire to experience oral sex (Santilla et al. 2008). Santilla and colleagues (2008) documented that men (relative to women) desired oral sex more often, and that men's (but not women's) relationship satisfaction was positively correlated with their actual frequency of experiencing oral sex (although the researchers did not distinguish between fellatio and cunnilingus). Notably, men's likelihood of orgasm at last copulation is unaffected by the addition of fellatio, but women are more likely to experience an orgasm when receiving cunnilingus than if they do not receive cunnilingus (Richters et al. 2006).

Previous research has documented a sex difference between the risk of a partner's infidelity and oral sex behaviors such that men whose partner is more attractive, a cue to partner infidelity (in men: Li and Kenrick 2006; and in women:

Table 4.1. Mate retention domains, categories, tactics, and sample items

Mate retention domain	Mate retention category	Mate retention tactic	Sample item
Cost-inflicting behaviors	Direct guarding	Vigilance	*Had my friends check up on my partner*
		Concealment of mate	*Refused to introduce my partner to my same-sex friends*
		Monopolization of time	*Would not let my partner go out without me*
	Intersexual negative inducements	Jealousy induction	*Flirted with someone in front of my partner*
		Punish mate's infidelity threat	*Became angry when my partner flirted too much*
		Emotional manipulation	*Threatened to harm myself if my partner ever left me*
		Commitment manipulation	*Asked my partner to marry me*
		Derogation of competitors	*Cut down the appearance of other men/ women*
	Intersexual negative inducements	Derogation of mate	*Told others my partner was "a pain"*
		Intrasexual threats	*Confronted someone who had made a pass at my partner*
		Violence against rivals	*Got my friends to beat up someone who was interested in my partner*
Benefit-provisioning behaviors	Positive inducements	Resource display	*Spent a lot of money on my partner*
		Sexual inducements	*Gave in to sexual pressure to keep my partner*
		Appearance enhancements	*Made sure that I looked nice for my partner*
		Love and care	*Told my partner "I love you."*
		Submission and debasement	*Gave in to my partner's every wish*
	Public signals of possession	Verbal possession signals	*Introduced my partner as my spouse or romantic partner*
		Physical possession signals	*Sat next to my partner when others were around*
		Possessive ornamentation	*Asked my partner to wear my ring*

Goetz et al. 2005; Schmitt and Buss 2001), reported greater interest in, and spent more time, performing oral sex on her (Pham and Shackelford 2013b). However, women whose partner was more attractive did *not* report more interest in, or spend more time, performing oral sex on their partner (Pham et al. 2013a). In addition, the relationships between partner infidelity risk and interest in, and time spent, performing oral sex were greater for men than women (Pham et al. 2013a).

Oral Sex and Mate Retention

Cunnilingus and Mate Retention in Men

Men may perform cunnilingus as a means of mate retention. Men at greater risk of a partner's infidelity report greater interest in, and spend more time, performing cunnilingus on their partner (Pham and Shackelford 2013b). In contrast, men do not typically perform cunnilingus on a woman during a casual, sexual encounter (i.e., "a one night stand"; Armstrong et al. 2009; Backstrom et al. 2012; Lewis et al. 2012; Reiber and Garcia 2010), a mating context that presents no risk of long-term partner infidelity. Pham and Shackelford (2013a) showed that men who report performing more mate retention behaviors, in general, also report greater interest in, and spend more time, performing cunnilingus on their partner.

Oral sex may be a benefit-provisioning mate retention behavior. Miner et al. (2009) documented that men at greater risk of partner infidelity performed more benefit-provisioning mate retention behaviors, but *not* more cost-inflicting mate retention behaviors. Women who receive cunnilingus from their partner, relative to those who do not, report greater relationship satisfaction (Kaestle and Halpern 2007; Santtila et al. 2008). Because greater partner relationship satisfaction is an outcome of benefit-provisioning mate retention, Pham and Shackelford (2013a) hypothesized and documented that men perform cunnilingus on their partner as part of a benefit-provisioning mate retention strategy. Specifically, men who report performing more benefit-provisioning mate retention behaviors, but *not* more cost-inflicting mate retention behaviors, also report greater interest in, and spend more time, performing cunnilingus on their partner. Further, men who report performing more cost-inflicting mate retention behaviors also report less interest in performing cunnilingus on their partner, and this is consistent with research documenting a negative correlation between the frequency with which men perform benefit-provisioning behaviors and cost-inflicting behaviors (Miner et al. 2009).

Fellatio and Mate Retention in Women

Women may perform fellatio on their partner as a means of mate retention. Sela et al. (2014) documented that women perform oral sex on their partner as a mate retention behavior. Specifically, women who report performing more benefit-provisioning mate retention behaviors also report greater interest in, and spend more time, performing oral sex on their partner. However, women's interest in, and time spent, performing oral sex was not related to *overall* mate retention behaviors. Women at greater risk of partner infidelity do *not* report greater interest in, or spend more time, performing oral sex on their partner (Pham et al. 2013a). Women, like men, are more likely to engage in oral sex within a committed relationship (vs. casual sexual encounter; Vannier and O'Sullivan 2012), a mating context that would elicit mate retention behavior.

Sex Similarities and Differences in Mate Retention Behaviors and Oral Sex

Overall Mate Retention and Domains of Mate Retention

Sela et al. (2014) found that men's *overall mate retention* behaviors are positively correlated with their interest in, and time spent, performing oral sex on their partner, but this relationship was nonsignificant in women. However, there were no sex differences in the magnitudes of these relationships. Both men's and women's benefit-provisioning mate retention behaviors, but *not* cost-inflicting mate retention behaviors, are positively related to their interest in, and time spent, performing oral sex on their partner, and there is no sex difference in the magnitudes of these relationships (Pham and Shackelford 2013a; Sela et al. 2014; see Tables 4.2 and 4.3). However, when comparing the unique effects of benefit-provisioning and cost-inflicting mate retention behaviors on men's and women's interest in, and time spent, performing oral sex, these effects are larger for men than women. There are no sex differences in the relationships between cost-inflicting mate retention behaviors and oral sex behaviors.

Men who perform more cost-inflicting mate retention behaviors report less interest in performing oral sex on their partner. This finding is consistent with the previous research documenting that the frequency with which men perform benefit-provisioning behaviors is correlated negatively with their cost-inflicting behaviors (Miner et al. 2009). Men who provision their partner with benefits must expend resources (e.g., "I bought my partner an expensive gift"). In contrast, men who inflict costs on their partner expend fewer resources, but the costs men inflict on their partner may lower her relationship satisfaction and cause her to terminate the relationship. Therefore, men who have the resources to provision their partner with benefits also tend to avoid the risks associated with inflicting costs on her.

Unlike men, women's cost-inflicting mate retention behaviors are not correlated with their oral sex behaviors. However, there are no sex differences in the magnitudes of these relationships. If oral sex functions to increase one's partner's relationship satisfaction, then these sex similarities are consistent with the previous research documenting that a spouse's cost-inflicting mate retention tactics (e.g., monopolization of time, jealousy induction, punishing or threatening to punishing infidelity, emotional manipulation) are negatively correlated with both husband's and wife's marital satisfaction (Shackelford and Buss 2000).

Table 4.2. Direction of zero-order correlations documented between target variables

Mate retention tactics	Oral sex variables			
	Interest in performing oral sex		Duration of oral sex	
	Women (fellatio)	Men (cunnilingus)	Women (fellatio)	Men (cunnilingus)
Vigilance	–	–	–	–
Concealment of mate	–	–	–	–
Monopolize mate's time	–	–	–	–
Threaten infidelity	–	–	–	–
Punish mate's threat to infidelity	–	–	–	–
Emotional manipulation	(+)	–	–	–
Commitment manipulation	–	(+)	–	(+)
Derogation of competitors	–	–	–	–
Resource display	–	(+)	–	(+)
Sexual inducements	–	(+)	–	(+)
Enhance physical appearance	–	–	–	(+)
Expressions of love and caring	(+)	(+)	(+)	–
Submission and debasement	–	–	–	–
Verbal signals of possession	(+)	(+)	(+)	(+)
Physical signals of possession	(+)	(+)	(+)	(+)
Possessive ornamentation	–	(+)	–	–
Derogation of mate to competitors	–	–	–	–
Intrasexual threats	–	(+)	–	(+)
Violence	–	–	–	–

Women's data from Sela et al. (2014). Men's data from Pham and Shackelford (2013a)
None of the relationships' magnitudes differed between men and women (Sela et al. 2014)
"–"=no significant correlation. "(+)"=positive correlation

Table 4.3. Direction of zero-order correlations documented between target variables

	Oral sex variables			
	Interest in performing oral sex		Duration of oral sex	
	Women (fellatio)	Men (cunnilingus)	Women (fellatio)	Men (cunnilingus)
Mate retention categories				
Direct guarding	–	–	–	–
Intersexual negative inducements	–	–	–	–
Intrasexual negative inducements	–	–	–	–
Positive inducements	–	(+)	(+)	(+)
Public signals of possession	(+)	(+)	(+)	(+)
Mate retention domains				
Benefit provisioning	(+)	(+)	(+)	(+)
Cost inflicting	–	–	–	–
Overall mate retention behaviors	–	(+)	–	(+)

Women's data from Sela et al. (2014). Men's data from Pham and Shackelford (2013a)
None of the relationships' magnitudes differed between men and women (Sela et al. 2014)
"–"=no significant correlation. "(+)"=positive correlation

Mate Retention Categories and Tactics

Men who report performing more mate retention with the categories of positive inducements and public signals of possession also reported greater interest in, and spent more time, performing oral sex on their partner (Pham and Shackelford 2013a). Sela et al. (2014) documented the same relationships for women (except interest in performing oral sex and positive inducement, which were unrelated), but there are no sex differences in the magnitudes of these relationships (see Tables 4.2 and 4.3).

Men who report greater *interest* in performing oral sex on their partner also report greater use of the mate retention tactics of intrasexual threats, resource display, sexual inducements, commitment manipulation, verbal and physical signals of possession, possessive ornamentation, and expressions of love and caring. Men who report spending more *time* performing oral sex on their partner also report greater use of the mate retention tactics of intrasexual threats, enhance physical appearance, commitment manipulation, resource display, sexual inducements, and verbal and physical signals of possession (Pham and Shackelford 2013a).

Women who report greater *interest* in performing oral sex on their partner also report greater use of the mate retention tactics of emotional manipulation, expressions of love and caring, and verbal and physical signals of possession. Women who report spending more *time* performing oral sex on their partner also report greater use of the mate retention tactics of expressions of love and caring, and verbal and physical signals of possession (Sela et al. 2014). However, there are no sex differences in the magnitudes of these relationships (Sela et al. 2014). Women may perform oral sex as a benefit-provisioning behavior, specifically through expressions of love and care, as documented by Sela et al. (2014). This is consistent with the research indicating that women in long-term, committed relationships who report that they love their partner "a lot" are also more likely to report performing fellatio on him (Kaestle and Halpern 2007). It is also consistent with the research indicating that women's second-most endorsed reason for engaging in oral sex (after a physical motive of pleasure and desirability) is an emotional motive of love and commitment, followed by an insecurity motive (which includes effort to retain their partner; Vannier and O'Sullivan 2012).

Limitations and Future Directions

A limitation of the two main studies (Pham and Shackelford 2013a; Sela et al. 2014) is exclusive reliance on self-reported mate retention behaviors. People may underreport the frequency with which they perform socially undesirable behaviors (e.g., "I told others of my same sex that my partner might have a sexually transmitted disease"). However, Shackelford et al. (2005) documented that both men's and women's self-reports of their mate retention behaviors are positively correlated with their partner's reports of these behaviors. Nevertheless, future research may benefit from securing data from both partners' reports of mate retention behaviors (their own, and their partners').

The two main studies rely on correlational data and, therefore, cannot offer clear statements about causality. Although we speculate that performing oral sex increases one's partner's relationship satisfaction and reduces their risk of infidelity, we cannot rule out the possibility that the direction of this relationship may work in reverse, or in both directions. Another limitation is that that both studies explain just a small amount of variance in individuals' reports of their mate retention behaviors and oral sex behaviors. Although these studies have documented relationships between these behaviors in both men and women, they offer only a modest first step towards identifying a function for oral sex, and only within a committed relationship.

The results of the two main studies might be explicable, in part, as a consequence of individual differences. Mate retention behaviors and oral sex behaviors, independently, are associated with individual differences such as personality (e.g., mate retention and oral sex; Holden et al. 2014), sociosexuality (mate retention; Kardum et al. 2006), and religiosity (e.g., oral sex; Mahoney 1980). The relationships between mate retention and oral sex behaviors may be further explained by personality traits. Individuals who are more altruistic and agreeable, for example, might be more likely to provision their partner with benefits, including sexually pleasuring their partner by performing oral sex. Indeed, Pham et al. (in press) have documented that men's benefit provisioning mate retention behaviors mediate the relationship between their agreeableness and their oral sex behaviors. Future research might profitably investigate the relevance of these and related individual differences on men's and women's oral sex behaviors and mate retention behaviors. Future research may test this explanation by assessing and statistically controlling the standings on personality traits that predict altruism and other individual difference correlates of partner-directed benefit-provisioning behaviors.

The Mate Retention Inventory (Buss 1988) assesses the frequency with which individuals perform various mate retention behaviors within the past month. Both studies asked about individual's oral sex behaviors during their most recent copulation to ensure that they remembered the details of, and therefore reported accurately, their oral sex behaviors. Future research investigating the relationship between mate retention behaviors and oral sex behaviors may consider securing reports of oral sex behaviors across multiple copulations within the past month, to ensure that measures of mate retention and oral sex assess behaviors that occur during the same time span.

A limitation of the literature on oral sex is reliance on problematic survey terminology, including the use of the term "oral sex" (e.g., Ashdown et al. 2011). Studies that assess oral sex with items such as "the number of partners you had oral sex with in the past year" (e.g., Prinstein et al. 2003) do not convey whether the respondent is the one performing oral sex, or oral sex is being performed on them. Future research must distinguish between fellatio and cunnilingus in heterosexual couples.

Other Hypothesized Functions of Oral Sex

Much evidence suggests a human evolutionary history of oral sex. Oral sex occurs in many cultures (e.g., Guadamuz et al. 2010; Iwawaki and Wilson 1983; Lurie et al. 1995; Leong 2012), including some preindustrial cultures (Hewlett and

Hewlett 2010), and is even depicted in ancestral cave paintings (Angulo and García 2005), indicating that oral sex is not a culture-specific practice.

Several functional hypotheses of oral sex have been suggested: infidelity detection, sperm retention, fertility detection, sexual arousal, sexual satiation, and mate retention (reviewed in Pham & Shackelford, in press). The infidelity detection hypothesis of oral sex has been supported for cunnilingus (Pham and Shackelford 2013b) but not fellatio (Pham et al. 2013a); men, but not women, at greater risk of their partner's infidelity report greater interest in, and spend more time, performing oral sex on their partner. The sperm retention hypothesis of oral sex has received mixed support. Men at greater recurrent risk of sperm competition are more likely to perform cunnilingus on their partner until she achieves orgasm; however, ejaculation usually does not occur in the brief time period during which orgasm might function to retain sperm (Pham et al. 2013c). The fertility detection hypothesis for cunnilingus has not been tested directly (to our knowledge), but there is preliminary support for the idea that men may detect a female's fertility status through scent cues. Men rate female genital odors that are produced during high fertility (relative to low-fertility) as more pleasant smelling (Cerda-Molina et al. 2013; Doty et al. 1975).

The sexual arousal hypothesis of oral sex has been supported for cunnilingus; men who spend more time performing cunnilingus on their regular partner also spend more time copulating with her, perform more semen-displacing copulatory behaviors, and report greater sexual arousal (e.g., more forceful ejaculation, greater orgasm intensity) (Pham et al. 2013b). Consistent with these findings, Cerda-Molina et al. (2013) found that men who smell vaginal odors produced at high fertility (relative to low fertility) also experience a surge in testosterone and report greater copulatory interest. The sexual arousal hypothesis of fellatio has not been tested, to our knowledge. The sexual satiation hypothesis of oral sex is generally supported because the occurrence of oral sex is positively associated with both men's and women's relationship satisfaction (e.g., Ashdown et al. 2011; Kaestle and Halpern 2007; Santtila et al. 2008). Finally, this chapter has reviewed recent evidence that supports the mate retention hypothesis of oral sex.

The proposed functions of oral sex are not necessarily mutually exclusive. Research on the function of oral sex is preliminary, and several directions exist for future research, as we have mentioned. The functional hypotheses for oral sex have all been informed by an evolutionary framework, and sperm competition theory, in particular. We suggest that researchers employ these approaches to guide future research on oral sex.

Conclusions

Men and women perform a diverse array of behaviors designed over evolutionary history to minimize the likelihood of their partner's infidelity. Both men and women may appease, threaten, conceal, or emotionally manipulate their partner to dissuade them from committing infidelity (Buss 1988). Together, the results from Pham and

Shackelford (2013a) and Sela et al. (2014) reviewed in this chapter suggest that both men and women perform oral sex on their partner to provision their partner with benefits, and thereby reduce the likelihood of their partner's infidelity. Men and women who perform more benefit-provisioning mate retention behaviors also report greater *interest* in, and more *time spent*, performing oral sex on their partner. These studies provide support for the hypothesis that oral sex is part of a broader benefit-provisioning mate retention strategy in men and women.

Acknowledgments This chapter is based on Pham and Shackelford (2013a), and on Sela, Y., Shackelford, T. K., Pham, M. N., & Euler, H. A. (2014).

References

Angulo, J., & García, M. (2005). Sex in stone: Sexuality, reproduction, and eroticism in the Paleolithic epoch. Madrid: Luzán, 5.

Armstrong, E. A., England, P., Fogarty, A. C. K., & Risman, B. J. (2009). *Families as they really are*. New York: Norton.

Ashdown, B. K., Hackathorn, J., & Clark, E. M. (2011). In and out of the bedroom: Sexual satisfaction in the marital relationship. *Journal of Integrated Social Sciences, 2,* 38–55.

Backstrom, L., Armstrong, E. A., & Puentes, J. (2012). Women's negotiation of cunnilingus in college hookups and relationships. *Journal of Sex Research, 49,* 1–12.

Brody, S., & Costa, R. M. (2009). Satisfaction (sexual, life, relationship, and mental health) is associated directly with penile-vaginal intercourse, but inversely with other sexual behavior frequencies. *The Journal of Sexual Medicine, 6,* 1947–1954.

Buss, D. M. (1988). From vigilance to violence: Tactics of mate retention in American undergraduates. *Ethology and Sociobiology, 9,* 291–317.

Buss, D. M., & Shackelford, T. K. (1997). From vigilance to violence. Mate retention tactics in married couples. *Journal of Personality and Social Psychology, 72,* 346–361.

Cerda-Molina, A. L., Hernández-López, L., de la O, C. E., Chavira-Ramírez, R., & Mondragón-Ceballos, R. (2013). Changes in men's salivary testosterone and cortisol levels, and in sexual desire after smelling female axillary and vulvar scents. *Frontiers in Endocrinology, 4,* 1–9.

Chambers, W. C. (2007). Oral sex: Varied behaviors and perceptions in a college population. *Journal of Sex Research, 44,* 28–42.

de Visser, R. O., Smith, A., Rissel, C. E., Richters, J., & Grulich, A. E. (2003). Sex in Australia: Heterosexual experience and recent heterosexual encounters among a representative sample of adults. *Australian and New Zealand Journal of Public Health, 27,* 146–154.

Doty, R. L., Ford, M., Preti, G., & Huggins, G. R. (1975). Changes in the intensity and pleasantness of human vaginal odors during the menstrual cycle. *Science, 190,* 1316–1318.

Goetz, A. T., Shackelford, T. K., Weekes-Shackelford, V. A., Euler, H. A., Hoier, S., Schmitt, D. P., & LaMunyon, C. W. (2005). Mate retention, semen displacement, and human sperm competition: A preliminary investigation of tactics to prevent and correct female infidelity. *Personality and Individual Differences, 38,* 749–763.

Guadamuz, T. E., Kunawararak, P., Beyrer, C., Pumpaisanchai, J., Wei, C., & Celentano, D. D. (2010). HIV prevalence, sexual and behavioral correlates among Shan, Hill tribe, and Thai male sex workers in Northern Thailand. *AIDS Care, 22,* 597–605.

Hewlett, B. S., & Hewlett, B. L. (2010). Sex and searching for children among Aka foragers and Ngandu farmers of central Africa. *African Study Monographs, 31,* 107–125.

Holden, C.J., Zeigler-Hill, V., Pham, M.P., & Shackelford, T.K. (2014). Personality features and mate retention strategies: Honesty-Humility and the willingness to manipulate, deceive, and exploit romantic partners. *Personality and Individual Differences, 57,* 31–36.

Iwawaki, S., & Wilson, G. D. (1983). Sex fantasies in Japan. *Personality and Individual Differences, 4,* 543–545.

Kaestle, C. E., & Halpern, C. T. (2007). What's love got to do with it? Sexual behaviors of opposite-sex couples through emerging adulthood. *Perspectives on Sexual and Reproductive Health, 39,* 134–140.

Kardum, I., Hudek-Knežević, J., & Gračanin, A. (2006). Sociosexuality and mate retention in romantic couples. *Psychological Topics, 15,* 277–296.

Leong, L. W. T. (2012). Asian sexuality or Singapore exceptionalism? *Liverpool Law Review, 33,* 11–26.

Lewis, M. A., Granato, H., Blayney, J. A., Lostutter, T. W., & Kilmer, J. R. (2012). Predictors of hooking up sexual behaviors and emotional reactions among U.S. college students. *Archives of Sexual Behavior, 41,* 1219–1229.

Li, N. P., & Kenrick, D. T. (2006). Sex similarities and differences in preferences for short-term mates: What, whether, and why. *Journal of Personality and Social Psychology, 90,* 468–489.

Lurie, P., Eugenia, M., Fernandes, L., & Hughes, V. (1995). Socioeconomic status and risk of HIV-1, syphilis and hepatitis B infection among sex workers in São Paulo State, Brazil. *AIDS, 9,* 31–37.

Mahoney, E. R. (1980). Religiosity and sexual behavior among heterosexual college students. *Journal of Sex Research, 16,* 97–113.

Miner, E. J., Starratt, V. G., & Shackelford, T. K. (2009). It's not all about her: Men's mate value and mate retention. *Personality and Individual Differences, 47,* 214–218.

Pham, M. N., & Shackelford, T. K. (2013a). Oral sex as mate retention behavior. *Personality and Individual Differences, 55,* 185–188.

Pham, M. N., & Shackelford, T. K. (2013b). Oral sex as infidelity-detection. *Personality and Individual Differences, 54,* 792–795.

Pham, M. N., Shackelford, T. K., & Sela, Y. (2013a). Women's oral sex behaviors and risk of partner infidelity. *Personality and Individual Differences, 55,* 446–449.

Pham, M.N., Shackelford, T.K., Welling, L.L.M., Ehrkel, A.D., Sela, Y., & Goetz, A.T. (2013b). Oral sex, semen displacement, and sexual arousal: Testing the ejaculate adjustment hypothesis. *Evolutionary Psychology, 11,* 1130–1139.

Pham, M. N., Shackelford, T. K., Sela, Y., & Welling, L. L. M. (2013c). Is cunnilingus-assisted orgasm a male sperm-retention strategy? *Evolutionary Psychology, 11,* 405–414.

Prinstein, M. J., Meade, C. S., & Cohen, G. L. (2003). Adolescent oral sex, peer popularity, and perceptions of best friends' sexual behavior. *Journal of pediatric psychology, 28,* 243–249.

Reiber, C., & Garcia, J. R. (2010). Hooking up: Gender differences, evolution, and pluralistic ignorance. *Evolutionary Psychology, 8,* 390–404.

Richters, J., de Visser, R., Rissel, C., & Smith, A. (2006). Sexual practices at last heterosexual encounter and occurrence of orgasm in a national survey. *Journal of Sex Research, 43,* 217–226.

Santtila, P., Wager, I., Katarina, W., Harlaar, N., Jern, P., Johansson, A., et al. (2008). Discrepancies between sexual desire and sexual activity: Gender differences and associations with relationship satisfaction. *Journal of Sex and Marital Therapy, 34,* 29–42.

Schmitt, D. P., & Buss, D. M. (2001). Human mate poaching: Tactics and temptations for infiltrating existing mateships. *Journal of Personality and Social Psychology, 80,* 894–917.

Sela, Y., Shackelford, T. K., Pham, M. N., & Euler, H. A. (2014). *Do women perform fellatio as a mate retention behavior?* Manuscript under editorial review.

Shackelford, T. K., & Buss, D. M. (2000). Marital satisfaction and spousal cost-infliction. *Personality and Individual Differences, 28,* 917–928.

Shackelford, T. K., Goetz, A. T., & Buss, D. M. (2005). Mate retention in marriage: Further evidence of the reliability of the Mate Retention Inventory. *Personality and Individual Differences, 39,* 415–425.

Vannier, S. A., & O'Sullivan, L. F. (2012). Who gives and who gets: Why, when, and with whom young people engage in oral sex. *Journal of Youth and Adolescence, 41,* 572–582.

Chapter 5
The French Connection: Sex and the English Revolution

Laura Betzig

One day in December of 1646, Henrietta Maria, who was the daughter of a king of France, sent a letter across the Channel to her husband, who was the king of England. "For as long as the parliament lasts, you are not king; and as for me, I shall not again set my foot in England." Henrietta wrote from Paris, in her native French; but her meaning was clear in English. She wanted Charles I to keep up the fight against his parliament, if it cost him his life (Henrietta Maria, *Letter* of 12/1646).

In January of 1649, that parliament took Charles I to the block, and lopped off his head with an ax. "For the advancement and upholding of a personal interest of will, power, and pretended prerogative to himself and his family, against the public interest, common right, liberty, justice, and peace of the people of this nation," their High Court of Justice had him executed (Gardiner 1906, no. 82).

That day, Henrietta Maria's son succeeded his father as Charles II of Scotland. It took another 12 years—all of Oliver Cromwell's and his own son, Richard's tenures—before he became Charles II of England. But he did. In the spring of 1660, at the age of 29, the Merrie Monarch crossed the Channel to Dover from Holland, and got to London on his 30th birthday, the 29th of May.

Charles II liked being a king, and he acted like one. The body of Oliver Cromwell was disinterred and decapitated; and nine of Cromwell's supporters were hanged, drawn and quartered. But many ladies were entertained. Andrew Marvell—who was John Milton's occasional amanuensis, and a fellow Cromwellian—put these words in Charles II's mouth: "Whatever it cost me, I'll have a French whore, … and when I am weary of her, I'll have more" (Marvell , The king's vowes, line 10).

It was not the first time English history had unfolded like that. The sons of powerful women—in particular, queens of England whose fathers were kings in their own right—tended to succeed to the throne young, and to have longer reigns. That was especially true of queens whose fathers—like Henrietta Maria's—were kings of France.

L. Betzig (✉)
Ann Arbor, Michigan, USA
http://laurabetzig.org
e-mail: lbetzig@gmail.com

T. K. Shackelford, R. D. Hansen (eds.), *The Evolution of Sexuality*,
Evolutionary Psychology, DOI 10.1007/978-3-319-09384-0_5

Almost every explanation in the book has been thrown at the conflict variously known as the Puritan Revolution, the English Revolution and the English Civil Wars of 1642–1649. They include the rise of Protestantism and decline of Catholicism; the decline of the aristocracy and rise of the middle class; and the difficulty of fighting domestic wars on three fronts (Gardiner 1883, 1901, 1906; Hill 1972; Russell 1990).

I favor of another cause here. Charles I was executed on the order of his parliamentary court, and his son was eventually restored, because his wife was a daughter of the king of France. The war was an effect, at least partly, of parent–offspring conflict.

As the evolutionary biologist Bob Trivers pointed out a generation ago, parents are genetically related to their offspring by half, but related to themselves by one. So offspring should be naturally selected to want twice as much from their parents as their parents are selected to give. Conflict over weaning and fledging, succession and inheritance ensue. And those conflicts are exacerbated or ameliorated by reproductive benefits and costs. To the extent that a parent's own reproductive potential is low—i.e., to the extent that he or she is unlikely to sire or bear more young—more of his or her investment should be made in offspring already born. But to the extent that a parent's own reproductive potential is high—i.e., to the extent that he or she is able to sire or bear more young—existing offspring can be expected to get less. Kings should often do better, genetically, by making efforts to father more children. But queens—in particular, mothers with their own royal fathers—should make efforts on behalf of their sons (Trivers 1974; Trivers and Willard 1973; Hartung 1982, 1985).

Henrietta Maria should have promoted Charles II, even if it cost Charles I his head.

Statistics

Altogether, 13 of the 33 kings of England who succeeded William the Conqueror had mothers who were born to the purple. Queens of England have included the daughters of kings of Castile, Scotland, Denmark and Norway, Bohemia, England, and France.

Queens whose fathers were kings of any country but France got their sons on the throne at a median age of 24—all of 7½ years younger than nonroyal queens' sons. But the sons of French queens did much better: they succeeded at a median age of just 16 (Welch's one-tailed T-test, $p=0.1684$ for sons of royal vs nonroyal queens, with $p=0.1863$ for sons of French vs nonroyal queens).

And queens who were the daughters of kings tended to have sons with longer reigns. Daughters of kings from any country but France had sons who were kings for a median of 23 years—a full 10½ years longer than the sons of nonroyal queens. And again, the daughters of French kings did better by far: their sons ruled for a median of 37 years (Kolomogorov-Smirnov one-tailed test, $p=0.1578$ for sons of royal vs nonroyal queens, with $p=0.0598$ for sons of French vs nonroyal queens).

In short, the daughters of kings had sons who got to the throne younger, and reigned longer, than any other English kings. And the daughters of French kings did best (Table 5.1; Figs. 5.1 and 5.2).

Table 5.1 The reigns of England's kings

King	King's mother	King's mother's father	Age reign began	Years reigned
William I				
William II	Matilda of Flanders	Baldwin V Count of Flanders	31	12
Henry I	Matilda of Flanders	Baldwin V Count of Flanders	32	35
Stephen	Adela of Blois	William I King of England	39	5
Henry II	Matilda of England	Henry I King of England	21	34
Richard I	Eleanor of Aquitaine	William X Duke of Aquitaine	31	9
John	Eleanor of Aquitaine	William X Duke of Aquitaine	32	17
Henry III	Isabella of Angouleme	Aymer Count of Angouleme	9	56
Edward I	Eleanor of Provence	Raymond IV Count of Provence	33	34
Edward II	Eleanor of Castile	Ferdinand III King of Castile	23	19
Edward III	Isabella of France	Philip IV King of France	14	50
Richard II	Joan of Kent	Edmond Earl of Kent	10	22
Henry IV	Blanche of Lancaster	Henry Duke of Lancaster	32	13
Henry V	Mary de Bohun	Humphrey Earl of Hereford	26	9
Henry VI	Catherine of Valois	Charles VI King of France	1	38
Edward IV	Cecily Neville	Ralph Earl of Westmoreland	18	9
Richard III	Cecily Neville	Ralph Earl of Westmoreland	30	2
Henry VII	Margaret Beaufort	John Duke of Somerset	28	23
Henry VIII	Elizabeth of York	Edward IV King of England	17	37
Edward VI	Jane Seymour	Sir John Seymour	9	6
James I	Mary Queen of Scots	James V King of Scotland	1	57
Charles I	Anne of Denmark	Frederick II King of Denmark	24	23
Charles II	Henrietta of France	Henry IV King of France	18	36
James II	Henrietta of France	Henry IV King of France	51	3
William III	Mary Princess Royal	Charles I King of England	38	13
George I	Sophia of Hanover	Frederick V Elector Palatine and King of Bohemia	54	12
George II	Sophia of Celle	George Duke of Brunswick-Luneberg	43	33
George III	Augusta of Saxe-Gotha	Frederick II Duke of Saxe-Gotha	22	59

Table 5.1 (continued)

King	King's mother	King's mother's father	Age reign began	Years reigned
George IV	Charlotte of Mecklinburg-Strelitz	Charles Duke of Mecklinburg-Strelitz	57	10
William IV	Charlotte of Mecklinburg-Strelitz	Charles Duke of Mecklinburg-Strelitz	64	6
Edward VII	Victoria Queen of England	Edward Duke of Kent	59	9
George V	Alexandra of Denmark	Christian IX King of Denmark	44	25
Edward VIII	Mary of Teck	Francis Duke of Teck	41	1
George VI	Mary of Teck	Francis Duke of Teck	40	14

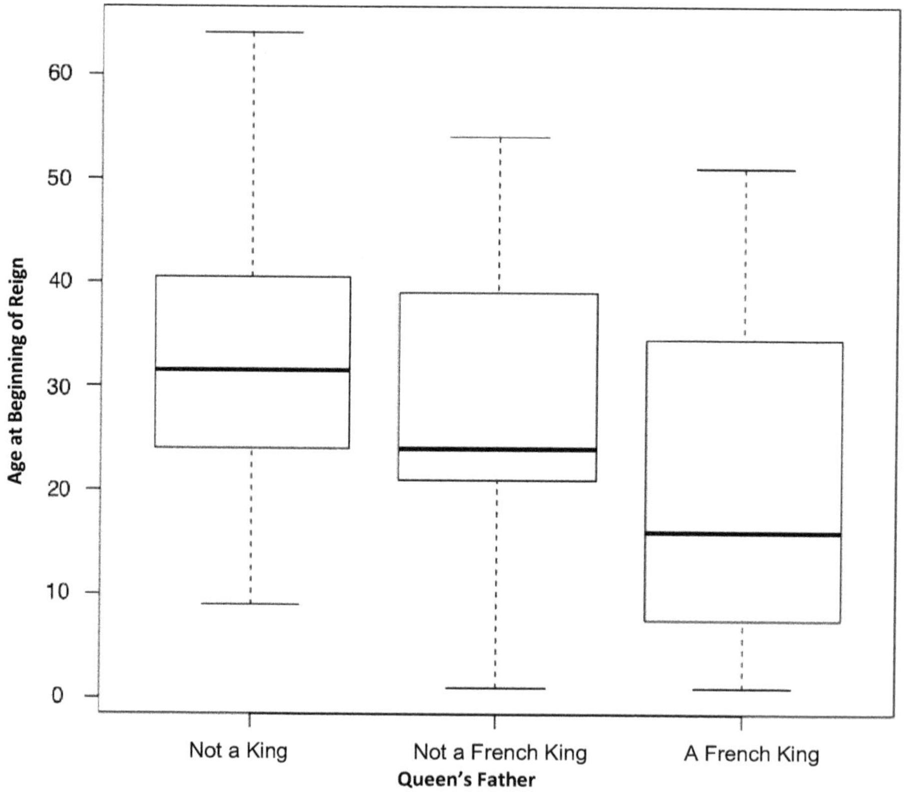

Fig. 5.1 The ages of kings of England when their reigns began

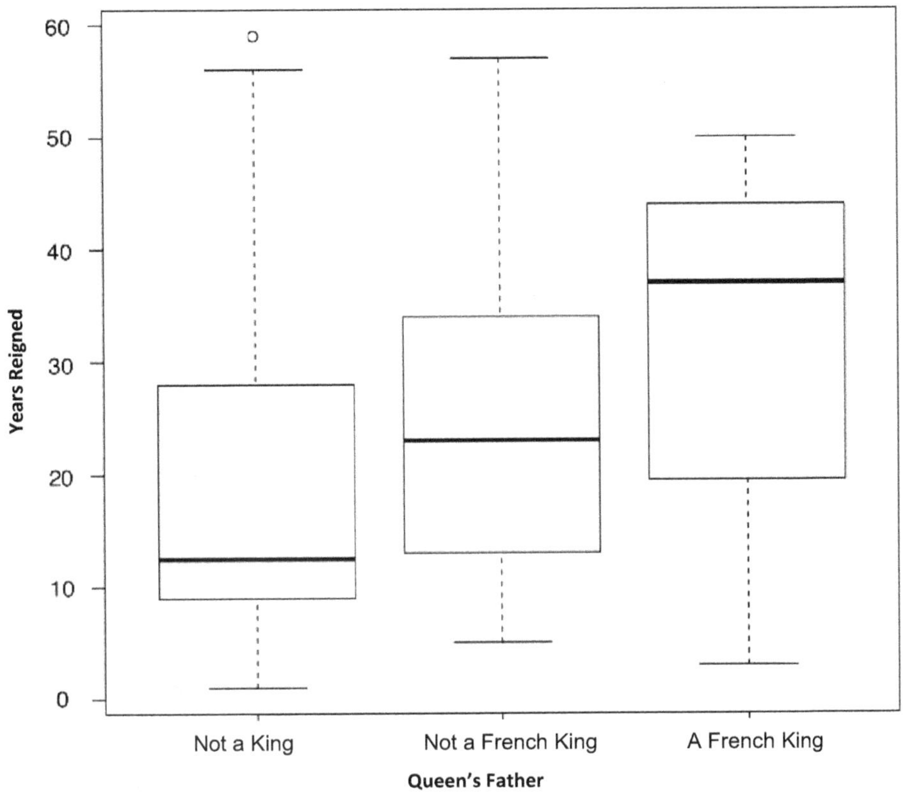

Fig. 5.2 The lengths of the reigns of English kings

Comparative Evidence

Many of those mothers made suspicious or conspicuous efforts to put their sons on the throne. Some were suspected of poisoning. Others raised armies (Strickland 1848).

William the Bastard became William the Conqueror in 1066, in part, because he was married to a strong wife. Matilda was the daughter of Baldwin V, the count of Flanders; and she was the granddaughter, through her mother Adéle, of Robert II of France. Matilda's oldest son, Robert, was made the Conqueror's heir to Normandy, his ancestral domains; and he fought to rule over the duchy, with autonomy, before his father died. He lost. But his mother abetted his cause. Her brother Robert I, the count of Flanders, and her cousin Philip I, the king of France, offered bases for his operations; and Matilda sent silver and gold, "recklessly squandered on jongleurs and parasites and courtesans." Robert unhorsed his father in a battle at Gerberoy in

1079—but let him go free; and he was captured in another battle at Tenchebrai in 1106, by his brother, Henry I—who eventually succeeded his father as England's king (Orderic Vitalis 1969, 3.103; Douglas 1964).

For the better part of two decades, another Matilda—who was a granddaughter of Matilda of Flanders, and the only legitimate daughter of Henry I—fought off usurpers from England's throne. She rode sidesaddle for 60 km from her captors in Winchester, and walked another 10 km out of an Oxford prison in the snow. But she was rewarded when, on 19 December 1154, her son Henry II was crowned king. His father, Geoffrey *Le Bel*, the 38-year-old count of Anjou who wore flowers (*Planta genista*) in his hair, had died of a 3-day fever in the summer of 1151—after a short interview with Eleanor of Aquitaine in Paris, followed by a ride home across the Loire with his 18-year-old son (Chibnall 1992).

Eleanor would go on to give her husband, Henry II, all of four adult sons. The oldest was crowned along with his father in 1170, at the age of 15; and when he married Margaret, a daughter of Louis VII of France, he was crowned again. The civil wars started a year later. And Young King Henry was backed by his mother. Eleanor was apprehended on a horse in men's clothes on her way to Louis VII's French court, and kept under guard in England for the next 16 years, as her gaoler romanced "fair" Rosamund Clifford and other wives and daughters of his vassals: "He became an open violator of the marriage bond" (Gerald of Wales, *Opera,* 5.303–305; Kelly 1950).

But it was especially dangerous to be married to the daughter of a king of France. Isabella the She Wolf, whose father was Philip IV, invaded England on her son's behalf. She married Edward II at the age of 12; but she went back to France in 1325, and was followed by her own 12-year-old son. That September, at a hunting lodge near Paris, he did homage to her brother, Charles IV of France; and the next August, he was engaged to a daughter of William I, the count of Hainault. Then in September of 1326—with help from count William, and from her brother Charles—Isabella brought a small army across the Channel, and had her husband deposed. *Incorrigible saunz esperance de amendement:* incorrigible and without hope of amendment, he abdicated in favor of Edward III. Edward II would be incarcerated soon afterward, and put to death in a few months; but his son would rule England for 50 years (*Articles of Accusation*, 1327; Phillips 2010).

Some queens were more subtle. Just 2 years after he signed the 1420 Treaty of Troyes that made him an heir to the crown and kingdom of France, Henry V met his wife Catharine of Valois near Paris. Catharine had given birth to the imminent Henry VI, at Windsor, on 6 December 1421; and she'd crossed the Channel to be with her husband the following May. The besieger of Harfleur ("What's it to me, when you yourselves are cause, if your pure maidens fall into the hand of hot and forcing violation?"), and the router of French forces at Agincourt ("God fought for us"), Henry took to his bed on the Seine that July, and was dead by the end of August. His death was variously attributed by contemporaries to dysentery, *infirmitas fluxus ventris,* and *un flux de ventre merveilleux.* There was witchcraft in Catherine's lips, as was eventually pointed out (Shakespeare, *Henry V,* 3.3, 4.7, 5.2; Wylie and Waugh 1929).

Other queens were quite obvious. In the summer of 1567, James VI succeeded his father on the throne of Scotland at the age of just 13 months. His mother, Mary Queen of Scots, may have had her husband blown up. Mary succeeded her father, James V of Scotland, at the age of just 6 days; and she was widowed by Francis II of France 3 days before she turned 18. She made her first cousin, Henry Darnley, her second husband; and a year later she presented him with an heir. Henry seems to have wanted more. In particular, he was after the Crown Matrimonial—which would have given him the right to rule Scotland alone if Mary died, and to pass the crown to his descendants by another queen. There would have been plenty of candidates. Lord Darnley was well proportioned, lusty and addicted to base pleasures; and he probably had venereal disease. When, at the beginning of the last century, the mathematician Karl Pearson examined his skull, he found it pitted with syphilis—"a melancholy proof of the effects of his incontinence." So, on 10 February 1567, with or without her consent, Mary's friends had her incontinent husband blown up. Her son became James I of Scotland that summer, and James VI of England 36 years later (Keith 1928; Warnicke 2006).

By bacterial, viral or human intervention, strong queens got their boys on the throne young. That happened again to Charles II, Henrietta Maria's son.

The Case of Charles II

Unlike other queens of England—from Matilda of Flanders, to Mary Queen of Scots—Henrietta Maria fought *for* her husband, not *against* him. She was his "she-generalissima," the head of his "Queen's Army," the dodger of Parliamentary bullets at Bridlington Bay. She worked hard to encourage Charles' allies, and to provide him with cash, ammunition, and armies. But without the queen, there might have been no war at all. It might not have gone on as long, and her husband might not have lost his head (Bone 1972; Hibbard 1991; Cust 2005; White 2006).

Because the mother of Charles II was the daughter of a king of France. Because, like other English queens with royal fathers, she promoted her young son—eventually, at her husband's expense. Because the Revolution was, at least partly, about parent–offspring conflict.

In October of 1638, Henrietta's widowed mother, Marie d'Medici, landed at Harwich with 600 attendants. The first casualties were inflicted within months on the Scots. And in the spring of 1639, Henrietta Maria appealed to "the Loialty and Affection of his Majesty's Catholick Subjects" for money to fund her husband's army. Under her direction, letters were disseminated from London; around £ 14,000 was raised (Rushworth, *Historical collections,* 2.820; Hibbard 1982). Then in the spring of 1640, when Charles I summoned and dissolved the Short Parliament after a hiatus of 11 years, she petitioned for funds from Rome. "She not only intimates the hope of obtaining large contributions from the Catholics but even of obtaining some help from the pope." After Charles I summoned his Long Parliament that fall, Henrietta reportedly looked for assistance from France. "The parliamentarians were warned by a person not yet divulged, that their majesties had concluded a treaty with France to introduce 10 regiments of that nation into that country, to join the troops." Ships were ready to sail from Dieppe, the Venetian ambassador wrote. Then after another rebellion broke out in Ireland, the Commons worried again about the French connection. "The parliamentarians have conceived some suspicion that the queen may have given some encouragement to these movements in Ireland, in secret ways" (*Calendar of state papers, Venice*, 25.52, 117, 150, 192, 241).

So help was fetched from the Dutch. Marie de' Medici left Dover for the Netherlands at the end of 1641; Henrietta Maria landed at Brill in March of 1642. With help from her oldest daughter, princess Mary, and her son-in-law, prince William II of Orange, no less than six shipments of ordnance, soldiers and horses were sent from Holland to England over the next 10 months. Henrietta brought another installment—no less than 300 officers, 1000 soldiers, 20,000 suits of armor, and £ 80,000—when she landed in February of 1643 near York, and other shipments would follow (*Calendar of state papers, Venice*, 26.252; *Calendar of state papers, Domestic,* 1644–1645, 469–470). Henrietta Maria had met with ambassadors to her husband's uncle in Holland, who reassured her "the King of Denmark would assist you in everything to that you might desire," though he offered no specifics (*Letters of queen Henrietta Maria,* 1/8/1643). Then in July of 1644, Henrietta Maria went back to France, and petitioned the Bourbons for help. She appealed to her sister-in-law Anne, the queen regent, who offered to "spend her last crown and devote the last soldier of France" to her cause, and sent £ 20,000; she lobbied her brother Gaston, the duke of Orléans, who hoped to offer 4 million *florins*; and she pressed for an "army to the king's assistance" from Gaston's brother-in-law, the duke of Lorraine, who promised 10,000 men (*Calendar of state papers, Venice*, 27.4, 40, 150; Rushworth, *Historical collections,* 6.131; *King's cabinet opened,* nos. 2, 3, 31, 39).

There's no question that Henrietta Maria encouraged her husband to take up arms against parliament. In the spring of 1641, months after the rumors about French ships, she was partly responsible for an army plot. The Lords had word from the Commons "that there have been, and still are, secret practices to discontent the army with the proceedings of parliament, and to engage them in some design of dangerous consequence." Then in the first days of 1642, months after the Irish rebellion, treason charges were brought against five members of parliament who opposed the queen. The Lords and the Commons appointed committees to determine the legality of those charges; so Charles I set out from Whitehall Palace and marched through the streets with Charles Louis, the Elector Palatine—his sister Elizabeth's son—and another 300 or 400 armed men. The five members got away on the river, and Henrietta Maria was mad. "Go you coward and pull those rogues out by the ears, or never see my face more," she reportedly said (*Lords journal,* 4.236; Gardiner 1883, 10.136).

There is no question that Henrietta Maria encouraged her husband to keep up the fight. She wrote often to her dear heart, with admonitions and advice. Days after she landed in Holland, in March of 1642, she said, "assuredly, God will assist us, and whatever may be said to you, do not break your resolution"; then in April, "my whole hope lies only in your firmness and constancy, and when I hear anything to the contrary, I am mad"; then in May, "continue to act with resolution, for now is the time to show that you will make good what you have undertaken, or you are lost"; then in June, "I am ever returning to the old point—lose no time" (*Letters of queen Henrietta Maria,* pp. 52, 61, 70, 80).

And there is no question that parliament considered Henrietta Maria a threat. In February of 1642, just weeks before the queen took ship for Brill, a declaration of causes and remedies was made in the House of Commons. "Your Majesty will be pleased not to entertain any advice, or mediation, from the queen, in matters of religion, or concerning the government of any of your majesty's dominions," it began. Then just months after she landed back in England, in May of 1643, they had her impeached. "Resolved, upon the question, *nemine contradicente,* that the queen bath levied war against the parliament, and kingdom. Resolved, upon the question, that Henrietta Maria, queen of England, shall be accused by this House, in the name of all the Commons of England, of High Treason" (*Commons journal,* 2.443, 3.98).

A year later, when she sailed to Brest, Henrietta had a couple of objectives. One was to get her son Charles restored. "Losing the title of queen, I have lost all my happiness in this life," she wrote to her son (*Letters of queen Henrietta Maria,* pp. 350–358). He made her a queen mother in short order: 20-year-old Charles II was crowned king of Scotland on 1 January 1651 at Scone—though he was back in Normandy that October, having eluded Oliver Cromwell's forces. But it took almost another decade to recover the second crown. On 29 May 1660, the day he turned 30 years old, and a year and a half after Oliver Cromwell died, Henrietta's son rode back into London. "All the world in a merry mood because of the king's coming," wrote an eventual Fellow of the Royal Society. The king's Proclamation against Drinking Swearing, and Debauchery was soon afterwards read out (Pepys, *Diary,* 5/31 and 6/4/1660).

Henrietta Maria's second objective was to find Charles a wife. Even before she left England in 1644, there were rumors of a marriage to the sister of the prince of Orange, or to a daughter of the duke of Orléans (*Letters of king Charles the first to queen Henrietta Maria,* no. 6; *Calendar of state papers, Venice,* 27.40–41, 131, 151). Those negotiations continued in Paris. But the winning candidate was Catharine of Braganza, John IV of Portugal's daughter, whose marriage contract was signed a year after Charles II got to London. Catharine miscarried several times, and none of her children survived.

Genetic Effects

There would be plenty of other children. It was obvious to John Fenn, who worked as a paymaster in the navy, that the court was never worse "for gaming, swearing, whoring, and drinking, and the most abominable vices that ever were in the world; so that all must come to naught"; and even better off civil servants like the Secretary of the Admiralty, Samuel Pepys, worried about "the vices of drinking, swearing, and loose amours, that I know not what will be the end of it" at court, and wondered that "the great bawdy-house at White Hall" had not been pulled down (Pepys, *Diary,* 8/31/1661, 7/27/1667, 3/25/1668). John Evelyn, who was Pepys' fellow Royal Society member, shared his friend's concerns, and suspected that Charles II "would doubtless have been an excellent prince had he been less addicted to women, which made him uneasy and always in want to supply their unmeasurable profusion, and to the detriment of many indigent persons" (Evelyn, *Diary,* 2/6/1685).

Many of the Merrie Monarch's women and children were well known. Months after his father lost his head, Lucy Walter, "a brown, beautiful, bold but insipid creature," gave the second Charles Stuart an Adonis of a son, James, who otherwise took after his father (Evelyn, *Diary,* 7/18/1649). A year later, Elizabeth Killigrew gave him a daughter, called Charlotte Jemima. Catherine Pegge—who took up with Charles II as an exile, around the time he was rumored to have 17 mistresses abroad—gave him another son, Charles, and Catherine, another daughter. Barbara Villiers gave the king a daughter, Anne, a year after he was restored, who was followed by three FitzRoys—Charles, Henry and George. There were bastards by a couple of actresses: "pretty, witty Nell" Gwyn gave birth to another Charles, and to a son who died young; and the "pretty girle" Moll Davis had a daughter named Mary (Pepys, *Diary,* 4/3/1665, 4/17/1666, 4/26/1667). In between, Louise de Quérouaille became the mother of another namesake, Charles Lennox. They were all very well provided for at the public expense (Hutton 1989).

Henrietta Maria's posterity did not end there. When her oldest son, Charles, died after a 25 year reign, he was succeeded by her second son, James. Who was already well known for his pursuit of other men's wives. Among others, he went after Jane Middleton—who was a relative of John Evelyn's; he went after Frances Stuart—who married the duke of Richmond; he went after Anne Carnegy when her husband, the earl of Southesk, was in Scotland; he went after Isabella Robartes, whose husband, the earl of Radnor, was offered revenues in Ireland; he went after sir John Denham's much younger bride; he went after Arabella Churchill, whose father was said to have prostituted his daughter; he went after Catherine Sedley, whose engagement to Arabella's brother was broken off; and he was smitten with Elizabeth Stanhope, lady Chesterfield, which made the earl of Chesterfield sad. "In England they looked with astonishment upon a man who could be so uncivil as to be jealous of his wife," was the opinion of Anthony Hamilton, who wrote in French (Hamilton, *Memoirs,* p. 329).

James is supposed to have been fond of his first wife, the mother of his two oldest daughters, Anne Hyde; but she seems to have been less than satisfied. As Anthony Hamilton went on: "The duchess of York was one of the highest feeders in England; as this was an unforbidden pleasure, she indulged herself in it, as an indemnification for other self-denials. It was really an edifying sight to see her at table. The duke, on the contrary, being incessantly in the hurry of new fancies, exhausted himself by his inconstancy, and was gradually wasting away" (Hamilton, *Memoirs,* p. 306; Miller 1977, 1997).

There would have been plenty of opportunities in house. By the summer of 1642, when the royal standard was raised at Nottingham, subjects of the first Charles Stuart and his French wife were supporting 580–620 servants in the king's chamber, 500 servants or servants' servants below stairs, 400–800 officers' servants, over 200 yeomen of the guard and over 250 in the stables, with more than 200 servants in the chambers of three princes (Charles, James and Henry) and two princesses (Mary and Elizabeth). There were others in Henrietta Maria's chambers (Aylmer 1974, 2002; Cruickshanks 2000). Those numbers would come down, but progress was slow.

The households of English kings had always been large. The *Constitutio domus Regis,* put together under the William the Conqueror's son, accounts for the cash and candle allowances of chief officers (chamberlain, chancellor, steward, butler and constable), and of the subordinates who worked for them: there were 20 serjeants and four hornblowers on the hunting staff alone. The Plantagenet kings who descended from Henry II supported households that numbered in the mid- to high-hundreds. The Tudors' Yorkist ancestors held up Solomon as an exemplar, whose great renowned riches and high largess "the noble quene of Saba doth wondyrly recommend:" hundreds of servants worked upstairs in the *domus magnificencie*—ushers of the chamber and gentlemen ushers, yeomen (Edward IV kept around 325) and grooms (Edward IV kept around 120)—with another 200 or 300 in the *domus providendae* below stairs (*Constitutio domus Regis*; *Liber Niger,* 4.2; Given-Wilson 1986). And the principal Tudor palace at Whitehall held around 2000 rooms, many of them put to use (Loades 1992).

Most inventories did not mention women, but there were of course women at court. A man by the name of Oin Purcell worked as a marshal of the *meretrices*, or whores, at the court of Henry I; a Ranulf de Broc, who inherited his lands and offices from Oin Purcell, marshaled the whores as *marscallus custodiendi meretrices de curia domini Regis* for Henry II; and a Henry de la Mere held the office of *serjentiam custodiendi meretrices sequentes curiam domini regis* for Henry III (Round 1911; Vincent 2007). Edward I made John de Windreshull his *serjantum ad custodiendum meretrices in exercitu domini regis,* or custodian of the whores in his army; Edward II let John de Warblinton succeed his father, Thomas, as *marescallus de mertricibus in hospito regis,* or marshal of the whores in his chamber; and Edward III made Ralph de Middleton his clerk of the market, and marshal of the meretricious women in his household—*dismembrare malefactores adjudicatos:* dismembering adjudged malefactors (Given-Wilson 1986; Prestwich 2006). Henry VIII had a Groom of the Stool, or chamberpot cleaner, William Compton, who sent offerings and gifts to some of Henry's conquests, and intercepted or quarreled with others. Compton's successor, Henry Norris, did the same. "Master Nores was bawd between the King" and one of Catherine of Aragon's ladies-in-waiting, Anne Boleyn (*Letters & papers of Henry VIII*, 6.923; Bernard 1981; Starkey 1987, 1991). At home or abroad, women were taken advantage of.

And like other kings, kings of England fathered a number of bastards (Given-Wilson and Curteis 1984; Betzig 1986, 2002). Matilda of Flanders may have failed to get her oldest son, Robert, on the Norman throne young; but William the Conqueror was eventually succeeded, in England, by two of her younger sons. The first, William II, died unmarried, but kept "troops of pathics, and droves of harlots" at court (William of Malmesbury, *Gesta,* 1093). And the second, Henry I, lost his only legitimate boy in the *White Ship* disaster, after which he promoted his bastards. There were two Roberts, two Williams, a Richard, a Reginald, a Gilbert, a Henry, a Fulk, three Matildas, a Juliana, a Mabel, a Constance, an Aline, an Isabel, two Sibyls, a Gundrada, a Rohese, an Emma and an Adeliza, as far as we know (Hollister and Keefe 1973).

With help from some of those bastards, Henry I's only legitimate daughter, Matilda, successfully promoted her oldest legitimate son. By Eleanor of Aquitaine, Henry II got four legitimate boys; and he had plenty of chances with others. One of Eleanor's secretaries, Peter of Blois, wrote about the actors, gamblers, buffoons and *candidatrices*, or whores, at his court; and Henry's martyred archbishop of Canterbury, Thomas Becket, is supposed to have been surrounded by adulterers and harlots. "His house very often emptied the king's, so much that the king began to be angry with it," says one of his many *Lives*. It was Henry's "pimp," Ranulf de Broc, who assaulted Becket in his cathedral, and scattered his brains on the pavement (Peter of Blois, *Letters,* 14; Guernes, *Life of Becket,* lines 396–430).

Eleanor of Aquitaine failed to advance the career of her oldest son, the Young King; but two of her younger sons, John and Richard, managed to succeed their father. *Coeur de Lion*, whose prowess was obvious, died in the arms of a whore. "He was strongly lubricious and burning with love for women, unwilling to curb his pleasures, belittling safe counsels, not worrying about his wound." Some of Softsword's amours would have been less well known. Anonymous women were sent chaplets of roses, or given *dolia* of good wine, or paid £ 30; and a handful of named women were provided for. There was a *domicella, amica domini Regis,* Suzanne, who was awarded a tunic for her services on a Misae Roll; there was a Hawise, whose son was pardoned a Pipe Roll relief payment on account of his mother's efforts; there was another Hawise, a widow, who was ordered a pension of a penny a day in a Memoranda Roll entry; and an Oblate Roll entry had the wife of the chief forester, Hugh Neville, promised 200 chickens, so that she might spend a night *with her husband* (Walter of Guisborough, *Chronicle,* 1199; Painter 1949).

Edward III, whose mother, the She Wolf of France, invaded England on his behalf, got 14 children on his queen consort, Philippa of Hainault. But there were other consorts, and parliament did not like it. Representatives in the House of Commons complained that a young lady named Alice Perrers had every year from the treasury 2000 or 3000 pounds in gold and silver, without any notable profit and to the great damage of the king. So in 1376, she was *bannitz hors du roialme,* or banished from the realm; and she was *forsclose de commune libertee,* or deprived of common liberty, in 1377. Members of parliament knew that Alice, like other women at court, pressed her advantage *par voie de meyntenance:* by way of maintenance. And they wanted those women to stay out (*Rotuli parliamentorum,* 2.329, 374; 3.12).

Catharine of Valois' husband was succeeded by her infant son, who was vulnerable to Commons petitions for most of his reign. They objected to the "continuell importable chargez" associated with his inability to live on his own, and set out to "establissh, apoynte and ordeign" good rule in his home. In 1445, a Household

Ordinance was drawn up: it asked that ushers of the chamber of Henry VI, and the marshals of his hall, come to his counting house every day with their expense accounts; the "excesse" was to be cut, and the "straungers" thrown out. Another Ordinance of 1454 put a lid on Henry VI's domestic servants. He might have kept as many as 800 by then, and the drafters wanted the numbers brought down to 120 for the queen, with another 424 for the king (*Rotuli parliamentoru,* 5.62–63; *Ordinance of 1445,* c. 2–4; *Ordinance of 1454,* preamble; Morgan 1973).

And James I, the son of Mary Queen of Scots, was continually had up for spending too much. His frugal predecessor, Elizabeth I, had kept her household expenses down around £ 40,000 a year; but James I would start spending at an annual rate of £ 87,000, and his costs would go up. The office of works budget went up from £ 4000 under Elizabeth Tudor to over £ 20,000 under the Stuarts; and that did not count new constructions—like the Banqueting House at Whitehall, which ran to around £ 15,000. And it did not count works of art. Under James' son, the exchequer laid out more than £ 17,500 for tapestries, £ 2500 for statues, £ 1400 apiece for court masques, and £ 7800 on paintings by the likes of Rubens and Van Dyck—£ 3000 of it on the Banqueting House ceiling alone. And that was a fraction of what went up in smoke: James I spent £ 36,000 on fireworks and festivities at his own coronation party, and another £ 100,00 when his daughter Elizabeth got hitched. He made vague promises to "reduce our Household," and to "abate the great and dayly Charge and Expense." To little effect. As the most eloquent Stuart put it in one of his many speeches to the Commons: "I was born to be begged of and not to beg" (*Ordinances of the household of king James I,* preamble; *Proceedings in parliament 1614,* 1.11f; Seddon 1980; Smuts 1987, 1999).

In some cases, the conflicts between kings and princes over sexual access was explicit. There were the rumors that Geoffrey Plantagenet competed with his son, Henry II, for Eleanor of Aquitaine's affections: a "charge secretly made" had Geoffrey slip into Eleanor's bed; but she only had eyes for his heir (Map, *Courtiers' trifles,* 5.6). And there were the rumors that Henry II competed with his son, Richard I: *Coeur de Lion* was intended for Alice, a daughter of Louis VII of France, but the marriage never came off—"inasmuch as the king of England, his own father, had been intimate with her, and had had a son by her." Many witnesses were produced to establish that fact (Roger of Hoveden, *Annals,* 1191). Other rumors pitted Henry VII against Henry VIII. Henry Tudor may have won the throne of England at Bosworth Field, but he kept it by his marriage to Elizabeth of York. In November of 1501, at the age of 15, their oldest son, Arthur, was married to Catherine, the daughter of Isabella of Castile and Ferdinand II of Aragon. But Arthur died the next year, and his mother died a year later. So a marriage treaty was drawn up between Arthur's widow and Arthur's younger brother, the eventual Henry VIII; but it took another 6 years for the marriage to come off. In the meantime, Henry VII may have considered a "monstrous proposal" to marry Catherine of Aragon himself (Gairdner 1889, p. 190; Chrimes 1972, p. 287).

In all cases, succession almost certainly had genetic effects. At home or abroad, kings had more favors to dispense. The sons of poor mothers were pardoned relief payments, and the husbands of prominent women were awarded peerages. Sex with a monarch paid off.

Conclusions

Shortly before Henrietta Maria's 6-month birthday, her father died. Her mother was implicated. "Very melancholy in the morning, upon some presages and tokens of his death," Henry IV of France was mortally wounded by the dagger of an obscure friar, just a day after Marie d'Medici was finally, formally crowned queen. Marie's son, who was Henrietta's older brother, Louis XIII, succeeded his father at the age of 8 (*Life and Death of Henrietta Maria de Bourbon,* p. 3).

Like Henry IV, Charles I was put to death by his subjects. Like Marie d'Medici, Henrietta Maria was implicated. And like Louis XIII, Charles II became king as a young man. Then he became a libertine.

Of course, parliamentary armies fought against the religious sympathies of Charles Stuart and his Catholic wife. They were suspicious of the treaty that promised "to forbear all manner of proceeding against His Majesties Catholic subjects," even before Henrietta got to England; they ordered the chapel she had built at Somerset House torched; they sent her Capuchin friars—"a scandal to our religion, and a danger to our peace"—back to France; and they abhorred her appeals to "the very Romish Antichrist himself" for help (*Calendar of state papers, Domestic,* 1.16; *Lords journal,* 4.285–287; *Lord George Digby's cabinet,* p. 8).

Of course, parliamentary armies fought against the absolutism of Charles Stuart and his French queen. They were unhappy with his 11 years of Personal Rule, when no parliament met; they were upset when he dissolved his Short Parliament after just 5 weeks; after the Long Parliament assembled, they were distressed by the army plot gossip and Charles' march to Whitehall with hundreds of armed troops; and they objected, often, to taxation without representation—in the form of forced loans, fines for forest encroachments, the granting of monopolies, and ship money payments used to build a fleet. "This whole kingdom most miserably distempered with levies of men and money, and imprisonments of those who denied to submit," said their Grand Remonstrance (Gardiner 1906, no. 43, item 74).

But it was Henrietta Maria who provoked those wars. And resisted peace. This "very little, plain old woman" who was "borne an Alien, bred up in a contrary Religion," cultivated Catholic allies even before she left England; she raised foreign arms and armies in Holland and France; she exhorted a husband who "saw with her eyes and determined by her judgment" to stand firm against parliament in a torrent of correspondence; and she sought and found a safe harbor on the Continent for her eldest son. The first Charles Stuart was expendable. Because the second Charles Stuart, all of 18-years-old when his father was brought to the block, was living with his mother in France. As John Milton—who was Andrew Marvel's employer, and worked for Oliver Cromwell as Secretary of Foreign Tongues—would put it soon after his friends had Charles' neck: "Examples are not farr to seek, how great mischief and dishonour hath befall'n to Nations under the Government of effeminate and Uxorious Magistrates" (Pepys, *Diary,* 11/22/1660; *King's cabinet opened,* p. 43; Clarendon, *Life,* 1.185; Milton, *Eikonoklastes,* pp. 64–65).

By stealth or by spectacle, queens of England—like strong mothers from other countries, and from other species—had fought on their sons' behalf, from Matilda of

Flanders to Mary Queen of Scots. Henrietta Maria was no exception. In the face of armed parliamentary opposition, she pushed Charles I to take risks—for the sake of her ultimately Catholic, and always absolutist, son (see Halpern 2001 on Absalom, and Surbeck et al. 2011 on bonobos).

Egalitarianism in England was not an effect of the English Revolution. Political and religious freedom in England advanced for generations before the Revolution began, and for generations after the Revolution was over. The Reformation on the Continent, and in England, anticipated the wars of the 1640s by over a century. And political representation grew up in fits and starts, from *Magna Carta* in 1215, to parliaments under the first three Edwards, to the deposition of Richard II in 1399, to the bloodless, Glorious Revolution that sent James II packing in 1688. The rhetoric of the Civil Wars was a bright spot in the clashes between monarchs, bishops and Commons; but it was not an end point. James II was deposed, but Charles II was restored.

Egalitarianism in England, as elsewhere, was an effect of finding new places, and new ways, to support a family. It followed the opening up of new habitats and new markets after Urban II made his 1095 speech to the first crusaders at Claremont, and after Columbus made landfall off the American shore in 1492. It came in the wake of the great ships that brought men and supplies across the Mediterranean to Outremer in the Near East, and across the Atlantic to New England and other parts of the Far West. Heresy and the Reformation, the rise of a middle class and political representation of the Commons were the forms it took (Betzig 2012, 2013, 2014a, b).

The Revolution of the 1640s was Henrietta Maria's war. It ended in the execution of her husband, and the Restoration—at the age of 20 in Scotland, and at 30 in England—of her oldest son. Charles I—like other English kings—was done in, in part, by his French wife. He was a casualty, more or less, of parent–offspring conflict.

Acknowledgements I am honored to thank eminent Stuart historians Richard Cust, Malcolm Smuts, and Michelle White for reading and commenting on this chapter. Elena Yudovina of the University of Michigan Department of Statistics offered patient and understanding quantitative advice.

References

Articles of accusation against Edward II, quoted from Adam of Orleton. *English Historical Review, 113,* 852–881, 1998.

Aylmer, G. E. (1974). *The king's servants: The civil service of Charles I.* London: Routledge.

Aylmer, G. E. (2002). *The crown's servants: government and service under Charles II.* Oxford: Oxford University Press.

Bernard, G. W. (1981). The rise of sir William Compton, early Tudor courtier. *Englsh Historical Review, 96,* 754–77.

Betzig, L. L. (1986). *Despotism and differential reproduction: A Darwinian view of history.* New York: Aldine.

Betzig, L. L. (2002). British polygyny. In M. Smith (Ed.), *Human biology and history* (pp. 30–97). London: Taylor and Francis.

Betzig, L. L. (2012). Means, variances and ranges in reproductive success: Comparative evidence. *Human Behavior and Evolution, 33,* 309–317.
Betzig, L. L. (2013). Darwin's question: How can sterility evolve? In K. Summers & B. Crespi (Eds.), *Human social evolution* (pp. 365–374). New York: Oxford University Press.
Betzig, L. L. (2014a). Eusociality: From the first foragers to the first states. *Human Nature, 23,* 1–5.
Betzig, L. L. (2014b). Eusociality in history. *Human Nature, 23,* 80–99.
Bone, Q. B. (1972). *Henrietta Maria: Queen of the cavaliers*. Urbana: University of Illinois Press.
Calendar of state papers, Domestic, edited by W. D. Hamilton. London: Public Record Office 1890.
Calendar of state papers, Venice, edited by A. B. Hinds. London: Public Record Office, 1924.
Chibnall, M. (1992). *The empress Matilda*. Oxford: Blackwell.
Chrimes, S. (1972). *Henry VII*. London: Methuen.
Clarendon, E. The life of Edward, earl of Clarendon. Oxford: Clarendon, 1827.
Commons journal (Journal of the House of Commons). London: House of Commons, 1802.
Constitutio domus Regis (Establishment of the royal household), translated by S. D. Church. Oxford: Oxford University Press, 2007.
Cruickshanks, E. (2000). *The Stuart courts*. Stroud: Sutton.
Cust, R. (2005). *Charles I: A political life*. London: Longman.
Douglas, D. C. (1964). *William the Conqueror*. Berkeley: University of California Press.
Essay of scandal. (1681). In J. H. Wilson, Court satires of the Restoration. Columbus: Ohio State University Press, 1976.
Evelyn, J. *Diary* (E. S. de Beer (Ed.)). Oxford: Clarendon, 1955.
Gairdner, J. (1889). *Henry the Seventh*. London: Macmillan.
Gardiner, S. R. (1883). *History of England from the accession of James I to the outbreak of the civil war*. London: Longmans.
Gardiner, S. R. (1901). *History of the great civil war, 1642–1649*. London: Longmans.
Gardiner, S. R. (1906). *The constitutional documents of the Puritan Revolution*. Oxford: Clarendon.
Gerald of Wales. *Opera* (J. S. Brewer, J. F. Dimock, & G. F. Warner (Eds.)). Cambridge: Cambridge University Press, 2012.
Given-Wilson, C. (1986). *The royal household and the king's affinity*. New Haven: Yale University Press.
Given-Wilson, C., & Curteis, A. (1984). *The royal bastards of medieval England*. London: Routledge and Kegan Paul.
Halpern, B. (2001). *David's secret demons: messiah, murderer, traitor, king*. Grand Rapids: Eerdman's.
Hamilton, A. *Memoirs of Count Grammont* (W. Scott (Ed.)). London: Chatto and Windus, 1876.
Hartung, J. (1982). Polygyny and the inheritance of wealth. *Current anthropology, 23,* 1–12.
Hartung, J. (1985). Polygyny, bride-price, inheritance, and Oedipus. In G. Volger & R. Weld (Eds.), *The bride, given away, sold, bartered, stolen* (pp. 794–797). Köln: Rautenstrauch-Joest Museum.
Hibbard, C. (1982). The contribution of 1639: Court and country Catholicism. *Recusant history, 16,* 42–60.
Hibbard, C. (1991). The role of a Queen Consort: the household and court of Henrietta Maria, 1625–1624. In R. Asch & A. Burke (Eds.), *Princes, patronage and the nobility* (pp. 393–414). New York: Oxford University Press.
Hill, J. E. C. (1972). *The world turned upside down: Radical ideas during the English revolution*. New York: Viking.
Hollister, C. W., & Keefe, T. K. (1973). The making of the Angevin Empire. *Journal of British Studies, 12,* 1–25.
Hutton, R. (1989). *Charles the Second*. Oxford: Clarendon.
Keith, A. (1928). The skull of Lord Darnley. *British Medical Journal, 2*(3531), 456–458.
Kelly, A. (1950). *Eleanor of Aquitaine and the four kings*. Cambridge: Harvard University Press.

King's cabinet opened. (1645). London: Robert Bostock.
Letters of king Charles the first to queen Henrietta Maria, edited by J. Bruce. London: Camden Society, 1856.
Letters & papers of the reign of Henry VIII, edited by J. Gairdner. London: Longman, 1910.
Letters of Queen Henrietta Maria, edited by M. A. E. Green. London: Richard Bentley, 1857.
Liber Niger (Black book) of the household of Edward IV. In A. R. Myers (Ed.), *The household of Edward IV*. Manchester: Manchester University Press, 1959.
Life and death of Henrietta Maria de Bourbon, reprinted In G. Smeeton, *Historical and hagiographical tracts*. London: St Martin's, 1820.
Loades, D. (1992). *The Tudor court*. London: Batsford.
Lord George Digby's cabinet. London: Edward Husband, 1646.
Lords journal (Journal of the House of Lords). London: House of Lords, 1830.
Map, W. *Courtiers' trifles* (*De nugis curialium*; trans: M. R. James). Oxford: Clarendon, 1893.
Marvell, A. The king's vows. In H. M. Margoliouth & P. Legouis (Eds.), *Poems and letters*. London: Oxford University Press, 1971.
Miller, J. *James II: a study in kingship*. Sussex: Wayland.
Miller, J. (1997). *The restoration and the England of Charles II*. London: Longman.
Milton, J. *Eikonoklastes* (S. Orgel & J. Goldberg (Eds.)). London: Oxford University Press, 2008.
Morgan, D. A. L. (1973). The king's affinity in Yorkist England. *Transactions of the Royal Historical Society, 23*, 1–25.
Orderic Vitalis. *Historicae ecclesiasticae* (*Ecclesiastical history*; trans: M. Chibnall). Oxford: Oxford University Press, 1969.
Ordinance of 1445, translated in A. R. Myers, *The household of Edward IV*. Manchester: University of Manchester Press, 1959.
Ordinance of 1454, in *A collection of ordinances and regulations for the government of the royal household*. London: Society of Antiquaries, 1790.
Ordinances of the household of king James I, 1604, in *A collection of ordinances and regulations for the government of the royal household*. London: Society of Antiquaries, 1790.
Painter, S. (1949). *The reign of King John*. Baltimore: Johns Hopkins.
Pepys, S. (1660–69). *Diary* (R. W. Latham, W. Matthews, & W. A. Armstrong (Eds.)). Berkeley: University of California Press, 2000.
Peter of Blois. *Letters*, edited by I. A. Giles. London: T. Rodd, 1847.
Phillips, S. (2010). *Edward II*. New Haven: Yale University Press.
Prestwich, M. (2006). The court of Edward II. In G. Dodd & A. Musson (Eds.), *The reign of Edward II: New perspectives* (pp. 61–75). York: Boydell.
Proceedings in parliament 1614 (House of Commons), edited by M. Jansson. Philadelphia: American Philosophical Society, 1988.
Roger of Hoveden. (1853). *Annals* (trans: H. T. Riley). London: H. G. Bohn.
Rotuli parliamentorum (The parliament rolls), translated by C. Given-Wilson and others. Woodbridge: Boydell Press, 2005.
Round, J. H. (1911). *The king's sergeants*. London: Oxford University Press.
Rushworth, J. *Historical collections*. London: Browne, 1722.
Russell, C. S. R. (1990). *The causes of the English civil wars*. Oxford: Clarendon.
Seddon, P. R. (1980). Household reforms in the reign of James I. *Bulletin of the Institute of Historical Research, 53*, 44–54.
Shakespeare, W. *Henry V*, in *Complete works*. London: Longman, 2013.
Smuts, R. M. (1987). *Court culture and the origins of a royalist tradition in early Stuart England*. Philadelphia: University of Pennsylvania Press.
Smuts, R. M. (1999). *Culture and power in England 1585–1685*. New York: St Martin's Press.
Starkey, D. (1987). *The English court: From the wars of the roses to the Civil War*. London: Longman.
Starkey, D. (1991). *Henry VIII: A European court in England*. London: Collins & Brown.
Strickland, A. (1848). *Lives of the queens of England*. London: Colburn.

Surbeck, M., R. Mundry & G. Hohmann. (2011). Mothers matter! Maternal support, dominance status and mating success in male bonobos (Pan paniscus). *Proceedings of the Royal Society of London, B, 278,* 590–598.
Trivers, R. L. (1974). Parent-offspring conflict. *American Zoologist, 14,* 249–264.
Trivers, R. L. & D. Willard. (1973). Natural selection of parental ability to vary the sex ratio of offspring. *Science, 179,* 90–92.
Vincent, N. (2007). The court of Henry II. In C. Harper-Bill & N. Vincent (Eds.), *Henry II: New interpretations* (pp. 278–334). Woodbridge: Boydell Press.
Walter of Guisborough. *Chronicle* (H. Rothwell (Ed.)). London: Camden Society, 1957.
Warnicke, R. M. (2006). *Mary queen of Scots*. New York: Routledge.
White, M. A. (2006). *Henrietta Maria and the English Civil Wars*. Aldershot: Ashgate.
William of Malmesbury. *Gesta regum Anglorum (Chronicle of the kings of England*; trans: J. A. Giles). London: Henry Bohn, 1847.
Wylie, J. H. & Waugh, W. T. (1929). *The reign of Henry the Fifth*. Cambridge: Cambridge University Press.

Chapter 6
An Evolutionary Functional Analysis of the Hormonal Predictors of Women's Sexual Motivation

James R. Roney

Introduction

What is the evolved function of women's sexual motivation? Recent research suggests that low sexual desire is a common condition among women, with rates of occurrence in the 20–40 % range even among premenopausal women (Gracia et al. 2007; Laumann et al. 1999; Stuckey 2008). The medical and scientific literatures have often treated low desire as a clinical disorder, with hypoactive sexual desire disorder codified as a recognized condition in the DSM-IV (4th ed., text rev.; *DSM-IV-TR*; American Psychiatric Association 2000). An important question from a functional perspective, however, is whether low desire is a disorder in the sense of brain mechanisms failing to operate as designed, or simply with respect to desired outcomes. It is not clear, for instance, that fairly constant, uninterrupted desire would have promoted women's reproductive success on average over the course of human evolution. Understanding both between-women and within-woman variance in desire would seem to require knowledge of the functional design of the brain mechanisms that regulate sexual motivation. Surprisingly, this issue of functional design has been largely ignored in both the medical and scientific literatures on human sexuality. This chapter will analyze women's sexual motivation from a functional perspective, with specific emphasis on the role of hormonal signals in the regulation of desire.

The chapter will also take a comparative approach to understanding women's sexual motivation. The physiology of the human menstrual cycle exhibits extensive parallels to that observed in the cycles of nonhuman primates, and broader homologies between human and nonhuman motivational systems (e.g., Roney and Maestripieri 2002) make it reasonable to suppose that the mechanisms that regulate human sexual motivation will be variations on designs found in nonhuman species.

J. R. Roney (✉)
Department of Psychological and Brain Sciences,
University of California, Santa Barbara, CA 93106, USA
e-mail: roney@psych.ucsb.edu

T. K. Shackelford, R. D. Hansen (eds.), *The Evolution of Sexuality*,
Evolutionary Psychology, DOI 10.1007/978-3-319-09384-0_6

This comparative perspective may help frame the issue of functional design in ways that may be obscured in the extant medical literature on human sexuality. As reviewed below, for instance, most nonhuman mammalian females are only sexually receptive on days of the estrous cycle when conception is possible, and thus by comparison the relevant functional question regarding women's sexuality is not why desire is so low but instead why it appears to be so much more frequent than in other species.

The plan for the chapter is to first review the endocrine events associated with estrous cycles in order to describe the types of functional information that may be carried by hormonal signals. A basic sketch of possible functions of sexual motivation will follow, which will then be used to generate predictions regarding which combinations of hormonal signals should predict variations in sexual motivation given specific functional hypotheses. A brief review of the nonhuman literature regarding hormonal regulation of female sexual behavior will then provide evidence regarding which functions are consistent with the extant data. Finally, the question of hormonal regulation of female sexual motivation will be applied to humans, with specific emphasis on a new study that addressed this question in natural menstrual cycles (Roney and Simmons 2013) and was the focus of my presentation at the 2013 Oakland University "Evolution of Sexuality" conference. Although this study suggests similarities between human and nonhuman females in the hormonal regulation of sexual motivation, there remain questions regarding uniquely human components of women's sexual desire, and the chapter will include a brief discussion of future research that may address those questions.

Cycle Phase Physiology and Hormones as Information

Physiology of the Human Menstrual Cycle

The hormonal events associated with mammalian female reproductive cycles are well-understood, and the dynamics of the human menstrual cycle will be summarized as an example here (for reviews, see Fauser and Van Heusden 1997; Hall 2004; Strauss and Williams 2004). Human females produce all of their gametes prenatally, which are stored in a resting pool as oocytes until recruited for final stages of maturation. "Follicles" refer to oocytes and the support cells that surround them. At a rate of approximately 1000 follicles per month, beginning even before birth, follicles are recruited out of the resting pool and begin to develop via the replication of the support cells. Follicle stimulating hormone (FSH) is a signal from the pituitary gland necessary for the continued development of follicles, and without this signal, follicles undergo a type of programmed cell death. FSH production at levels high enough to stimulate continued follicle development occurs at puberty with maturation of the hypothalamic–pituitary–gonadal axis. When FSH exceeds a threshold, follicle support cells replicate and eventually begin to express aromatase, the enzyme that produces estrogen from androgen precursors. The follicle

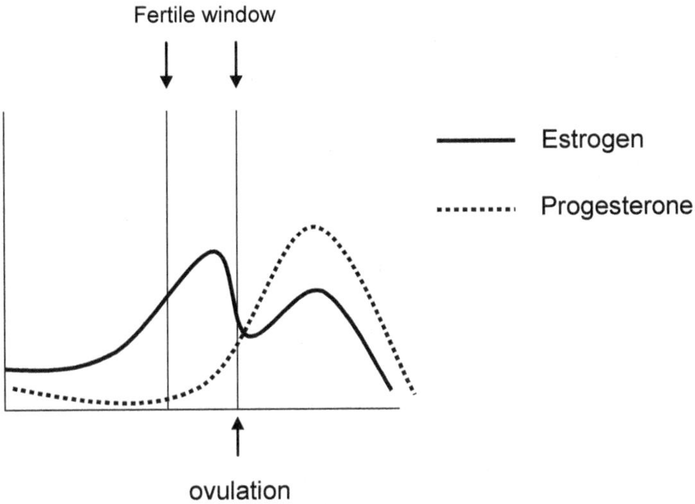

Fig. 6.1 Depiction of prototypical estrogen and progesterone concentrations across days of ovulatory menstrual cycles

that reaches a replication stage at which estrogen production begins also increases expression of luteinizing hormone (LH) receptors, which facilitate continued follicular development even absent FSH. Because estrogen has a negative feedback effect on FSH release, estrogen production from the first follicle that reaches this stage reduces the FSH signal to other developing follicles, and this in combination with other inhibitory signals leads the other follicles to die such that one oocyte and its surrounding support cells become the "dominant" follicle.

As cell replication continues in the dominant follicle, its support cells produce a steep rise in estrogen concentrations. This rise in estrogen in turn activates a surge in LH production from the pituitary, which causes rupture of the follicle and subsequent release of the ovum for possible fertilization. The support cells from the follicle then become a new structure called the corpus luteum, which secretes estrogen and progesterone in the second half of the cycle after ovulation. Rising estrogen before ovulation stimulates development of the uterine lining for possible implantation of a zygote, whereas corpus luteum production of estrogen and especially progesterone is necessary for the uterine lining to provide a nutritive environment.

The above sequence of events produces a prototypical pattern of estrogen and progesterone production in human ovulatory cycles that is summarized in Fig. 6.1. Progesterone is typically low in the follicular phase (the region of the cycle preceding ovulation) but then exhibits large peaks via corpus luteum production in the luteal phase (the region of the cycle after ovulation but preceding the next menstruation). Estrogen also begins low early in the follicular phase but then rises sharply with dominant follicle growth, peaks just before ovulation, falls sharply, and then rises again to a secondary but usually smaller peak in the luteal phase. Testosterone is a third hormone produced by the ovary as it serves as a precursor to estrogen production via the aromatase enzyme. About half of testosterone production comes from the ovary in ovulatory cycles (with the other half derived from adrenal androgen production; see Abraham 1974; Burger 2002), and several studies have reported midcycle testosterone peaks associated with the preovulatory estrogen surge, followed by fairly low testosterone production in the luteal phase (e.g., Abraham 1974; Campbell and Ellison 1992; Roney and Simmons 2013; Van Goozen et al. 1997). Importantly, the patterns depicted in Fig. 6.1 apply to ovulatory cycles, but when dominant follicle development is inhibited (as occurs during lactational amenorrhea, for example), estrogen and progesterone concentrations remain consistently low across time.

Information Content of Hormones in Human Menstrual Cycles

Although estrogen and progesterone have important signaling functions related to endometrial development within the reproductive tract, they are also released into the general circulation whereby they can reach brain mechanisms and thus provide information that can be used by psychological mechanisms that regulate behavior. What information do these hormones carry? Information related to fecundity (i.e., the probability of successful conception and gestation given unprotected intercourse) may be especially important, at two broad timescales. First, at a *within-cycle* timescale, estrogen and progesterone can signal time in the cycle when conception is possible. Conception is only possible within a narrow window from about 5 days before ovulation through the day of ovulation itself (e.g., Wilcox et al. 1998), which is a time period characterized by steeply rising estrogen but also fairly low progesterone (Fig. 6.1 provides a schematic depiction of the fertile window relative to hormone production). High estrogen combined with low progesterone may therefore signal elevated conception risk. Conception appears to be impossible during the luteal phase when progesterone exhibits pronounced peaks within ovulatory cycles, and thus high progesterone concentrations may provide an especially strong signal of low immediate conception risk. The information carried by testosterone is more ambiguous. Although testosterone on average exhibits a midcycle peak and reduced luteal phase production, the substantial adrenal sources of this hormone may cloud its signal value with respect to fecundity. As such, a priori, one would expect brain mechanisms to be designed to rely primarily on estrogen and progesterone as clearer signals of conception risk, though it is possible that testosterone elevations might supplement information carried by estrogen if larger dominant follicles produce greater amounts of both hormones.

Ovarian hormones may also carry information regarding fecundity at what could be thought of as a *between-cycle* (or lifespan) time scale. Although premenopausal women in industrialized countries tend to experience frequent ovulatory cycles due to both late marriage and the use of contraception, women in natural fertility populations thought to be more similar to human ancestral environments cycle much more rarely due to suppression of ovulation associated with events like lactation or food shortage (see Ellison 2001; Strassmann 1997). As such, throughout most of human evolution, women likely spent most of their reproductive lives with low ovarian hormone concentrations, with fecund cycles occurring only rarely between the end of lactational amenorrhea and the conception of the next child. Even within ovulatory cycles, however, evidence suggests that fecundity varies across cycles within the same women, with higher probabilities of conception in cycles with higher estrogen production (e.g., Lipson and Ellison 1996; Venners et al. 2006). In Fig. 6.1, then, imagine a second estrogen curve superimposed higher than the one depicted: for the same woman, fecundity would on average be greater in the cycle associated with the higher curve.

Throughout most of human history, therefore, elevated ovarian hormone concentrations may have signaled that a woman was experiencing one of the rare ovulatory cycles that occurred between longer stretches of anovulation (estrogen and progesterone are also highly elevated during pregnancy, but it seems likely that brain mechanisms are designed to distinguish pregnancy from ovulatory cycles). The information content of progesterone is especially interesting in this context since elevated progesterone would have provided opposite signals of fecundity at the within- and between-cycle timescales: whereas high progesterone concentrations signaled luteal timing and thus zero immediate fecundity, they also signaled that a woman was experiencing fecund cycles in which behaviors related to mating and sexuality may have assumed greater functional importance due to elevated probability of conception in near-future cycles. In other words, in humans, progesterone may act as a within-cycle signal of low fecundity but a between-cycle signal of high fecundity. Estrogen, on the other hand, should signal greater fecundity at both the within- and between-cycle timescales.

Functional Hypotheses Regarding the Endocrine Regulation of Sexual Motivation

Within-Cycle Predictions

A basic functional expectation is that female sexual motivation will be higher on cycle days for which conception is possible than on cycle days with negligible conception risk. This idea is predicated on the assumption that sexual behavior has recurrently entailed costs to females in terms of time, energy, risk of injury, risk of sexually transmitted disease and, in group-living species, perhaps also social costs in the form of harassment from other individuals (see Wallen 2001). Conception would have provided a countervailing fitness benefit. Other things equal, then, a first-order prediction is that females will be designed to avoid the costs of sex when the benefit of conception is absent but then exhibit enhanced sexual motivation (in the form of heightened receptivity and/or proceptivity) when conception is possible. In species in which males provide no benefits to females other than genetic material carried in sperm—which characterizes most mammals—this logic predicts that females will exhibit essentially zero sexual motivation outside of cycle days when conception is possible. In species in which males provide nongenetic material benefits, however, other things may often not be equal, and female sexual behavior may have functions related to the acquisition of these material benefits that makes predictions about the determinants of sexual motivation more difficult.

Considering first species in which males do not provide nongenetic benefits, the expectations regarding the endocrine regulation of sexual motivation are straightforward in that sexual behavior should be promoted by hormones that positively predict current fecundity but inhibited by hormones that negatively predict it. In all nonhuman mammals, estrogen is associated with follicle development and ovulation and thus estrogen should be a consistent positive predictor of sexual motivation in species without extended sexuality. In most nonhuman mammals, an extended diestrus similar in biology to the above described human luteal phase is associated with both zero immediate fecundity and enhanced progesterone production, leading to the prediction of an inhibitory effect of progesterone on sexual motivation. Most rodents are an exception, however, in which a fully formed corpus luteum occurs only in the case of conception (or pseudopregnancy), and progesterone peaks before ovulation and relatively soon after the estrogen peak, at a time when conception is possible (for reviews, see Carter 1992; Blaustein 2008). As such, rodents provide an interesting test case for the informational signaling content of hormones since progesterone (at least when not pregnant) should facilitate rather than inhibit sexual behavior. Finally, the expected effects of testosterone on sexual motivation are ambiguous due to its uncertain information content.

In species in which males provide nongenetic benefits to females—such as food, protection, grooming, and paternal care—females often exhibit "extended sexuality," which refers to sexual receptivity and/or proceptivity on cycle days for which conception is not possible (for an extensive and insightful discussion of extended sexuality, see Thornhill and Gangestad 2008). In chimpanzees, for instance, females mate promiscuously on follicular phase days outside of the fertile window, which has been explained as a paternity confusion device that decreases the probability that males will mistreat a female's offspring (for a review, see Emery Thompson 2009). Human females exhibit pronounced extended sexuality, with sexual behavior occurring at all times of the menstrual cycle. Various theories have proposed that extended sexuality in conjunction with concealed ovulatory timing may have evolved within the context of human pair-bonding as a means of promoting men's continued proximity to and material investments in their partners, since males would likely eschew such investment and instead compete to inseminate currently fertilizable females under conditions in which ovulatory timing could be precisely detected (see Alexander and Noonan 1979; Strassmann 1981; Symons 1979). If some version of these theories is correct, then one expects women's extended sexuality to be designed to have promoted the flow of nongenetic benefits from male partners within the context of relationship initiation and maintenance. Precisely what this design might be has been a neglected topic in the study of human sexuality. Grebe et al. (2013) presented evidence that women were more proceptive (i.e., initiated sex more often) during the luteal phase when they were more invested in their relationship than their long-term partners, which the authors interpreted as a means of motivating greater male investment. Overall, however, little is understood regarding the predictors of extended sexuality.

Although extended sexuality complicates the story considerably, I predicted that estrogen would be a positive predictor of women's sexual motivation at a within-cycle timescale. Women's extended sexuality could be regulated by nonhormonal mechanisms that are not linked to fecundity, such that hormonal and nonhormonal mechanisms operate concurrently. Whatever the determinants of extended sexuality, then, the arguments for greater sexual motivation during fecund cycle days still applies to humans, as the costs of sex should, other things equal, produce design for reduced sexual motivation on days when conception is not possible. My expectations regarding the effects of progesterone on women's sexual motivation were less certain: whereas progesterone clearly signals low fecundity within-cycles, it might also act as a positive signal of fecundity at the between-cycle timescale (see below).

Between-Cycle Predictions

Motivational systems function to allocate attention and behavioral effort to those adaptive problems that are currently most pressing. Many of the conditions under which ovarian hormones are reduced in humans at the between-cycle (or lifespan) timescale involve adaptive problems that demand immediate behavioral effort. During intensive lactation, for instance, a shift in attentional and behavioral resources away from sexuality and toward maternal care appears especially functional, and indeed evidence suggests that lactation is associated with a substantial decrease in women's libido (e.g., Avery et al. 2000; Forster et al. 1994; Rupp et al. 2013). As reviewed above, conditions such as lactation, negative energy balance secondary to food shortage, as well as menopause, are all associated with decreases in ovarian hormones such that elevated estrogen and progesterone could signal that a woman is experiencing a life stage during which sexuality takes on greater importance and should thus be allocated increased attention and motivation.

Whether calibration of sexual motivation to between-cycle fluctuations in ovarian hormones should extend to days outside of the fertile window is unclear. At issue is whether, for instance, a luteal phase day in a high-estrogen cycle should be treated differently by brain mechanisms than a luteal phase day in a low-estrogen cycle: immediate fecundity is zero on both days, but the general problem of mating and reproduction may be more important in a life stage sense in the higher estrogen cycle with greater fecundity. Increased sexual motivation across days of more fecund cycles could have functions related to mate-seeking or mate-evaluation during time periods when conception is more likely to occur in the near future, for example, or could signal paternity confidence to a partner through increased sexual behavior near the time of conception. The possibility of between-cycle effects of ovarian hormones even outside of the fertile window led me to hypothesize that estrogen will be the primary endocrine regulator of women's sexual motivation, because estrogen is positively correlated with fecundity at both within- and between-cycle timescales. If there were benefits to increased sexual motivation in more fecund cycles, then inhibitory effects of progesterone might disrupt between-cycle calibration of libido to fecundity, which led to the expectation that progesterone may not be a consistent regulator of sexual motivation in humans. Thus, although sexual desire may decrease in the luteal phase relative to the fertile window, my expectation was that this effect would be mediated by estrogen rather than progesterone.

Empirical Evidence for Hormonal Predictors of Female Sexual Motivation

Nonhuman Mammals

Estradiol and Progesterone

In domestic ruminants (e.g., sheep, cows, horses, pigs) in which males typically provide no nongenetic benefits and females experience a nonfecund diestrus phase associated with elevated progesterone, sexual activity is mostly confined to a behavioral estrous period during which conception is possible, estradiol (the dominant circulating form of estrogen in mammals) is necessary for and promotes female sexual behavior, and progesterone inhibits the current expression of sexual behavior (for reviews, see Beach 1976; Crowell-Davis 2007; Fabre-Nys and Gelez 2007; Katz 2007; Pedersen 2007). In rodents such as rats and mice, estradiol also promotes sexual receptivity (for reviews, see Carter 1992; Pfaff et al. 2002), but progesterone likewise facilitates the expression of sexual behavior after obligatory estrogen priming; in fact, the standard procedure for inducing sexual receptivity in ovariectomized rodents is injection of estradiol about 48 h prior to sexual testing followed by a progesterone injection within a few hours of such tests (e.g., Blaustein 2008; Powers 1970; Whalen 1974). Because elevated progesterone signals impending ovulation in rodents but not in ruminants, this contrast in the effects of progesterone on sexual motivation provides comparative evidence that brain mechanisms use hormones as signals of fecundity for the purpose of regulating sexual behavior. The differing effects of progesterone likewise demonstrate evolutionary plasticity in the signaling function of the same chemical in different species, which leaves open the possibility that progesterone could either promote or inhibit sexual motivation in women, depending on its information content.

Females of many nonhuman primate species, in contrast to other mammals, are sometimes sexually receptive outside of fecund cycle days and may even mate when ovariectomized (at least in captive conditions), leading to the idea that primate sexual motivation has been released from strict hormonal control (for reviews, see Dixson 1998; Wallen 2001). Nonetheless, extensive evidence supports greater sexual motivation during cycle days when estradiol is elevated and progesterone is low. In rhesus macaques—the most extensively studied nonhuman primate—sexual behavior under group housing conditions correlates strongly and positively with estradiol concentrations, but strongly and negatively with progesterone concentrations (Wallen et al. 1984). Likewise, manipulations of estradiol promote female sexual receptivity and proceptivity (e.g., Michael et al. 1978; Zumpe et al. 1983), and positive effects on female sexual initiation have been demonstrated even with unresponsive males tested during the nonbreeding season, thus ruling out changes in male behavior as the cause of increased sexual interactions (Zehr et al. 1998). Experimental manipulations of progesterone are less common in nonhuman primates, and as such decreases in luteal phase sexual behaviors could be caused by decreases in estradiol rather than by correlated increases in progesterone. Kendrick and Dixson (1985), however, demonstrated in marmosets that progesterone administration increased refusals of male mount attempts and decreased receptive and proceptive tongue-flicking displays relative to ovariectomized females without progesterone, whereas the opposite outcomes were obtained via administration of estradiol.

In ape species, extended sexuality is the most pronounced among nonhuman mammals, though evidence likewise implicates positive and negative effects of estradiol and progesterone, respectively, on female sexual behavior (for a review, see Emery Thompson 2009). Sexual behavior in chimps and bonobos generally coincides with expression of sex skin swellings, which appear to be promoted by estrogen but inhibited by progesterone (see Deschner et al. 2004; Emery and Whiten 2003; Heisterman et al. 1996); indeed, copulations in chimps tend to cease after the luteal phase rise in progesterone triggers sex skin detumescence (e.g., Deschner et al. 2004). These patterns are difficult to interpret in terms of female motivation in chimps given high rates of male sexual coercion in this species (Muller et al. 2007), but such coercion appears absent in bonobos in which females nonetheless accept more male copulation attempts when swollen than when not, which suggests greater receptivity when estradiol is higher (Furuichi and Hashimoto 2004). Female proceptivity in gorillas and orangutans appears to peak near the cycle peak in estradiol and testosterone, but ends abruptly with the rise in luteal phase progesterone (reviewed in Emery Thompson 2009).

Overall, then, the same hormonal correlates of sexual behavior are found in mammalian species with and without extended sexuality, namely positive associations with estradiol and negative associations with progesterone. The main difference between these groups may be in the extended time-course of estrogen production (or sensitivity) during the follicular phase in many primate species. Female chimpanzees and bonobos are in estrogen-dependent, highly swollen states for 10–12 days per cycle on average, and are partially swollen for even longer, despite the fact that maximum fecundity is probably restricted to the few days preceding and including the day of ovulation (Deschner et al. 2004; Heisterman et al. 1996). Thus, although extended sexual receptivity associated with prolonged sexual swellings must have nonconceptive functions, it appears that the endocrine mechanisms for it have been conserved from species with more transient periods of behavioral estrous.

Testosterone

In contrast to the clear roles of estradiol and progesterone in female sexual motivation among nonhuman mammals, the role of testosterone is less certain. Receptivity can be primed in ovariectomized females of many species without testosterone administration, which demonstrates that ovarian testosterone is not a necessary signal. Nonetheless, some evidence in rodents suggests that testosterone may produce synergistic effects with estradiol and other signals to produce higher levels of proceptivity than found without testosterone administration (e.g., Fernandez-Guasti et al. 1991). An early literature provided evidence for positive effects of androgens on the sexual receptivity of ovariectomized female macaques in captive pair tests (reviewed in Wallen 2001), but subsequent research provided strong evidence for estradiol as the primary positive regulator of sexual motivation in female macaques. Studies that induced artificial menstrual cycles in ovariectomized females via cyclic administration of estradiol, progesterone, and testosterone, for instance, demonstrated that estradiol promoted female sexual behavior but that omission of testosterone from the artificial cycles had no effects on sexual motivation (Michael et al. 1978; Zumpe et al. 1983). In addition, positive effects of androgens on sexual receptivity appear to require aromatizable androgens (e.g., Wallen and Goy 1977), suggesting that androgens may exhibit positive effects via conversion to estrogens.

At most, then, testosterone may supplement the positive effects of estradiol in the regulation of some components of mammalian female sexual motivation. That estradiol and progesterone appear to be the more important signals makes functional sense given that these hormones provide clear information regarding fecundity via their links to follicle development and corpus luteum formation. As explained earlier, it is less clear what functional information is provided by testosterone over the course of female reproductive cycles.

Humans

Between-Cycle (Lifespan) Timescale

Extant evidence for the role of ovarian hormones in women's sexual motivation has addressed effects of hormones at various time-scales. Perhaps the strongest evidence for hormonal regulation of libido has been at the lifespan timescale with respect to menopause and the partial reversal of its effects via hormone replacement therapy. Several studies have provided convergent evidence that sexual motivation decreases after natural or surgical menopause (Alexander et al. 2004; Dennerstein et al. 1977, 2005; Gracia et al. 2007), and additional evidence supports the efficacy of hormone replacement therapy in increasing sexual desire in menopausal women via use of estrogen (Dennerstein et al. 1980; Nathorst-Boos et al. 1993; Sherwin 1991; Wiklund et al. 1993), testosterone combined with estrogen (Braunstein et al. 2005; Floter et al. 2002; Sherwin et al. 1985), or testosterone alone (Davis et al. 2008). Although positive effects of both estrogen and testosterone provide ambiguous evidence regarding which signal may be most important in natural cycles (especially since testosterone can be converted to estrogen), these studies provide clear support for hormonal influences on the regulation of women's sexual motivation.

Other lines of evidence suggest that declining estrogen may be the key hormonal determinant of menopausal decreases in sexual motivation. First, in natural menopause, testosterone does not usually change significantly (beyond the gradual decrease with age), and in fact the decrease in sex hormone-binding globulin (SHBG) that occurs at menopause leads to an increase in free androgens (see Burger 2002); thus, decreases in sexual motivation at natural menopause occur in an endocrine context characterized by reduced estrogen but increased free testosterone. Second, studies that have followed women longitudinally across the menopausal transition have produced evidence that changes in sexual functioning are predicted by changes in estradiol but not by changes in androgens (Dennerstein et al. 2002, 2005; Freeman et al. 2007; McCoy 1990). Dennerstein et al. (2005) demonstrated that estradiol had a positive effect on sexual responsiveness that was independent of relationship factors such as feelings for the partner and acquisition of a new partner, both of which were positive predictors of sexual motivation. They concluded from their study: "…the effect of the menopausal transition on sexual function is overwhelmingly caused by the marked decline in E_2" (p. 179).

Decreases in sexual motivation attributable to menopausal declines in estradiol are consistent with the hypothesis that estradiol calibrates sexuality at between-cycle or lifespan timescales. The large and extended decline in estradiol may signal a nonreproductive lifestage during which the benefits of sex over evolutionary time were reduced on average, thus increasing the cost–benefit ratio of sexual behavior. Reduced sexual motivation at this time may not only have reduced the costs of sexual behavior but also functioned to reallocate attention and motivation toward those adaptive problems that were more fitness relevant at this lifestage: in particular, a shift in attention from mating and sexuality to investment in kin. Benefits of nonconceptive sex associated with extended sexuality may still have been present after menopause (if sex had positive effects on pair-bond maintenance and thus promoted male investment in descendants, for instance), although the mechanisms regulating this are unclear and may include nonhormonal signals.

Although the data on menopause are consistent with between-cycle hormonal regulation of sexual motivation, menopause is an extreme case of extended hormone deprivation and it is unclear whether between-cycle fluctuations in estradiol would calibrate sexual motivation in premenopausal women. Chemical suppression of ovarian hormone concentrations over an 8-week period led to large decreases in self-reported sexual functioning in a sample of healthy premenopausal women (Schmidt et al. 2009), which indicates that minimum hormone concentrations are necessary for normal sexual motivation in premenopausal women. In that study, 4 weeks of estradiol or progesterone replacement had not fully restored sexual function, though there was a positive trend for effects of estradiol and mechanisms related to sexual motivation could have "fast off, slow on" design properties such that positive effects of estradiol might have been found had treatment continued longer. No previous studies have followed premenopausal women across multiple natural cycles to test whether sexual desire is higher in cycles with higher estradiol concentrations, however, and this was thus one of the goals of the research presented later in this chapter.

Within-Cycle Timescale

A second line of evidence for the role of ovarian hormones in women's sexual motivation comes from studies that have measured changes in sexual behavior or subjective desire associated with phases of the menstrual cycle. A large number of studies have provided evidence that measures of sexual motivation are higher near ovulation (i.e., inside the fertile window) than at other times of the cycle (e.g., Adams et al. 1978; Bullivant et al. 2004; Dennerstein et al. 1994; Diamond and Wallen 2011; Harvey 1987; Matteo and Rissman 1984; Pillsworth et al. 2004; Stanislaw and Rice 1988; Wilcox et al. 2004). A number of these studies have detected increased sexual initiation by women or increased subjective desire, suggesting that these effects cannot be explained simply by greater male interest near ovulation (in fact, rates of male initiation appear to be relatively constant across the cycle; e.g., Van Goozen et al. 1997). Despite this evidence, there have been some failures to replicate the midcycle increases in sexual motivation (for a review, see Regan 1996), though studies that have more precisely verified the timing of ovulation via frequent hormone sampling or LH tests appear to be fairly consistent in demonstrating periovulatory peaks (e.g., Bullivant et al. 2004; Dennerstein et al. 1994; Diamond and Wallen 2011; Wilcox et al. 2004).

Increased sexual motivation near ovulation supports the calibration of women's libido to within-cycle fluctuations in fecundity, but does not precisely address which combination of hormonal signals might produce these effects. Only a small number of studies have measured hormone concentrations across broad regions of the cycle in order to test for associations with measures of sexual motivation in natural menstrual cycles (Dennerstein et al. 1994; Morris et al. 1987; Persky et al. 1978a, b; Van Goozen et al. 1997). None of these studies reported significant within-cycle relationships between fluctuations in hormone concentrations and fluctuations in sexual desire or behavior. A number of studies did report that women with higher average or midcycle testosterone exhibited higher frequencies of sexual behavior (Morris et al. 1987; Persky et al. 1978b; Van Goozen et al. 1997; c.f., Bancroft et al. 1983), which provides evidence for a between-women effect of androgens (but see Wallen 2001, for a reanalysis of the data in Persky et al. 1978a, b that suggests positive between-women associations between estradiol and measures of women's sexual initiation). The null effects in these studies at the within-cycle timescale leave unspecified the physiological signals that may regulate cyclic shifts in libido.

Various methodological issues may have limited the conclusiveness of the studies that tested within-cycle hormonal correlates of sexual motivation. Sample sizes were small, raising issues of power, and most of the articles were published before the widespread use of multilevel regression modeling, which both increases power in within-subject analyses (relative to some alternative approaches) and correctly accounts for correlated error terms given nested data. In addition to small numbers of subjects, hormone sampling was typically infrequent; only univariate analyses were performed, with no studies having tested the effects of particular hormones while controlling for the effects of other hormones; and only same-day associations between hormones and outcome variables were tested despite the possibility of time delays for the genomic effects of ovarian hormones. Given these limitations, my lab implemented a large study designed to provide further evidence regarding the hormonal predictors of sexual motivation in natural menstrual cycles.

New Research on Hormonal Predictors of Women's Sexual Motivation

This section summarizes the methods and findings from a recent study on the hormonal predictors of women's sexual motivation (Roney and Simmons 2013). We attempted to secure data regarding both within- and between-cycle hormonal predictors of sexual desire and behavior by collecting daily measures from women across two different menstrual cycles. This allowed more comprehensive tests of within-cycle relationships between ovarian hormones and sexuality, as well as the first tests in premenopausal women of whether sexual motivation is higher in cycles with higher vs. lower average concentrations of particular hormones.

Summary of Methods

Our final sample with available hormone data included 43 undergraduate women who self-reported being heterosexual and naturally cycling (mean age = 18.76 years). Thirty-six women collected daily saliva samples across two full menstrual cycles (separated by 1–2 months), while seven completed data collection for only the first cycle. In addition to collecting saliva samples, participants logged on to a secure website each day to complete survey measures.

Two primary dependent variables were the focus of the present report, one assessing subjective sexual desire and the other assessing sexual behavior. The desire item read: "How much did you desire sexual contact?" and was assessed on a 1–7 scale. The behavior item was a yes/no assessment of whether sex occurred that day, with sex defined as "intercourse or other forms of genital stimulation with another person." Additional items inquiring whether the woman or her partner initiated sex, as well as whether masturbation occurred, are described in Roney and Simmons (2013).

Saliva samples were assayed for estradiol, progesterone, and testosterone (intra- and interassay CVs were below 10 % for each hormone). All samples in a 9-day window centered on an initial estimate of the day of ovulation were sent for assay, as were samples from alternating days outside of this window (3621 total assays). Once hormone values were available, the day of ovulation was re-estimated based on the conjunction of the midcycle estradiol drop and the beginning of the luteal phase increase in progesterone.

Data analyses employed multi-level regression models, which allowed tests of within-cycle (Level-1) predictors of sexual motivation (e.g., do day-to-day fluctuations in estradiol predict within-cycle fluctuations in desire?); within-women, between-cycle (Level-2) predictors (e.g., does change in average estradiol across the same woman's two cycles predict change in her average sexual motivation?); and between-women (Level-3) predictors (e.g., do women with higher estradiol on average report higher desire on average?). We predicted that estradiol would positively predict our measures of sexual motivation at all three levels, based on the positive links between estradiol and fecundity at both within- and between-cycle timescales. All three hormones were entered simultaneously into the regression models, giving each equal opportunity to predict the dependent variables. At Level-1 (within-cycle timescale), we constructed separate regression models testing the effects of current day, 1-day lag, and 2-day lag hormone concentrations to account for possible time delays in the effects of ovarian hormones. As mentioned earlier, estrogen administration primes sexual receptivity in female rodents at a time lag of approximately 48 h, and based on those findings we expected the strongest effects at a 2-day lag.

Summary of Results

At the within-cycle (Level-1) level of analysis, estradiol measured 2 days earlier was a positive predictor of current day subjective sexual desire, $b=0.16$, $p=0.01$, whereas current day estradiol had a marginally significant effect, $b=0.09$, $p=0.096$. Unlike estradiol, we had not predicted significant effects of progesterone, but found strong evidence for negative associations with sexual desire at all three timescales: two-day lag, $b=-0.20$, $p=0.0001$; one-day lag, $b=-0.11$, $p=0.04$; current day: $b=-0.13$, $p=0.01$. There were no significant effects of testosterone at any timescale.

Figure 6.2 plots estimated sexual desire against estimated day of the cycle, with cycles aligned on the day of ovulation as day zero. Progesterone concentrations are also plotted on the secondary y-axis. It can be seen, first, that there was a visible decrease in sexual desire just as progesterone was reaching its highest luteal phase values. Second, although less obvious visually, there was a significant within-cycle effect of fertile window timing (days −5 to 0) on sexual desire, with higher desire on average inside the fertile window (mean = 3.74) than on other days (mean = 3.48), $b=0.26$, $p=0.023$. Furthermore, the rise in luteal progesterone mediated the decrease in desire when moving from the fertile window to the luteal phase, whereas estradiol and testosterone did not mediate this pattern (for statistical details, see Roney and Simmons 2013). These results are consistent with progesterone acting as a within-cycle stop signal that truncates the midcycle rise in sexual motivation.

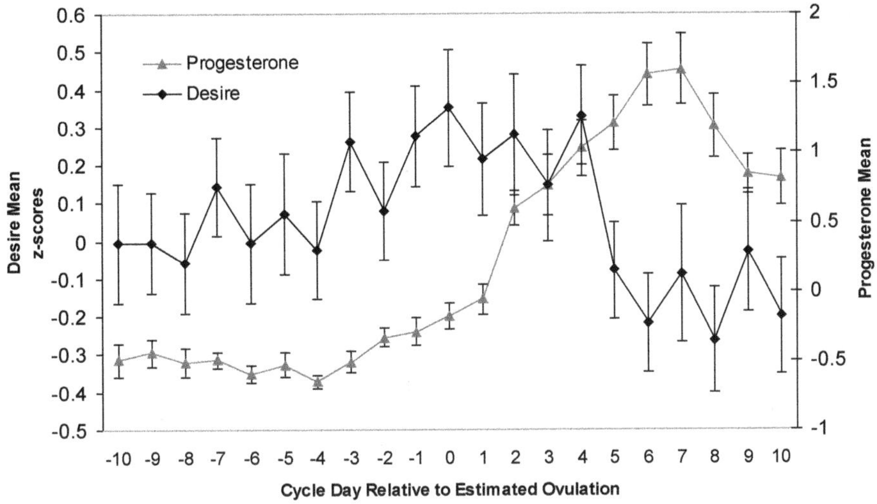

Fig. 6.2 Mean desire for sex and mean progesterone concentrations aligned against estimated day of cycle, where day zero represents the estimated day of ovulation. Values are standardized within-cycles such that zero points on the y-axes represent the mean values within a given cycle. *Error bars* are SEM

One other variable was a within-cycle predictor of self-reported desire: weekend timing. Figure 6.3 plots average sexual desire against day of the week. It can be seen, first, that Monday was a poor day for desire. Friday and Saturday were associated with pronounced increases in desire, on average, and a binary weekend variable comparing Friday and Saturday to other days of the week was a significant within-cycle predictor of desire, $b=0.40$, $p<0.0001$. The weekend timing and hormone variables were independent of one another and did not interact in the prediction of sexual desire.

There were no significant effects of any hormone at the within-woman, between-cycle or between-women levels of analysis. Thus, contrary to our prediction, among participants with two cycles of data, women did not experience higher desire on average in the cycle with higher mean estradiol. Variance in hormone concentrations was restricted at the between-cycle level, however, as hormone values were similar on average within-women across their two cycles. Thus, although we found no evidence for between-cycle calibration of desire to hormone concentrations, more rigorous tests of such calibration may require following women across cycles with larger differences in hormone concentrations than those observed in this study.

For sexual behavior, logistic mixed regression models were used to model the probability of sex on given response days. Two variables were significant predictors at the within-cycle level of analysis. Current day estradiol positively predicted the probability of sex, with a one standard deviation increase in estradiol associated with a 34% increase in the odds of sex relative to other days in the same cycle ($p=0.02$). Weekend timing was also a positive predictor, with the odds of sex approximately three times greater on weekend days relative to weekdays ($p<0.0001$).

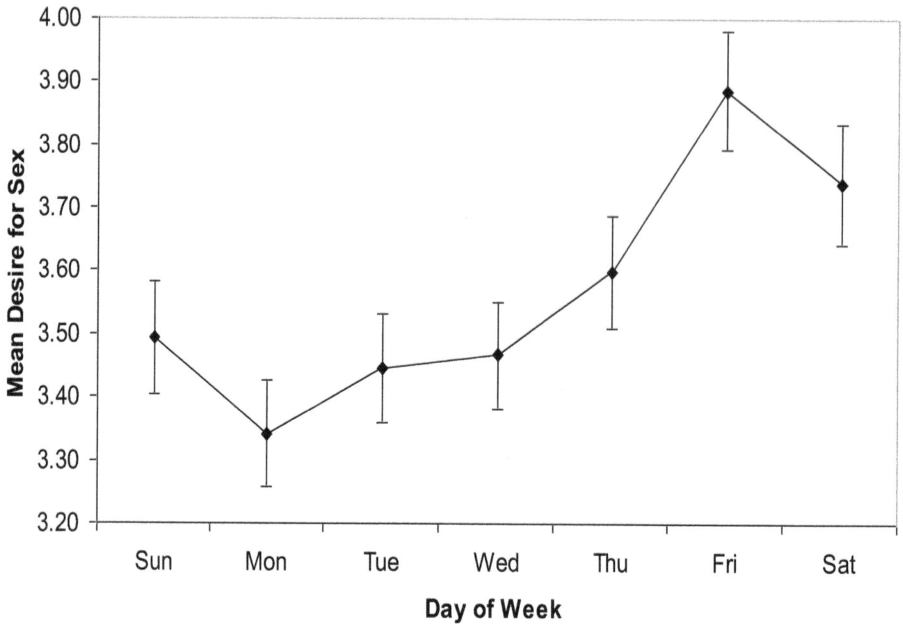

Fig. 6.3 Mean desire for sex aggregated across all women and plotted against day of the week. *Error bars* are SEM

Unlike sexual desire, there was no evidence of a periovulatory peak in sexual behavior. At higher levels, the only significant finding was a negative Level-2 effect of progesterone, meaning that for those women with two cycles of data, sexual frequency tended to be lower in the cycle with higher average progesterone.

Implications of the New Data and Directions for Future Research

The patterns reported in Roney and Simmons (2013) are consistent with those found in females of many nonhuman primates: estradiol was a positive predictor of indices of sexual motivation, whereas progesterone was a negative predictor. This similarity argues for homologous brain mechanisms in the regulation of human and nonhuman sexual motivation, although the extent of modification of those mechanisms for new functions in humans is an open question. At the within-cycle timescale, functions may be similar across species: avoidance of the costs of sexual behavior when conception is absent as a countervailing fitness benefit, as well as relative allocation of attention and motivation to other adaptive problems during non-fecund regions of the cycle. With respect to the latter, for instance, Fessler

(2003) has argued that attention to feeding is downregulated during the follicular phase of ovulatory cycles when mating takes on greater relative importance, but then upregulated in the luteal phase. Trade-offs in the amount of attention and motivation that can be allocated to different tasks may thus help explain the luteal phase decrease in sexual desire, as this region of the cycle may involve an increase in attention to problems unrelated to mating.

One way in which we anticipated modification of regulatory mechanisms in humans was with respect to the role of progesterone. In a nonpregnant state, progesterone is a positive signal of between-cycle fecundity. Because we reasoned that higher sexual motivation may have been functional across days of more fecund cycles in pair-bonding humans, we did not expect progesterone to have the strong inhibitory effects that it is has in most nonhuman mammals. That expectation was not supported, as progesterone was the most consistent negative predictor of desire in our study. Our prediction of a between-cycle positive association between estradiol and sexual motivation was also not supported, as there was no evidence for greater desire or frequency of sexual behavior in the cycle with higher mean estradiol among those women with two cycles of data.

Although we found no evidence for between-cycle effects of hormones, other lines of evidence suggest that they do occur. Dennerstein et al. (2005) found that estradiol was the only endocrine variable to predict longitudinal changes in sexual functioning across the menopausal transition. Likewise, although we found that progesterone was the strongest predictor of sexual desire at the within-cycle timescale, progesterone cannot be the primary regulator of sexual desire or else desire should increase at menopause when progesterone is consistently lower. Instead, the most sensible model appears to be one in which estrogen priming maintains neural networks in a state in which sexual motivation is upregulated, progesterone produces relative declines in such motivation against the background of estrogen priming, and extended estrogen deprivation then produces a general decrease in sexual motivation. Estrogen, then, may have effects at different timescales, with longer-term effects on baseline sexual motivation but also more acute effects associated with rapid increases in estrogen near ovulation. At a between-cycle timescale, sexual motivation may not respond to modest changes in estradiol from cycle to cycle, but likely does respond to more extended periods of estrogen deprivation.

Despite the evidence for within-cycle endocrine predictors of sexual motivation, sexual desire and behavior occurred at all times of the cycle and nonhormonal factors must also be important in explaining variance in libido. Weekend timing—which was uncorrelated with hormone concentrations—was a strong and consistent predictor of both desire and behavior in our study. Given the likely exposure to social stimuli that occurs differentially on weekends in an undergraduate sample, this result suggests that sexual motivation is calibrated simultaneously by exogenous social stimuli and endogenous endocrine signals. The weekend effect was independent of and did not interact with the hormone effects, which suggests that these endogenous and exogenous influences may act separately, at least within the hormonal milieu that characterizes premenopausal women. Retaining the capacity to respond with sexual desire to social stimuli at any point in the cycle (i.e.

independent of hormone concentrations) may be functional given long-term pair-bonds in humans, because desirable long-term partners could be met at any time, and expressions of sexual desire to current partners could have signaling functions that are uncorrelated with current fecundity. Thus, although sex hormones may have main effects in modulating sexual motivation up or down across specific time periods, other variables associated with relationship initiation and maintenance are likely to be important determinants of variance in women's libido.

Calibration of sexual motivation to relationship factors intersects with the broad issue of extended sexuality, and determination of the variables that predict extended sexuality is an important direction for future research on women's sexual motivation. Women's sexual desire tends to decline with increasing relationship length (e.g., Dennerstein et al. 2005; Murray and Milhausen 2012; Pillsworth et al. 2004), which is consistent with the possibility that high sexual motivation has functions related to relationship initiation and pair-bond establishment, since frequent sexual behavior may have been a strong signal of commitment to male partners over most of human history given both the absence of contraception and the large physiological costs associated with human gestation. If true, then high sexual motivation (even on nonfecund days) early in relationships may have functioned as a mate acquisition tactic for women. Once pair-bonds were established and offspring produced, however, attention and motivation may have undergone relative shifts toward more pressing adaptive problems related to parental investment. Even within established reproductive relationships, though, maintenance of some nonreproductive sexual motivation may have functioned to promote continued male investment in the relationship. These speculations regarding the functions and patterns of women's extended sexuality are consistent with available evidence, but more theoretical and empirical work is necessary to understand the mechanisms that regulate nonreproductive sexual motivation.

Conclusion

The mechanisms that regulate sexual motivation should be designed to increase the desire for sex under circumstances in which such desire had higher fitness benefits than costs (including the opportunity costs of alternative behaviors) on average over the course of evolution. In general, the benefit-to-cost ratio of sexual behavior was likely higher on fecund days than on days with little or no conception risk, leading one to expect higher sexual motivation during fecund regions of the cycle. That expectation is borne out across all mammalian species that have been examined. Hormones produced by the ovary can carry information to the brain regarding fecundity, and one can therefore predict based on that information which hormonal signals should promote and inhibit female sexual motivation. Consistent with a fecundity-signaling role, estradiol promotes but progesterone inhibits sexual receptivity and proceptivity in the vast majority of nonhuman mammalian females. Despite this phylogenetic pattern—as well as the physiological homologies between the reproductive cycles of human and many nonhuman species—it has long been thought that testosterone is the primary regulator of women's libido (for reviews, see Wallen 2001, 2013). As reviewed here, however, both findings regarding the endocrine predictors of changing sexual motivation at menopause (e.g., Dennerstein et al. 2005) and new data on the hormonal correlates of sexual desire in natural cycles (Roney and Simmons 2013) support estradiol as the primary positive regulator of women's sexual motivation. The Roney and Simmons (2013) study likewise identified progesterone as the primary negative predictor of women's desire. It therefore appears that conserved brain mechanisms in women use specific hormonal signals to upregulate sexual motivation during fecund relative to subfecund periods of time.

This chapter has focused on the role of hormonal signals in women's sexual motivation, but there is a need for additional functional analyses of nonhormonal predictors of libido. Theorists have persuasively argued that women's nonreproductive, extended sexuality likely evolved to promote the acquisition of nongenetic, material resources from male partners (e.g., Thornhill and Gangestad 2008). However, the design of the brain mechanisms that govern such sexual behavior—including both the proximate variables that activate increases in sexual motivation as well as the physiological signals that respond to those variables—has not been extensively investigated. Thus, although there is now strong evidence that women have inherited from nonhuman ancestors hormonal mechanisms that calibrate sexual motivation to fluctuations in fecundity, this is only one part of the story, and future research is necessary to systematically test functional hypotheses regarding additional predictors of variance in libido.

References

Abraham, G. E. (1974). Ovarian and adrenal contribution to peripheral androgens during the menstrual cycle. *Journal of Clinical Endocrinology and Metabolism, 39,* 340–346.

Adams, D. B., Gold, A. R., & Burt, A.D. (1978). Rise in female-initiated sexual activity at ovulation and its suppression by oral contraceptives. *New England Journal of Medicine, 299,* 1145–1150.

Alexander, J. L., Kotz, K., Dennerstein, L., Kutner, S. J., Wallen, K., & Notelovitz, M. (2004). The effects of postmenopausal hormone therapies on female sexual functioning: a review of double-blind, randomized controlled trials. *Menopause, 11,* 749–765.

Alexander, R. D., & Noonan, K. M. (1979). Concealment of ovulation, paternal care, and human social evolution. In N. A. Chagnon & W. G. Irons (Eds.), *Evolutionary biology and human social behavior: An Anthropological Perspective* (pp. 436–453). Scituate: North Duxbury.

American Psychiatric Association. (2000). *Diagnostic and statistical manual of mental disorders* (4th ed., text rev.).

Avery, M. D., Duckett, L., & Frantzich, C. R. (2000). The experience of sexuality during breastfeeding among primiparous women. *Journal of Midwifery and Women's Health, 45,* 227–237.

Bancroft, J., Sanders, D., Davidson, D., & Warner, P. (1983). Mood, sexuality, hormones, and the menstrual cycle. III. Sexuality and the role of androgens. *Psychosomatic Medicine, 45,* 509–516.

Beach, F. A. (1976). Sexual attractivity, proceptivity, and receptivity in female mammals. *Hormones and Behavior, 7,* 105–138.

Blaustein, J. D. (2008). Neuroendocrine regulation of feminine sexual behavior: Lessons from rodent models and thoughts about humans. *Annual Review of Psychology, 59,* 93–118.

Braunstein, G. D., Sundwall, D. A., Katz, M., Shifren, J. L., Buster, J. E., & Simon, J. A., et al. (2005). Safety and efficacy of a testosterone patch for the treatment of hypoactive sexual desire disorder in surgically menopausal women. *Archives of Internal Medicine, 165,* 1582–1589.

Bullivant, S. B., Sellegren, S. A., Stern, K., Spencer, N. A., Jacob, S., Mennella, J. A., et al. (2004). Women's sexual desire during the menstrual cycle: Identification of the sexual phase by noninvasive measurement of luteinizing hormone. *Journal of Sex Research, 41,* 82–93.

Burger, H. G. (2002). Androgen production in women. *Fertility and Sterility, 77,* S3–S5.

Campbell, B. C., & Ellison, P. T. (1992). Menstrual variation in salivary testosterone among regularly cycling women. *Hormone Research, 37,* 132–136.

Carter, C. S. (1992). Neuroendocrinology of sexual behavior in the female. In J. B. Becker, S. M. Breedlove, & D. Crews (Eds.), *Behavioral endocrinology* (pp. 71–95). Cambridge: MIT Press.

Crowell-Davis, S. L. (2007). Sexual behavior of mares. *Hormones and Behavior, 52,* 12–17.

Davis, S. R., Moreau, M., Kroll, R., Bouchard, C., Panay, N., Gass, M., et al. (2008). Testosterone for low libido in postmenopausal women not taking estrogen. *New England Journal of Medicine, 359,* 2005–2017.

Dennerstein, L., Wood, C., & Burrows, G. D. (1977). Sexual response following hysterectomy and oophorectomy. *Obstetrics & Gynecology, 49,* 92–96.

Dennerstein, L., Burrows, G. D., Wood, C., & Hyman, G. (1980). Hormones and sexuality: Effect of estrogen and progesterone. *Obstetrics & Gynecology, 56,* 316–322.

Dennerstein, L., Gotts, G., Brown, J. B., Morse, C. A., Farley, T. M. M., & Pinol, A. (1994). The relationship between the menstrual cycle and female sexual interest in women with pms complaints and volunteers. *Psychoneuroendocrinology, 19,* 293–304.

Dennerstein, L., Randolph, J., Taffe, J., Dudley, E., & Burger, H. (2002). Hormones, mood, sexuality, and the menopausal transition. *Fertility and Sterility, 77,* S42–S48.

Dennerstein, L., Lehert, P., & Burger, H. (2005). The relative effects of hormones and relationship factors on sexual function of women through the natural menopausal transition. *Fertility and Sterility, 84,* 174–180.

Deschner, T., Heistermann, M., Hodges, K., & Boesch, C. (2004). Female sexual swelling size, timing of ovulation, and male behavior in wild West African chimpanzees. *Hormones and Behavior, 46,* 204–215.

Diamond, L. M., & Wallen, K. (2011). Sexual minority women's sexual motivation around the time of ovulation. *Archives of Sexual Behavior, 40,* 237–246.

Dixson, A. F. (1998). *Primate sexuality: Comparative studies of the prosimians, monkeys, apes, and humans*. Oxford: Oxford University Press.

Ellison, P. T. (2001). *On fertile ground.* Cambridge: Harvard University Press.

Emery Thompson, M. (2009). The endocrinology of intersexual relationships in the apes. In P. T. Ellison & P. B. Gray (Eds.), *Endocrinology of social relationships* (pp. 196–222). Cambridge: Harvard University Press.

Emery, M. A., & Whitten, P. L. (2003). Size of sexual swellings reflects ovarian function in chimpanzees (Pan troglodytes). *Behavioral Ecology and Sociobiology, 54,* 340–351.

Fabre-Nyes, C., & Gelez, H. (2007). Sexual behavior in ewes and other domestic ruminants. *Hormones and Behavior, 52,* 18–25.

Fauser, B. C. J. M., & Van Heusden, A. M. (1997). Manipulation of human ovarian function: Physiological concepts and clinical consequences. *Endocrine Reviews, 18,* 71–106.

Fernandez-Guasti, A., Vega-Matuszczyk, J., & Larsson, K. (1991). Synergistic action of estradiol, progesterone and testosterone on rat proceptive behavior. *Physiology & Behavior, 50,* 1007–1011.

Fessler, D. M. T. (2003). No time to eat: An adaptationist account of periovulatory behavioral changes. *Quarterly Review of Biology, 78,* 3–21.

Floter, A., Nathorst-Boos, J., Carlstrom, K., & von Schoultz, B. (2002). Addition of testosterone to estrogen replacement therapy in oophorectomized women: Effects on sexuality and well-being. *Climacteric, 5,* 357–365.

Forster, C., Abraham, S., Taylor, A., & Llewellyn-Jones, D. (1994). Psychological and sexual changes after the cessation of breast-feeding. *Obstetrics & Gynecology, 84,* 872–873.

Freeman, E. W., Sammel, M. D., Lin, H., Gracia, C. R., Pien, G. W., Nelson, D. B., et al. (2007). Symptoms associated with menopausal transition and reproductive hormones in midlife women. *Obstetrics & Gynecology, 110,* 230–240.

Furuichi, T., & Hashimoto, C. (2004). Sex differences in copulation attempts in wild bonobos at Wamba. *Primates, 45,* 59–62.

Gracia, C. R., Freeman, E. W., Sammel, M. D., Lin, H., & Mogul, M. (2007). Hormones and sexuality during the transition to menopause. *Obstetrics & Gynecology, 109,* 831–840.

Grebe, N. M., Gangestad, S. W., Garver-Apgar, C. E., & Thornhill, R. (2013). Women's luteal-phase sexual proceptivity and the functions of extended sexuality. *Psychological Science, 24,* 2106–2110.

Hall, J. E. (2004). Neuroendocrine control of the menstrual cycle. In J. F. Strauss & R. L. Barbieri (Eds.), *Yen and Jaffe's reproductive endocrinology* (pp. 195–211). Philadelphia: Elsevier.

Harvey, S. M. (1987). Female sexual behavior: Fluctuations during the menstrual cycle. *Journal of Psychosomatic Research, 31,* 101–110.

Heistermann, M., Mole, U., Vervaecke, H., van Elsacker, L., & Hodges, J. K. (1996). Application of urinary and fecal steroid measurements for monitoring ovarian function and pregnancy in the bonobo (*Pan paniscus*) and evaluation of perineal swelling patterns in relation to endocrine events. *Biology of Reproduction, 55,* 844–853.

Katz, L. S. (2007). Sexual behavior of domesticated ruminants. *Hormones and Behavior, 52,* 56–63.

Kendrick, K. M., & Dixson, A. F. (1985). Effects of oestradiol 17B, progesterone and testosterone upon proceptivity and receptivity in ovariectomized common marmosets (*Callithrix jacchus*). *Physiology & Behavior, 34,* 123–128.

Laumann, E. O., Paik, A., & Rosen, R. C. (1999). Sexual dysfunction in the United States: Prevalence and predictors. *Journal of the American Medical Association, 281,* 537–544.

Lipson, S. F., & Ellison, P. T. (1996). Comparison of salivary steroid profiles in naturally occurring conception and non-conception cycles. *Human Reproduction, 11,* 2090–2096.

Matteo, S., & Rissman, E. F. (1984). Increased sexual activity during the midcycle portion of the human menstrual cycle. *Hormones and Behavior, 18,* 249–255.

McCoy, N. L. (1990). Estrogen levels in relation to self-reported symptoms and sexuality in perimenopausal women. *Annals of the New York Academy of Sciences, 592,* 450–452.

Michael, R. P., Richter, M. C., Cain, J. A., Zumpe, D., & Bosnall, R. W. (1978). Artificial menstrual cycles, behavior and the role of androgens in female rhesus monkeys. *Nature, 275,* 439–440.

Morris, N. M., Udry, R. J., Khan-Dawood, F., & Dawood, M. Y. (1987). Marital sex frequency and midcycle female testosterone. *Archives of Sexual Behavior, 16,* 27–37.

Muller, M. N., Kahlenberg, S. M., Emery Thompson, M., & Wrangham, R. W. (2007). Male coercion and the costs of promiscuous mating for female chimpanzees. *Proceedings of the Royal Society of London B, 274,* 1009–1014.

Murray, S. H., & Milhausen, R. R. (2012). Sexual desire and relationship duration in young men and women. *Journal of Sex & Marital Therapy, 38,* 28–40.

Nathorst-Boos, J., Wiklund, I., Mattsson, L. A., Sandin, K., & von Schoultz, B. (1993). Is sexual life influenced by transdermal estrogen therapy? A double blind placebo controlled study in postmenopausal women. *Acta Obstetrics Gynecology Scandanavia, 72,* 656–660.

Pedersen, L. J. (2007). Sexual behaviour in female pigs. *Hormones and Behavior, 52,* 64–69.

Persky, H., Charney, N., Lief, H. I., O'Brien, C. P., Miller, W. R., & Strauss, D. (1978a). The relationships of plasma estradiol level to sexual behavior in young women. *Psychosomatic Medicine, 40,* 523–535.

Persky, H., Lief, H. I., Strauss, D., Miller, W. R., & O'Brien, C. P. (1978b). Plasma testosterone level and sexual behavior of couples. *Archives of Sexual Behavior, 7,* 157–173.

Pfaff, D., Frohlich, J., & Morgan, M. (2002). Hormonal and genetic influences on arousal—sexual and otherwise. *Trends in Neuroscience, 25,* 45–50.

Pillsworth, E. G., Haselton, M. G., & Buss, D. M. (2004). Ovulatory shifts in female sexual desire. *Journal of Sex Research, 41,* 55–65.

Powers, J. B. (1970). Hormonal control of sexual receptivity during the estrous cycle of the rat. *Physiology & Behavior, 5,* 831–835.

Regan, P. C. (1996). Rhythms of desire: The association between menstrual cycle phases and female sexual desire. *Canadian Journal of Human Sexuality, 5,* 145–156.

Roney, J. R., & Maestripieri, D. (2002). The importance of comparative and phylogenetic analyses in the study of adaptation. *Behavioral and Brain Sciences, 25,* 525.

Roney, J. R., & Simmons, Z. L. (2013). Hormonal predictors of sexual motivation in natural menstrual cycles. *Hormones and Behavior, 63,* 636–645.

Rupp, H. A., James, T. W., Ketterson, E. D., Sengelaub, D. R., Ditzen, B., & Heiman, J. R. (2013). Lower sexual interest in postpartum women: Relationship to amygdala activation and intranasal oxytocin. *Hormones and Behavior, 63,* 114–121.

Schmidt, P. J., Steinberg, E. M., Negro, P. P., Haq, N., Gibson, C., & Rubinow, D. R. (2009). Pharmacologically induced hypogonadism and sexual function in healthy young women and men. *Neuropsychopharmacology, 34,* 565–576.

Sherwin, B. B. (1991). The impact of different doses of estrogen and progestin on mood and sexual behavior in postmenopausal women. *Journal of Clinical Endocrinology and Metabolism, 72,* 336–343.

Sherwin, B. B., Gelfand, M. M., & Brender, W. (1985). Androgen enhances sexual motivation in females: A prospective, crossover study of sex steroid administration in the surgical menopause. *Psychosomatic Medicine, 47,* 339–351.

Stanislaw, H., & Rice, F. J. (1988). Correlation between sexual desire and menstrual cycle characteristics. *Archives of Sexual Behavior, 17,* 499–508.

Strauss III, J. F., & Williams, C. J. (2004). The ovarian life cycle. *Reproductive Endocrinology. Physiology, Pathophysiology, and Clinical Management, 4,* 213–253.

Strassmann, B. I. (1981). Sexual selection, paternal care, and concealed ovulation in humans. *Ethology and Sociobiology, 2,* 31–40.

Strassmann, B. I. (1997). The biology of menstruation in Homo sapiens: total lifetime menses, fecundity, and nonsynchrony in a natural-fertility population. *Current Anthropology, 38,* 123–129.

Stuckey, B. G. A. (2008). Female sexual function and dysfunction in the reproductive years: The influence of endogenous and exogenous sex hormones. *Journal of Sexual Medicine, 5,* 2282–2290.

Symons, D. (1979). *The evolution of human sexuality*. Oxford: Oxford University Press.

Thornhill, R., & Gangestad, S. W. (2008). *The evolutionary biology of human female sexuality*. Oxford: Oxford University Press.

Van Goozen, S. H. M., Wiegant, V. M., Endert, E., Helmond, F. A., & Van de Poll, N.E. (1997). Psychoneuroendocrinological assessment of the menstrual cycle: The relationship between hormones, sexuality, and mood. *Archives of Sexual Behavior, 26,* 359–382.

Venners, S. A., Liu, X., Perry, M. J., Korrick, S. A., Li, Z., Yang, F., et al. (2006). Urinary estrogen and progesterone metabolite concentrations in menstrual cycles of fertile women with non-conception, early pregnancy loss or clinical pregnancy. *Human Reproduction, 21,* 2272–2280.

Wallen, K. (2001). Sex and context: Hormones and primate sexual motivation. *Hormones and Behavior, 40,* 339–357.

Wallen, K. (2013). Women are not as unique as thought by some: Comment on "Hormonal predictors of sexual motivation in natural menstrual cycles," by Roney and Simmons. *Hormones and Behavior, 63,* 634–635.

Wallen, K., & Goy, R. W. (1977). Effects of estradiol benzoate, estrone, and propionates of testosterone or dihydrotestosterone on sexual and related behaviors of rhesus monkeys. *Hormones and Behavior, 9,* 228–248.

Wallen, K., Winston, L. A., Gaventa, S., Davis-DaSilva, M., & Collins, D. C. (1984). Periovulatory changes in female sexual behavior and patterns of ovarian steroid secretion in group-living rhesus monkeys. *Hormones and Behavior, 18,* 431–450.

Whalen, R. E. (1974). Estrogen-progesterone induction of mating in female rats. *Hormones and Behavior, 5,* 157–162.

Wiklund, I., Karlberg, J., & Mattsson, L. A. (1993). Quality of life of postmenopausal women on a regimen of transdermal estradiol therapy: A double-blind placebo-controlled study. *American Journal of Obstetrics and Gynecology, 168,* 824–830.

Wilcox, A. J., Baird, D. D., Dunson, D. B., McConnaughey, D. R., Kesner, J. S., & Weinberg, C. R. (2004). On the frequency of intercourse around ovulation: Evidence for biological influences. *Human Reproduction, 19,* 1539–1543.

Wilcox, A. J., Weinberg, C. R., & Baird, D. D. (1998). Post-ovulatory ageing of the human oocyte and embryo failure. *Human Reproduction, 13,* 394–397.

Zehr, J. L., Maestripieri, D., & Wallen, K. (1998). Estradiol increases female sexual initiation independent of male responsiveness in rhesus monkeys. *Hormones and Behavior, 33,* 95–103.

Zumpe, D., Bosnall, R. W., & Michael, R. P. (1983). Behavior of rhesus monkeys during 28-day cycles of estrogen treatment. *Behavioral Neuroscience, 97,* 615–623.

Chapter 7
Evolutionary Science of Female Orgasm

John R. Wheatley and David A. Puts

Introduction

Despite decades of scientific study, the functional significance of the human female orgasm remains unsettled. Whereas male orgasm is usually coupled with ejaculation, there is no parallel association between women's orgasm and a physiological process critical to reproduction. Indeed, even in a culture in which women's orgasm was reportedly unknown, women managed to conceive without noticeable difficulty (Messenger 1971). It strikes many as curious that an event of such remarkable psychological import that it has been called *la petite mort* ("the little death") would have no obvious reproductive function. This apparent paradox has inspired a number of scientists to offer hypotheses about the utility of the female orgasm, resulting in a heated and ongoing debate. As we discuss, some researchers have suggested that orgasm in women is a nonfunctional by-product of orgasm in men, whereas others suggest that women's orgasm has been shaped by selection for its own function—in other words, that orgasm is an adaptation in women. In this chapter, we outline the debate between these viewpoints and review evidence for several functional hypotheses that are among the most plausible.

Adaptations and By-products

An adaptation is a morphological, physiological, psychological, or behavioral trait that has been shaped by natural selection to perform some function that contributed to the inclusive fitness in its ancestral bearers. The more precisely and efficiently a trait performs its putative function, the more probable its status as an adaptation for this function (Buss et al. 1998; Williams 1966). It is not

J. R. Wheatley (✉) · D. A. Puts
Department of Anthropology, Pennsylvania State University, State College, PA, USA
e-mail: jrw5492@psu.edu

T. K. Shackelford, R. D. Hansen (eds.), *The Evolution of Sexuality*,
Evolutionary Psychology, DOI 10.1007/978-3-319-09384-0_7

the case, however, that traits that appear to serve some function are necessarily adaptations for that apparent function. For example, it is possible for a preexisting trait to acquire a new fitness-enhancing effect without any modification by natural selection (Andrews et al. 2002). Such a trait is termed an *exaptation* (Gould and Vrba 1982).

Traits that are adaptations in one sex are not necessarily adaptations in the other. Males and females commonly have different phenotypic optima for the same trait, resulting in sexually antagonistic selection pressures. Selection is often able to disrupt the expression of a sex-specific adaptation in the opposite sex via the regulation of the associated genes by sex steroids. For example, antler growth is regulated by testosterone (Suttie et al. 1995), so that female elk, moose, and white-tailed deer, whose testosterone levels are low, do not grow antlers. Disruption of sex-specific adaptations in the opposite sex is not always complete, however, so that the ontogenetic mechanisms that produce an adaptation in one sex may result in nonadaptive *by-products* in the other. For example, nipples are functional in mammalian females, but of no apparent utility in males. Nipples seem to exist in males only because males and females share genes and some developmental pathways, and the pathways necessary for the development of female nipples are incompletely disrupted in males (Symons 1979). To take another example, female leopard geckos appear to possess hemipenes only because of selection for these structures in males. Hemipenes are paired, bilateral copulatory organs, either of which may be used during copulation (Holmes et al. 2005). Much like male nipples in mammals, female hemipenes in leopard geckos have no apparent function and are reduced in their expression, being less than one twentieth the cross-sectional area of males' and with much smaller associated musculature.

When strong selection for a trait in one sex maintains its vestiges in the other sex, this is known as *sexually antagonistic pleiotropy*. Both male nipples in mammals and female hemipenes in leopard geckos demonstrate two important characteristics of sexually antagonistic by-products. First, by-products are notable for the reduction of their expression in one sex relative to their expression in the other. Second, they do not appear to have been shaped by selection to efficiently perform any function. These two characteristics are central to our consideration of the evolution of the human female orgasm.

Occasionally, selection may modify a by-product to serve a new function, producing what is known as a secondary adaptation (Gould and Vrba 1982). Although antlers evolved in ancestral deer in the service of male contests (Clutton-Brock 1982), antlers also develop in females of one extant deer species, caribou or reindeer (*Rangifer tarandus*). Female antlers differ in shape from male antlers, grow at a different time of year, and are used in competition over feeding sites (Henshaw 1968) and perhaps in defense of young (Espmark 1971), but not in competition for mates. Antlers thus appear to be secondary adaptations in female caribou. We note here again that, despite having originated as by-products, secondary adaptations do not appear vestigial or functionless.

Is Female Orgasm Vestigial?

Most women do not reliably experience orgasm during penile–vaginal intercourse without the additional stimulation provided by hands, vibrators, etc. A review of 33 studies on women's orgasms revealed that only about 25 % of women report "reliably" achieving orgasms during intercourse, while perhaps 5 % of women report never having achieved orgasm through any type of stimulation (Lloyd 2005). In addition, the glans clitoridis is much smaller than its anatomical homologue in the male, the glans penis, and therefore might be deemed a vestige of the development of male genitalia. The diminished size of the glans clitoridis, the greater difficulty of inducing orgasm in women relative to men, and the ability of women to conceive in the absence of orgasm have led several researchers to conclude that female orgasm is a nonfunctional by-product of male orgasm (Gould 1987; Lloyd 2005; Symons 1979). Yet, several lines of evidence challenge this hypothesis.

The Phenomenology of Female Orgasm

First, the phenomenology of female orgasm does not appear to be a diminished version of male orgasm. Meston et al. (2004, p. 174) describe women's orgasm as a "variable, transient peak sensation of intense pleasure, creating an altered state of consciousness, usually with an initiation accompanied by involuntary, rhythmic contractions of the pelvic striated circumvaginal musculature, often with concomitant uterine and anal contractions, and myotonia that resolves the sexually induced vasocongestion (sometimes only partially), generally with an induction of well-being and contentment." Similarly, Mah and Binik's (2002). Two-dimensional model characterizes orgasm as an experience that has both physical-sensory (e.g., swelling, flushing, shooting, throbbing) and cognitive-affective (e.g., pleasure, relaxation, euphoria) components.

One study demonstrated that a group of male and female gynecologists, psychologists, and medical students were unable to sex-identify men's and women's descriptions of their own orgasms, implying that the phenomenology of orgasm is quite similar between the sexes (Vance and Wagner 1976). Other researchers suggest that women's orgasms are psychologically more complex and perhaps even more intense than men's (Mah and Binik 2001, 2002).

Subjective descriptions of orgasm, such as this one provided by a woman from the United States, allow us to more fully appreciate the intensity of its physical-sensory aspect: "Orgasm feels like an intense drawing together sensation, located in my genital area (I can't differentiate in feeling between my clitoris and vagina at that point), then my whole body tenses and the sensation is one of total involvement without any 'will' or thought involved. '*It*' takes over completely. The physiological sensation is best described by the word 'outrageous' in terms of its devastating total effect. It's over within seconds, but fantastic when it occurs. The only awareness I can state is a certain stiffening all over, in addition to the intense 'implosion' in the undifferentiated genital area" (Hite 1976, p. 83). Another description provided by a Sudanese woman expresses the intensity of the cognitive–affective aspect of orgasm: "I feel as if I am losing all consciousness, and I love him most intensely at that moment. I tremble all over. My vagina contracts strongly and I have a feeling of great joy. Then I relax all over, and I am so happy to be alive and to be married to my husband" (Lightfoot-Klein 1989, p. 387). Orgasm in women, then, can be described as a singular event that is often characterized by strong physical sensations, feelings of contentment, and sometimes, when experienced with a partner, feelings of love and intimacy. Women are also far likelier than men to experience multiple orgasms (Masters and Johnson 1966), with one study reporting that 43 % of female respondents usually experience multiple orgasms (Darling et al. 1991). The experience of orgasm appears to be at least as intense in women as it is in men.

Genital Anatomy

The anatomical structure most frequently associated with female orgasm, the clitoris, also does not appear vestigial relative to its homologue in men. Perhaps because it is superficially distinct from the vagina, the glans clitoridis (the external part of the clitoris) is frequently mistaken to be the only part of the clitoris relevant to orgasm. However, the clitoris is a larger, more complex structure than its external aspect would suggest; it is composed internally of bulbs, corpora, and crura, all of which are attached via extensive supportive tissues to the mons pubis, labia, urethra, and vagina (O'Connell et al. 2008; O'Connell et al. 1998; O'Connell et al. 2005). These separate but intimately interrelated structures form a tissue cluster—a "clitoral complex"—that is an integrated and functional entity that represents the locus of female orgasm (Buisson et al. 2010; Colson 2010; O'Connell et al. 2008).

As a result, orgasms may be induced either through stimulation of the glans clitoridis or the vagina, and simultaneous stimulation of both has mutually reinforcing effects on sexual arousal and orgasm (Hoch 1986; Ingelman-Sundberg 1997). Hoch (1986) proposed a "clitoral/vaginal sensory arm of orgasmic reflex" that includes the clitoral glans, the anterior vaginal wall, and associated deep tissues. He suggested that rather than being pure types, vaginal and clitoral orgasms could most parsimoniously be conceptualized simply as a single "genital orgasm." Although some evidence suggests psychological differences between women who report having "vaginal" versus "clitoral" orgasms (e.g., Brody 2007; Brody and Costa 2008), the responses of women who were asked to describe where their orgasms originated indicate that women may be unable to reliably differentiate clitorally-and vaginally-induced orgasms (Clifford 1978; Prause 2011). Given the lack of controlled physiological studies investigating possible differences in the two putative types, there is insufficient evidence to conclude that stimulation of particular areas of women's genitals results in distinct types of orgasm. In fact, the subjective descriptions and objective physiological measures associated with orgasms achieved through fantasy alone are not significantly different from those of orgasms induced by other means (Whipple et al. 1992). This suggests that, although orgasms may vary in their intensity and specific sensory qualities (e.g., "deep" versus "surface" orgasm as described by King and Belsky 2012), the experience of orgasm is essentially the same phenomenon regardless of the specific area stimulated.

Though clitoral anatomy and its relation to orgasm are not suggestive of by-products, variability in clitoral anatomy has been interpreted as evidence for an absence of function in women's orgasm. Wallen and Lloyd (2008) reported that clitoral length (specifically, the length of the glans clitoridis) is more variable than penile length, taking this as evidence of weaker selection on orgasmic potential in women than in men. To Wallen and Lloyd, the greater variability of female orgasm compared to male orgasm suggests that female orgasm has been under less selective pressure than male orgasm. Consequently, "one would expect that the genital structures primarily responsible for triggering orgasm in women, the clitoris… and in men, the penis, would demonstrate a similar difference in variability" (Wallen and Lloyd 2008, p. 1). There are a variety of problems with this interpretation. First, there is little reason to assume that clitoral (much less glans) size is systematically related to women's orgasm frequency or quality, and therefore little reason to believe that variability in clitoral size reflects differences in orgasmic potential. Second, Wallen and Lloyd treat the glans clitoridis as the clitoris in its entirety, implying that only stimulation of this aspect of female genital anatomy is relevant to orgasm. As described above, the clitoris is far more extensive than its external aspect (O'Connell et al. 2008; O'Connell et al. 1998; O'Connell et al. 2005), and orgasm may be most efficiently induced by simultaneous stimulation of the glans clitoridis and its associated internal tissues (Hoch 1986; Ingelman-Sundberg 1997). Even women who have undergone removal of the glans clitoridis remain capable of experiencing orgasms through penile–vaginal intercourse (Alsibiani and Rouzi 2010; Lightfoot-Klein 1989). Third, neither the study from which Wallen and Lloyd derived clitoral variability (Lloyd et al. 2005) nor the study from which they derived penile variability (Spyropoulos et al. 2002) reported intra- or intermeasurer reliability. Thus, much and perhaps all of the reported difference in variability between clitorises and penises may have been due to the greater difficulty of precisely measuring smaller structures (clitoral length was 16 % of penile length). Fourth, penises, unlike clitorises, also function in urination and as intromittent organs necessary for insemination. These additional roles mean that whatever selection pressures operated on clitorises and penises for orgasmic potential, overall selection on these two organs necessarily differed. Thus, variability in penile and clitoral dimensions simply cannot shed light on the relative strengths of selection specifically on male and female orgasmic potential. Additional complications with this study have been discussed elsewhere (Hosken 2008; Lynch 2008; Puts et al. 2012a).

The Neural Basis of Orgasm

The neural correlates of orgasm in women also differ from men's in important respects. Although neuroimaging studies have found that men's and women's orgasms share a common distributed cerebral network involving activation of the insula;

putamen; temporal, parietal, and prefrontal cortices; septal region; and cerebellum (reviewed in Bianchi-Demicheli and Ortigue 2007), sex differences have also been observed. Specifically, women's orgasms involve activity in the nucleus accumbens, anterior cingulate, hippocampus, hypothalamus, and preoptic area, while in men orgasms involve the ventral tegmental area, thalamus, and visual cortex (Bianchi-Demicheli and Ortigue 2007). A recent study also demonstrated sex differences in the pituitary during orgasm, with female orgasm being uniquely associated with increased activity in this brain region (Huynh et al. 2013). Higher pituitary activation in women was interpreted by the authors to signify greater plasma concentrations of oxytocin, a hormone that is released from the pituitary (Du Vigneaud 1954) and is found in much higher plasma concentrations in women than in men after orgasm (Exton et al. 1999; Kruger et al. 2003). Thus, at the neural and endocrine levels, female orgasm is not a diminished version of male orgasm, but is instead elaborated and distinct from male orgasm in some respects.

Female Orgasm May Promote Conception

Like antlers in caribou, orgasm may have arisen as a male adaptation that was subsequently shaped as a secondary adaptation in females. Although females of some other primate species exhibit signs of orgasm (Puts et al. 2012a), most do not, so the presence of orgasm in males and its absence in females seems likely to be the ancestral state. If orgasm arose in females as a by-product of male orgasm and was subsequently modified for a special function in females, what might that function be?

Some of the strongest evidence concerns the potential role of women's orgasm in increasing the likelihood of conception. Two areas of increased brain activation during orgasm in women are the cingulate cortex and medial amygdala (Komisaruk et al. 2004). Electrical stimulation of these brain areas in animals induces peristaltic uterine contractions (Beyer et al. 1961; Setekleiv 1964) which transport sperm through the reproductive tract in humans (Zervomanolakis et al. 2007; Zervomanolakis et al. 2009) and nonhuman animals (Fox and Fox 1971; Singer 1973). As noted above, orgasm releases the hormone oxytocin into the bloodstream (Blaicher et al. 1999; Carmichael et al. 1994; Carmichael et al. 1987) from the pituitary, an event which itself is probably initiated by activation of the paraventricular nucleus (PVN) of the hypothalamus during orgasm (Cross and Wakerley 1977; Komisaruk et al. 2004). Like stimulation of brain regions involved in orgasm, oxytocin induces peristaltic contractions of the uterus and oviducts (Knaus 1950; Wildt et al. 1998), which transport semen-like fluid from the vagina to the oviducts (Kunz et al. 2007; Wildt et al. 1998; Zervomanolakis et al. 2007). Significantly, fluid is transported to the uterus throughout the cycle, but into the oviducts only during the follicular phase (Wildt et al. 1998; Zervomanolakis et al. 2007). Most importantly, during the periovulatory (fertile) phase of the menstrual cycle, transport is directed with increasing frequency to the oviduct of the dominant follicle as the follicle grows and ovulation approaches (Wildt et al. 1998; Zervomanolakis et al. 2007). The uterus

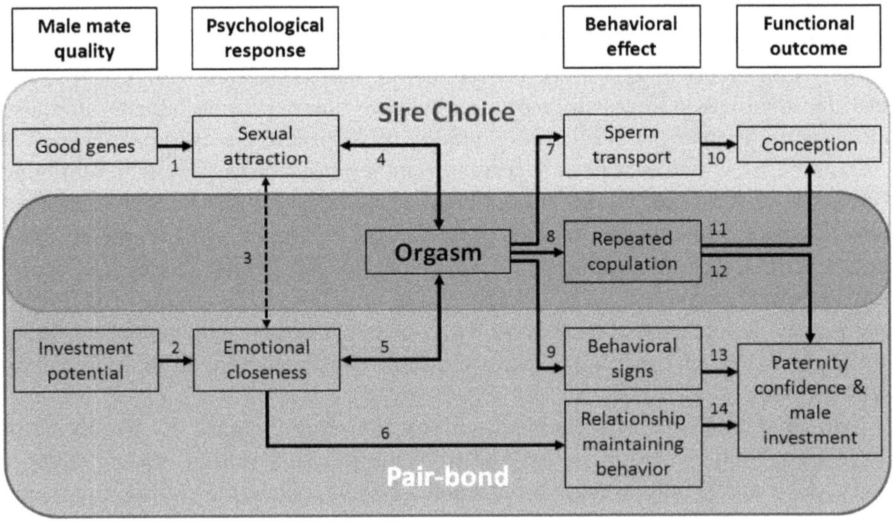

Fig. 7.1 Predicted causal pathways between men's mate quality, women's orgasm, and functional outcomes according to the sire choice hypothesis (*top region*), the pair-bond hypothesis (*bottom region*), as well as where these hypotheses overlap (*center region*). Numbers indicating each pathway are referred to in the text. The *dashed line* indicates a possible relationship that is not a necessary prediction of either model

and fallopian tubes may therefore be conceptualized as a "functionally united peristaltic pump" that is controlled hormonally by the follicle-bearing ovary (Zervomanolakis et al. 2007, p. 1) and whose activity is enhanced by oxytocin and thus perhaps orgasm.

Thus, the neural activation and endocrine changes that occur with orgasm appear to cause contractions of the uterus and oviducts that may facilitate fertilization by increasing the proximity between sperm and ovum (Fig. 7.1, path 7). Interaction between sperm and oviductal epithelium may also prolong sperm longevity, increase the number of capacitated sperm (sperm capable of fertilizing an ovum), and lengthen the interval over which at least some sperm in an ejaculate are capacitated (Smith 1998; Suarez 1998). Although early studies failed to show movement of semen-like substances through the cervix following orgasm (Gräfenberg 1950; Masters and Johnson 1966), these studies placed a cap over the cervix (Fox et al. 1970). With the cervix unobstructed, the uptake of a semen-like substance into the uterus occurs spontaneously, regardless of orgasm, oxytocin treatment, or menstrual cycle phase (Zervomanolakis et al. 2007). It is the directed transport of sperm from the uterus toward the follicle-bearing ovary that is increased by oxytocin treatment and, therefore, probably orgasm. Directed transport into the oviduct with the dominant follicle (vs. transport into both oviducts) is associated with a higher probability of pregnancy (Zervomanolakis et al. 2007).

In addition to causing the muscular contractions that facilitate sperm transport, orgasm (Fox et al. 1970) and oxytocin (Wildt et al. 1998) reverse uterine pressure from outward to inward, which may prevent sperm loss from "flowback" and aid sperm in reaching the oviducts. Baker and Bellis (1993) found that female orgasm within 1 min before and 45 min after ejaculation was associated with higher sperm retention than was no orgasm or orgasm at other times. Female orgasm may also allow sperm to enter the cervix more quickly by resolving the "vaginal tenting" of sexual arousal, which elevates the cervix from the posterior vaginal wall, removing it from the semen pool (Levin 2002). Earlier entry of sperm into the cervix may reduce flowback and help sperm reach the oviducts (Fox and Fox 1971) thereby promoting fertilization, though Levin (2011) has noted that greater numbers of sperm near the ovum may result in polyspermy (and thus nonviable ovum) or increased sperm enzyme release that may cause ovum degeneration. It remains to be demonstrated, however, that increased sperm transport is associated with these nonadaptive outcomes.

Two final mechanisms associated with orgasm may increase the likelihood of conception. First, prolactin is secreted from the pituitary during orgasm (Krüger et al. 2002), and if able to enter the vaginal, cervical, or uterine environment, may capacitate sperm (Meston et al. 2004). Second, the vaginal contractions associated with female orgasm may excite male ejaculation (Fox and Fox 1971; Meston et al. 2004), which could coordinate ejaculation with the various conception-enhancing processes associated with orgasm in women.

If female orgasm promotes conception through neural, endocrine, and physiological mechanisms, then one might predict that orgasms would be most likely to occur during the late follicular phase of the ovulatory cycle when conception is possible. Some evidence supports this prediction. Udry and Morris (1968) reported on a total of between 911 and 997 menstrual cycle samples of 40 and 80 normally cycling women. In both samples, women reported significantly more orgasms near ovulation, with the peak rate occurring 14 days before the onset of the next menstrual cycle, the approximate date of peak fertility. Several subsequent studies reported significant or nonsignificant trends toward more frequent orgasms during the fertile ovulatory phase (Clayton et al. 1999; Matteo and Rissman 1984; Worthman 1978). These cyclic changes in the female orgasmic response are likely to be mediated by hormones such as estradiol and progesterone (van Anders and Dunn 2009; Zumpe and Michael 1968). It is noteworthy that rates of copulation also increased near ovulation in some but not all of these studies, so cyclic changes in orgasm frequency are partly confounded by changes in copulation frequency.

Zietsch and Santtila (2013) recently found in a large twin sample that women's rates of self-reported orgasm positively predicted their number of offspring. The correlation between intrapair differences in orgasm frequency and fertility was stronger in fraternal twins than in identical twins, indicating that variation in the strength of this correlation was due primarily to environmental differences. Although the authors interpret the lack of genetic correlation to imply that orgasm does not promote conception, their reasoning is unclear. As the authors acknowledge, if orgasm promotes conception, then both genetic and environmental factors would be expected to influence the correlation between orgasm frequency and fertility. Given that most (approximately 70%; Dawood et al. 2005; Dunn et al. 2005) of the variation in coital orgasm frequency is due to environmental effects, one might expect that most of the correlation between orgasm frequency and fertility would be due to environmental effects. Women have been found to report higher orgasm frequency during the fertile phase of the ovulatory cycle with more compatible mates (Garver-Apgar et al. 2006), so differences in the compatibility of women's mates is an environmental variable that may influence the correlation between orgasm frequency and fertility. While Zietsch and Santtila's study provides modest support for a positive association between orgasm and fertility, any such association would have been obscured by contraceptive use. In addition, the measure of orgasm frequency obtained (orgasm from "sexual stimulation or intercourse" over the past 4 weeks) is not ideal for testing whether female orgasm contributed to women's total number of offspring. The relevant measure would be frequency of coital orgasm during the fertile window of the ovulatory cycle throughout the participants' sexually active adult lives.

Overall, the evidence for a potential conception-enhancing effect of orgasm is indirect and incomplete (Levin 2011; Lloyd 2005). Because past studies have administered supraphysiological doses of oxytocin to participants, it is not clear that the observed effects of exogenous oxytocin administration duplicate those that would occur naturally after orgasm (Levin 2011). Moreover, these studies examined sperm transport in women who were in a sexually relaxed state; the actions of oxytocin may differ in sexually aroused women (Levin 2011). Yet these points undermine neither the brain stimulation research reviewed above suggesting that female orgasm increases sperm transport, nor the modest association reported between fertility and orgasm rate. More importantly, satisfactory alternative explanations of the evidence are lacking. Why, for example, would *exogenous* oxytocin induce sperm transport toward the oviducts if the high levels of *endogenous* oxytocin released at orgasm did not trigger a similar response? Why do oxytocin-induced contractions transport sperm selectively to the oviduct containing the dominant follicle specifically during the fertile phase of the ovulatory cycle, if not to promote conception? And why would women who exhibit such directional (vs. bilateral) transport following oxytocin administration have a higher probability of later pregnancy (Zervomanolakis et al. 2007)?

To summarize, the available evidence suggests that female orgasm increases the likelihood of conception. At orgasm, vaginal contractions may promote male ejaculation (if it has not yet occurred), and the vaginal tenting that occurs during sexual arousal is resolved, allowing the cervix to contact the semen pool in the posterior vagina. The decreased pressure in the uterus following orgasm draws the semen upward where it is subsequently transported to the oviducts and directed toward the ovum (if one is present) by peristaltic contractions of the uterus and oviducts themselves. Directed transport toward the ovum predicts later probability of conception. The release of prolactin from the pituitary at orgasm may capacitate sperm and therefore facilitate fertilization of the ovum. Orgasm may be more likely to occur during the fertile phase of the ovulatory cycle, suggesting that it is a facultative adaptation that occurs most reliably at the time when its effects have the best chance of promoting conception. Finally, some evidence suggests an association between the frequency of women's orgasms and their fertility, even in a society with widespread contraceptive use.

Mate Choice

Of the many functional hypotheses advanced for human female orgasm (Lloyd 2005), those that posit a mate choice function (Alcock 1980; Smith 1984; Thornhill et al. 1995) are among the most plausible. Two mate choice hypotheses in particular have received considerable attention and empirical support: the sire choice hypothesis and the pair-bond hypothesis (Fig. 7.1). The *sire choice hypothesis* suggests that female orgasm is a mechanism whereby women increase their chances of conceiving with men of superior genetic quality. The *pair-bond hypothesis*, in contrast, holds that female orgasm functions to bond women psychologically and behaviorally to males with high investment potential, thereby promoting male investment in the woman and/or her offspring.

We have already reviewed evidence suggesting that female orgasm increases the probability that a coital act will result in fertilization. But as suggested at the outset of the chapter, orgasm is not merely physiological; it is also a psychological event that, however brief, is of rare intensity. Therefore, in exploring the possible functions of female orgasm, it is necessary to examine both its physiological aspects and its psychological correlates. Consider why male orgasm has positive affective correlates (intense pleasure) in addition to its physiological features (e.g., ejaculation). Affect may evolve to elicit adaptive behavior (Plutchik 1982)—fear diverts us from danger, pain prevents our using damaged body parts, and pleasure motivates behaviors that likely augmented fitness ancestrally. The affective reward value of orgasm would seemingly motivate whatever sexual behavior elicited it, which for men would presumably mean copulating until ejaculation. Orgasm is clearly also salient to women, and the prospect of orgasmic pleasure may function to motivate women to continue copulating until orgasm is achieved or perhaps to copulate again with males with whom they experienced orgasm (Fig. 7.1, path 8).

If female orgasm is a mate choice mechanism, then orgasm should be triggered less easily in women than it is in men (Allen and Lemmon 1981; Hosken 2008; Puts 2007). This follows from the idea that only some males will meet females' mate selection criteria and that women have evolved to be choosier than men about mates. The latter is true specifically in mating contexts such as purely sexual relationships when male investment is minimal, and the woman may end up gestating, nursing, and caring for a child (Clark and Hatfield 1989; Daly and Wilson 1983; Kenrick et al. 1993; Schmitt 2005; Symons 1979; Trivers 1972). Indeed, only about 60 % of Western women report experiencing orgasm most of the time (> 50 %) during copulation, and only about 25 % report that they always experience orgasm (Lloyd 2006; Puts 2007). The frequency for men, of course, is much greater. The lower frequency of coital orgasm in women compared to men is thus consistent with what one would predict if female orgasm functions as a mate choice mechanism.

The Sire Choice Hypothesis

If a woman's orgasm serves as a mate choice mechanism, then the likelihood of its occurrence might depend, in part, on the genetic quality of her mate (Alcock

1980; Baker and Bellis 1993; Smith 1984; Thornhill et al. 1995). After all, women's lifetime reproductive potential is limited, and every pregnancy and child requires a substantial investment of time and resources. It stands to reason—and much evidence suggests—that women possess psychological and physiological mechanisms that have evolved to increase the genetic quality of their offspring via mate choice (Gangestad and Thornhill 2008; Roberts and Little 2008). If female orgasm enhances the probability of conception, then it might be more reliably induced by partners of superior genetic quality (Fig. 7.1, paths 1 and 4) whose beneficial genes will be passed down to a woman's offspring (Fig. 7.1, paths 7, 8, 10, and 11). This prediction is known as the sire choice hypothesis.

Though evolutionary biologists have no ideal metric for genetic quality, physical attractiveness is one putative measure (Andersson 1994; Grammer et al. 2003). As predicted by the sire choice hypothesis, several studies provide evidence that men's attractiveness is positively associated with their female partners' orgasm frequency. Thornhill et al. (1995) found that men's attractiveness as assessed by independent raters was marginally significantly related to the proportion of copulations that resulted in their female partner achieving orgasm, while attractiveness was unrelated to the frequency of noncoital orgasms. This suggests that male sire quality increases the likelihood of female orgasm specifically during sexual behaviors that might result in conception. Another study found that when several variables, including relationship satisfaction, relationship duration, and a woman's rating of her partner's attractiveness were entered into a multiple regression, only her partner's attractiveness predicted a woman's probability of orgasm during her last copulation (Shackelford et al. 2000). A more recent study reported that women mated to more physically attractive men (assessed both independently and via self-ratings) reported that their coital orgasms more frequently occurred during or after male ejaculation (Puts et al. 2012b). This relative timing of female orgasm corresponds approximately to the window of greatest sperm retention, according to one study (Baker and Bellis 1993).

Another commonly employed measure of genetic quality that is also related to physical attractiveness is fluctuating asymmetry (FA). FA refers to asymmetry in anatomical traits such as facial features that are normally bilaterally symmetric. FA may negatively indicate genetic quality, because it results from perturbations in growth and development, such as those caused by mutations or parasitic infection (Møller and Pomiankowski 1993; Parsons 1990, 1992; Van Valen 1962), and is moderately heritable in several species (Møller and Thornhill 1998). Thornhill et al. (1995) found that when controlling for several other variables, women's reported orgasm frequencies were significantly higher if their mates were more symmetrical. In addition, women reported achieving orgasm more frequently with extra-pair males relative to their in-pair partners (Baker and Bellis 1993), and women's extra-pair partners have been found to exhibit low FA (Gangestad and Thornhill 1997).

Masculinity is a third proxy measure for a man's genetic quality. The degree of development in androgen-dependent, masculine characteristics (e.g., muscularity) may indicate heritable disease resistance, because androgens may be produced in proportion to inherited resistance to infectious disease (Folstad and Karter, 1992). Alternatively, immune system activation may suppress testosterone production

(Boonekamp et al. 2008). If a male's immune system more quickly and efficiently deals with immune threats, then testosterone production might be suppressed less frequently, less severely, or for shorter durations, and a more masculine phenotype would emerge. To the extent that such immune efficiency was heritable, ancestral women may have produced healthier offspring by mating with phenotypically masculine men (Puts et al. 2012). Males with fewer deleterious mutations may also be able to produce and maintain more well-developed masculine traits (Zahavi and Zahavi 1977). Furthermore, because masculine traits tend to be costly to produce and are frequently tested by competitors in dominance contests, they should provide accurate information about male quality to potential mates (Berglund et al. 1996).

In a test of the possible association between women's orgasms and their partners' masculinity, Puts et al. (2012b) found that scores on a principle component defined by men's independently rated facial masculinity, objectively measured facial masculinity, partner-rated overall masculinity, and partner-rated dominance were positively related to a principle component defined by the frequency of their female partners' coital orgasms, as well as their partners' tendency to achieve orgasm before male ejaculation. Earlier-timed orgasms are associated with greater sexual pleasure (Darling et al. 1991), which may stimulate elevated oxytocin release (Carmichael et al. 1994; Carmichael et al. 1987) leading to sperm transport (Wildt et al. 1998). The findings of Puts et al. (2012b) thus suggest that possible conception-promoting correlates of female orgasm may be especially effective or likely when copulation occurs with masculine males. It is relevant to note that neither men's masculinity nor their attractiveness predicted their partners' frequency of orgasm from self-masturbation or noncoital partnered sexual behavior (Puts et al. 2012b).

The research reviewed above thus indicates that women's orgasms are more frequent both near ovulation and during copulation with males who are putatively of high genetic quality. However, if female orgasm functions in sire choice, these variables might also be expected to interact in predicting a woman's orgasm frequency. That is, women should be especially likely to achieve copulatory orgasm with a high-quality male when they are nearer to ovulation. The one study to test this prediction found that women reported more orgasms if their partner had discordant (i.e., compatible) major histocompatibility (MHC) complex genes, but only during the fertile phase of the ovulatory cycle (Garver-Apgar et al. 2006). The MHC is the main genomic region mediating disease resistance, and preferences for MHC-discordant mates would ancestrally have produced offspring whose immune systems are capable of coping with a greater variety of infectious diseases (Potts and Wakeland 1993). Olfactory preferences for MHC-discordant mates have been observed across vertebrate taxa, including humans (reviewed in Milinski 2006). Given that Garver-Apgar et al. utilized genotype data rather than proxies for genetic quality, their findings are perhaps the most convincing demonstration of the association between a man's genetic quality and his ability to induce orgasm in his mate (Fig. 7.1, paths 1 and 4). It is difficult to explain why this link was observed only during ovulation if female orgasm has not been shaped to recruit high-quality genes for increased offspring viability.

The Pair-Bond Hypothesis

In species such as ours in which both parents cooperate in rearing offspring, male mate quality is not limited to the possession of beneficial genes. It can also be measured by investment that the male is willing and able to commit to his partner and any offspring resulting from their sexual union. For ancestral humans, most reproduction probably occurred in the context of socially monogamous pairs (Mellen 1982), with a small proportion of polygynous men investing less in their offspring (Gavrilets 2012). It would therefore have been adaptive for a woman to attract and establish an affectionate bond with a man—and assure him of his likely paternity of her subsequent offspring—to promote long-term, cooperative parenting relationships (Buss and Kenrick 1998; Fisher 1998; Mellen, 1982). The pair-bond hypothesis asserts that female orgasm functions in these roles (Barash 1977; Beach 1974; Eibl-Eibesfeldt 1975; Hamburg 1978; Morris 1967).

Women's orgasm could promote pair-bonds and male investment in several ways. First, as noted above, the pleasure and relative difficulty of achieving orgasm may encourage women to seek and be receptive to future copulations with the same male (Fig. 7.1, path 8). In a survey of 202 women, 76 % said that having an orgasm with a partner was somewhat important to very important, compared with only 6 % saying that it would be somewhat unimportant to very unimportant (Eschler 2004). Repeated copulation would presumably also elevate a man's perceived probability of siring a woman's future offspring, which would tend to promote his investment (Fig. 7.1, path 12), although data on this are lacking.

Second, though sometimes faked (Ellsworth and Bailey 2013; Muehlenhard and Shippee 2010), the outward expression of female orgasm (e.g., clutching, vocalizations; Fig. 7.1, path 9) may demonstrate to the male that his partner is bonded to him and that she is likely to conceive his child, thereby encouraging his investment (Fig. 7.1, path 13). Evidence suggests that women sometimes manipulate men's perceptions of their romantic relationships by consciously vocally advertising their orgasms (Brewer and Hendrie 2011). Ellsworth and Bailey (2013) found that the intensity of women's orgasmic behaviors (e.g., vocalizations) did not significantly predict women's sexual fidelity in their current relationship or men's perceptions of female sexual fidelity. However, women's self-reported orgasmic behavioral intensity was positively related to their perceptions of their partners' investment in the relationship, and men's reports of their partners' orgasmic behavioral intensity positively predicted men's relationship satisfaction (Ellsworth and Bailey 2013).

Finally, if emotional closeness is enhanced by orgasm (Fig. 7.1, path 5), then this might promote behaviors targeted toward ensuring the male's investment, such as mate monitoring by the woman or displays of her affection and fidelity (Fig. 7.1, path 6). Some studies have reported no association between whether women achieved orgasm at their most recent copulation and their relationship satisfaction (Shackelford et al. 2000) or orgasm frequency and professed love for their partner (Thornhill et al. 1995). However, in most studies women's orgasmic pleasure and satisfaction predict their relationship satisfaction and emotional intimacy with their partner (King et al. 2011; Mah and Binik 2005; Ortigue et al. 2007). Frequency of orgasm from penile–vaginal intercourse has also been positively associated with global measures of relationship satisfaction (Brody and Weiss 2011; Costa and Brody 2007; Ellsworth and Bailey 2013; Gebhard 1966), as well as measures of intimacy, passion, and love (Costa and Brody 2007; Haning et al. 2007).

Oxytocin, a hormone released at high levels during orgasm (Blaicher et al. 1999; Carmichael et al. 1987, 1994), and which plays a key role in mammalian social behavior (Carter et al. 2008), may mediate the relationship between orgasm and pair-bonding. Oxytocin is colloquially known as the "love hormone" because of its apparent role in the formation of pair-bonds in humans and nonhuman mammals (reviewed in Carter and Porges 2013). In rodents, mating facilitates the initiation of pair-bonding (Williams et al. 1992), probably via increased oxytocin (Carter 1998; Sanchez et al. 2009; Williams et al.1994), especially in females (Sanchez et al. 2009). Oxytocin has similar prosocial effects in humans, including increased trust (Andari et al. 2010; Kosfeld et al. 2005), empathy (Carter et al. 2009), and co-operation (Andari et al., 2010; Hurlemann et al. 2010; Rilling et al. 2012). Interestingly, oxytocin administration also selectively facilitates the recognition of positive sex and relationship words (e.g., kissing), but not other words (e.g., safety/threat words) (Unkelbach et al. 2008). Oxytocin levels are generally higher in women in relationships than in single women (Schneiderman et al. 2012; Taylor et al. 2006; Turner et al. 1999). Plasma oxytocin levels 3 months into a relationship positively predicted couples' likelihood of remaining together by 6 months (Schneiderman et al. 2012). In addition, women who reported more partner support (Grewen et al. 2005) and hugging (Light et al. 2005) had higher baseline oxytocin levels. Finally, couples given oxytocin prior to a conflict discussion exhibited more instances of positive communication compared to placebo-treated couples (Ditzen et al. 2009). Thus, though some studies have found associations between oxytocin and negative emotions, anxiety, and relationship distress in couples (Marazziti et al. 2006; Tabak et al. 2011; Taylor et al. 2010), likely because the effects of oxytocin depend on situational and personality variables (Bartz et al. 2011; Graustella and MacLeod 2012), available evidence generally suggests that oxytocin is associated with positive aspects of interpersonal relationships, particularly for women.

Although research on oxytocin suggests a mechanism whereby women's orgasm may facilitate pair-bonding, emotional closeness may also facilitate women's orgasm (Fig. 7.1, path 5). Eschler (2004) found that 82 % of women agreed that emotional closeness was an important factor in vaginal intercourse, and women in relationships who have difficulty becoming aroused and achieving orgasm most commonly cite emotional and relationship difficulties as the cause (King et al. 2007). Women in long-term relationships were roughly 50 % more likely to have achieved orgasm in their last sexual encounter if they reported it was somewhat likely that they would eventually marry their partner compared to women who did not believe they would marry their partner (Armstrong et al. 2012). Orgasm was more than twice as likely when women reported they were very likely to marry their partners. These results, along with qualitative data from one-on-one interviews with 85 individuals, led the authors to conclude that commitment and affection are likely to enhance orgasm and sexual satisfaction (Armstrong et al. 2012). If emotional closeness facilitates orgasm, then women with less secure romantic attachment styles should be less likely to achieve orgasm in partnered intercourse. (For a review of adult romantic attachment styles, see Holmes and Johnson 2009) Indeed, women with avoidant (Cohen and Belsky 2008) and anxious (Birnbaum 2007; Costa and Brody 2011) attachment styles have reported lower orgasmic responsivity than women with more secure attachment styles.

Again, oxytocin is a likely mediator of relationships between emotional closeness and orgasm. Affectionate contact with a partner increases oxytocin levels (Grewen et al. 2005), is sexually arousing (van Anders et al. 2013), often precedes sexual intercourse (van Anders et al. 2013), and predicts orgasmic function (Fugl-Meyer et al. 2006). A rise in oxytocin following affectionate touch may explain findings that duration of foreplay is associated with orgasm frequency and consistency (Brody and Weiss 2010; Gebhard 1966; Singh et al 1998). Furthermore, oxytocin induces and maintains sexual receptivity in rodents (e.g., Arletti et al. 1990; Benelli et al. 1994; Caldwell 1986; Cushing and Carter 1999; Pedersen and Boccia 2002), and may facilitate sexual arousal and orgasm in humans as well (Anderson-Hunt and Dennerstein 1994; IsHak et al. 2010; MacDonald and Feifel 2012). Thus, though the oxytocin surge experienced at orgasm likely increases the emotional closeness a woman feels with her partner, oxytocin also appears to facilitate sexual responsivity and orgasm. Oxytocin rises in response to affectionate touch, increases throughout the sexual arousal phase, peaks at orgasm, then quickly returns to baseline levels. The ease with which this rise in oxytocin is produced, and therefore the ease with which orgasm is induced, may depend upon the degree to which a woman already feels emotionally close to her sexual partner.

Associations between women's orgasm and their relationship satisfaction and emotional closeness to their partner suggest a link between orgasm and male investment, but few data directly assess this link. Thornhill et al. (1995) reported no relationship between women's orgasm frequency and their mates' reported love, nurturance, commitment, or exclusivity. However, Ellsworth and Bailey (2013) found that men's relationship satisfaction was positively related to their reports of their partners' orgasm frequency and orgasm behavioral intensity and negatively related to their reports of their partners' frequency of faking orgasm. In addition, women's ratings of their own orgasm frequency and behavioral intensity positively predicted their ratings of their partners' investment.

Female Orgasm in Context

The conditions under which women's orgasm is most likely to occur can provide further information about its function. Women generally report higher orgasm frequencies within long-term relationships than in short-term mating contexts such as one-night stands (Armstrong et al. 2012; Eschler 2004; Garcia et al. in submission). For example, among 6591 unmarried women at 21 universities who had both (a) engaged in at least one sexual hookup and (b) been in at least one romantic relationship with a man that lasted 6 months or more, 11 % of women had achieved orgasm during their last hookup, while 67 % reported having had an orgasm during their last sexual event within a long-term relationship (Armstrong et al. 2012).

These data might appear to support the pair-bond hypothesis and contradict the sire-choice hypothesis. However, given women's relative reticence over uncommitted sex (Clark and Hatfield 1989), many women may not feel entirely comfortable having sex with a man whom they have recently met, or with whom they are otherwise relatively unfamiliar. Over most of human evolution, people lived in small groups in which individuals would have been intimately familiar. Having sex with a stranger or near-stranger probably occurred mainly when women had little influence over their mating circumstances, such as in wife exchange between villages or abduction and sexual coercion (e.g., Walker and Bailey 2013; Walker et al. 2011). Yet, across cultures, women sometimes choose to have sexual liaisons outside of committed relationships, such as in extramarital sexual affairs. A study of 75 normally-ovulating, polyandrous women reported significantly elevated rates of copulatory orgasm with extra-pair males relative to orgasm rates with in-pair mates, and visual inspection of summary data suggests that in-pair orgasm frequencies did not differ between monogamous and polyandrous women (Baker and Bellis 1993: Fig. 7.1). In addition, women who reported having more open attitudes toward casual sex were particularly likely to have had an orgasm during their last hookup (Armstrong et al. 2012). Therefore, apart from any affection she might feel for her partner, the likelihood of a woman achieving orgasm may also be directly related to her role in initiating sexual intercourse, and how relaxed and receptive she feels.

Women's state of relaxation and receptivity during partnered sexual activity may thus function to facilitate orgasm across a variety of mating contexts, depending on the woman's sexual strategies (Buss 1998; Buss and Schmitt 1993). Orgasm could therefore serve both sire choice and pair-bonding functions. One possibility is that women's orgasm promotes conception within the fertile window of the menstrual cycle with good-genes males and promotes pair-bonding, perhaps via increases in oxytocin levels, outside of the fertile window when greater partner attentiveness is required for its induction. This hypothesis explains elevated orgasm rates in the context of long-term, emotionally close relationships, as well as in uncommitted sexual contexts such as extra-pair sex or when a sexual hookup is desired. It also helps explain the conspicuous behavioral signs of orgasm, which would seemingly be most effective at increasing male investment if orgasm also promotes conception; behavioral signs would increase paternity confidence not only by signaling a woman's arousal and comfort with her mate but also her elevated probability of conceiving his offspring. During the fertile phase of the ovulatory cycle, and in the context of purely sexual relationships such as extra-pair sex, a man's attractiveness, symmetry, masculinity, and other measures of genetic quality should especially increase women's orgasm rates. Orgasm may promote pair-bonding to a greater extent in some women, such as those less likely to benefit from extra-pair copulations, while promoting conception to a greater extent in others, such as those pursuing a short-term rather than a long-term mating strategy.

Alternatively, orgasm may function to promote conception with both good-genes males and investing males. Pregnancy in the context of a long-term relationship may induce investing males to remain with the woman and continue investing (Betzig 1989). Moreover, men are sensitive to cues to paternity in allocating investment (Daly and Wilson 1982; Regalski and Gaulin 1993; Welling et al. 2011), so it may often have benefitted ancestral females to promote fertilization from males who demonstrated investment potential. Fatherhood may also redirect reproductive effort from mating toward parenting (e.g., Storey et al. 2000), even in males who were initially poor investors. This hypothesis predicts positive effects of both male genetic quality and investment potential on women's orgasm rates at midcycle, and perhaps a positive interaction, such that orgasm rates are particularly elevated at midcycle with men who are high on both dimensions of mate quality. The latter could occur if emotional closeness increases women's sexual attraction to men, sexual attraction increases emotional closeness, or both (Fig. 7.1, path 3). Outside the fertile phase, female orgasm may indicate to men that their partner is both sexually satisfied and likely to conceive, thereby encouraging his investment.

Thus, orgasm may be a contextually dependent adaptation that increases the odds of conception during the fertile phase of the ovulatory cycle, perhaps especially with good-genes males, while also serving to emotionally bond women to their partners and encourage male investment. Future research may demonstrate that its occurrence is dependent upon women's sexual strategies (Buss 1998; Buss and Schmitt 1993). Orgasm may be most responsive to a mate's genetic quality in women seeking uncommitted sex, and to signs of male investment in women seeking a committed relationship. That is not to say that a man's genetic quality is unimportant to orgasm in committed relationships, perhaps especially during the fertile phase of the ovulatory cycle. Indeed, most studies (reviewed above) that reported relationships between women's orgasm and measures of their mates' genetic quality were conducted on dating couples. Nor do we imply that male investment potential plays no role outside of committed relationships. A man's genetic quality and a woman's perception of his commitment to her may have independent, as well as interactive, effects on the likelihood of orgasm in women. For example, attractive women, who can be most demanding when choosing a mate, desire partners who are both loving and possess indicators of genetic quality (Buss and Shackelford 2008).

Conclusion

Symons (1979) argued that female orgasm is too difficult to induce, and its expression too variable among women, to be an adaptation. Lloyd (2005) echoed these sentiments: "very wide variability of rates of orgasm with intercourse suggests that there is no selection on female orgasm with intercourse. If there was, then we would expect a high, consistent expression of the trait of female orgasm with intercourse…." (p. 134). The reasoning is that if female orgasm is not reliably induced within and across women, then this challenges how efficiently and precisely it can perform any hypothetical function.

However, evidence reviewed above indicates that female orgasm may selectively promote conception with partners who possess specific orgasm-facilitating qualities. If so, then relatively difficult induction and relatively high variability in expression are expected. Because women are generally choosier than men over mates (Clark and Hatfield 1989; Schmitt 2005; Trivers 1972), a certain partner-specific contingency in orgasm induction is predicted by a mate choice hypothesis. This "choosiness" in orgasmic response should, and does (Lloyd 2005), apply especially to sexual intercourse, as opposed to less reproductively consequential behavior, such as masturbation. Much of the variation in orgasm frequency likely results from the facultative nature of orgasm—not all sexual stimulation is equal. Thus, women vary in orgasm frequency because they differ in the propitiousness of their mating circumstances, as well as in the response pattern relating those circumstances to their sexual responsiveness. Selection could reduce variation only in this response pattern. The relevant issue of how much orgasmic variation results from differences in women's response patterns is presently unknown, but variability among women's orgasm frequency *per se* does not warrant rejecting a functional hypothesis. In other words, the locus of selection and hence the focus of evolutionary analysis should be on the response pattern rather than the overall frequency. Moreover, even traits that have probably experienced strong selection, such as menstrual cycle length, stature, cognitive abilities, running speed, and facial attractiveness, are highly variable among individuals. Facultative adaptations (i.e. adaptations that are functional only in certain contexts) should exhibit even greater phenotypic variation (Hosken 2008; Puts 2007).

We may never be able to definitively state that female orgasm is an adaptation, as opposed to an exaptation or by-product. However, in the last several decades researchers have produced a great deal of evidence that has advanced our understanding of the phenomenon. Taken together, the data strongly suggest that female orgasm serves a reproductive purpose. Neither the phenomenology of female orgasm nor its underlying genital anatomy is vestigial, in contradiction with predictions of the by-product hypothesis. In fact, both female genital anatomy and the neural representation of orgasm in the brain are elaborated in some ways relative to homologous characteristics in the male, suggesting that female orgasm has been modified to serve a female-specific function. Neurological, endocrinological, and physiological studies suggest that female orgasm functions to increase the likelihood of conception. Oxytocin appears to play a dual role in orgasm, first facilitating its occurrence, then surging at the orgasmic peak to activate the mechanisms that transport sperm toward the ovum, and perhaps to facilitate or maintain a pair-bond. Sex differences in human orgasm mirror sex differences in choosiness over mates, and within-female variation in orgasm frequency and physiology tracks conception risk across the cycle, a key indicator that female orgasm is related to conception. Between-female variation in orgasm frequency is related to both the genetic quality of a woman's mate and the quality of their relationship. Finally, orgasms appear to occur most frequently in contexts where women are most receptive, such as within established romantic relationships, during extra-pair sexual liaisons, and in brief sexual encounters when women hold more positive attitudes toward casual sex.

Much work remains to be done. For example, naturalistic studies of sexual intercourse, oxytocin, and the physiology of sperm transport are needed to more decisively determine if female orgasm enhances the probability of conception. For example, future research should explore sperm-sized particulate transport during sexual arousal with and without naturally induced orgasm (Levin 2011). Future research should also explore a possible association between women's orgasm and fertility in a longitudinal study with couples attempting to conceive. Data should be collected on the timing of each coital act in the ovulatory cycle and whether each coital act led to orgasm. If orgasm promotes conception, then orgasm should increase the probability of conception from an act of intercourse, after daily variation in conception risk is statistically controlled. Additional research is also necessary to determine if female orgasm enhances pair-bonding, and if pair-bonding enhances orgasm. Although oxytocin provides a plausible mechanism, existing data linking orgasm to emotional closeness are correlational, so the direction of any causality is unclear. Researchers should explore whether other putative indicators of a man's genetic quality, such as his body odor, influence a woman's likelihood of orgasm. And very little research has directly examined the link between women's sexual strategies and the context in which their orgasms are most likely to occur, an area that is ripe for future investigation. Because so many of these variables (e.g., genetic quality, relationship quality, oxytocin levels, sexual strategies, context, and orgasmic frequency) are likely interrelated in complex ways, we encourage researchers to adopt an integrative approach and examine multiple variables simultaneously. This would help to unify what is currently a relatively fragmented corpus of evidence and would present a clearer picture of the determinants and consequences of the female orgasm.

References

Alcock, J. (1980). Beyond the sociobiology of sexuality: predictive hypotheses. *Behavioral and Brain Sciences, 3,* 181–182.

Allen, M. L., & Lemmon, W. B. (1981). Orgasm in female primates. *American Journal of Primatology, 1,* 15–34.

Alsibiani, S. A., & Rouzi, A. A. (2010). Sexual function in women with female genital mutilation. *Fertility and Sterility, 93,* 722–724.

Andari, E., Duhamel, J.-R., Zalla, T., Herbrecht, E., Leboyer, M., & Sirigu, A. (2010). Promoting social behavior with oxytocin in high-functioning autism spectrum disorders. *Proceedings of the National Academy of Sciences, 107,* 4389–4394.

Anderson-Hunt, M., & Dennerstein, L. (1994). Increased female sexual response after oxytocin. *BMJ: British Medical Journal, 309,* 929.

Andersson, M. (1994). *Sexual selection.* Princeton: University Press.

Andrews, P. W., Gangestad, S. W., & Matthews, D. (2002). Adaptationism, exaptationism, and evolutionary behavioral science. *Behavioral and Brain Sciences, 25,* 534–547.

Arletti, R., Benelli, A., & Bertolini, A. (1990). Sexual behavior of aging male rats is stimulated by oxytocin. *European Journal of Pharmacology, 179,* 377–381.

Armstrong, E. A., England, P., & Fogarty, A. C. K. (2012). Accounting for women's orgasm and sexual enjoyment in college hookups and relationships. *American Sociological Review, 77,* 435–462.

Baker, R. R., & Bellis, M. A. (1993). Human sperm competition: Ejaculate manipulation by females and a function for the female orgasm. *Animal Behaviour, 46,* 887–909.
Barash, D. P. (1977). *Sociobiology and behavior*: Elsevier: North-Holland.
Bartz, J. A., Zaki, J., Bolger, N., & Ochsner, K. N. (2011). Social effects of oxytocin in humans: Context and person matter. *Trends in Cognitive Sciences, 15,* 301–309.
Beach, F. (1974). Human sexuality and evolution. In W. Montagna & W. A. Sadler (Eds.), *Reproductive behavior* (pp. 333–365). New York: Plenum Press.
Benelli, A., Poggioli, R., Luppi, P., Ruini, L., Bertolini, A., & Arletti, R. (1994). Oxytocin enhances, and oxytocin antagonism decreases, sexual receptivity in intact female rats. *Neuropeptides, 27,* 245–250.
Berglund, A., Bisazza, A., & Pilastro, A. (1996). Armaments and ornaments: An evolutionary explanation of traits of dual utility. *Biological Journal of the Linnean Society, 58,* 385–399.
Betzig, L. L. (1989). Causes of conjugal dissolution: A cross-cultural study. *Current Anthropology, 30,* 654–676.
Beyer, C., Anguiano, G., & Mena, F. (1961). Oxytocin release in response to stimulation of cingulate gyrus. *American Journal of Physiology, 200,* 625–627.
Bianchi-Demicheli, F., & Ortigue, S. (2007). Toward an understanding of the cerebral substrates of woman's orgasm. *Neuropsychologia, 45,* 2645–2659.
Birnbaum, G. E. (2007). Attachment orientations, sexual functioning, and relationship satisfaction in a community sample of women. *Journal of Social and Personal Relationships, 24,* 21–35.
Blaicher, W., Gruber, D., Bieglmayer, C., Blaicher, A. M., Knogler, W., & Huber, J. C. (1999). The role of oxytocin in relation to female sexual arousal. *Gynecologic and Obstetric Investigation, 47,* 125–126.
Brewer, G., & Hendrie, C. (2011). Evidence to suggest that copulatory vocalizations in women are not a reflexive consequence of orgasm. *Archives of Sexual Behavior, 40,,* 559–564.
Brody, S. (2007). Vaginal orgasm is associated with better psychological function. *Sexual and Relationship Therapy, 22,* 173–191.
Brody, S., & Costa, R. M. (2008). Vaginal orgasm Is associated with less use of immature psychological defense mechanisms. *The Journal of Sexual Medicine, 5,* 1167–1176.
Brody, S., & Weiss, P. (2010). Vaginal orgasm is associated with vaginal (not clitoral) sex education, focusing mental attention on vaginal sensations, intercourse duration, and a prefernceprefference for longer penises. *The Journal of Sexual Medicine, 7,* 2774–2781.
Brody, S., & Weiss, P. (2011). Simultaneous penile–vaginal intercourse orgasm is associated with satisfaction (sexual, life, partnership, and mental health). *The Journal of Sexual Medicine, 8,* 734–741.
Buisson, O., Foldes, P., Jannini, E., & Mimoun, S. (2010). Coitus as revealed by ultrasound in one volunteer couple. *The Journal of Sexual Medicine, 7,* 2750–2754.
Buss, D. M. (1998). Sexual strategies theory: Historical origins and current status. *Journal of Sex Research, 35,* 19–31.
Buss, D. M., & Kenrick, D. T. (1998). Evolutionary social psychology. In D. T. Gilbert, S. T. Fiske, & G. Lindzey (Eds.), *The handbook of social psychology: Vol. 2* (pp. 982–1026). New York: McGraw-Hill.
Buss, D. M., & Schmitt, D. P. (1993). Sexual strategies theory: An evolutionary perspective on human mating. *Psychological Review, 100,* 204–232.
Buss, D. M., & Shackelford, T. K. (2008). Attractive women want it all: Good genes, economic investment, parenting proclivities, and emotional commitment. *Evolutionary Psychology, 6,* 134–146.
Buss, D. M., Haselton, M. G., Shackelford, T. K., Bleske, A. L., & Wakefield, J. C. (1998). Adaptations, exaptations, and spandrels. *American Psychologist, 53,* 533–548.
Caldwell, J. D., Prange Jr, A. J., & Pedersen, C. A. (1986). Oxytocin facilitates the sexual receptivity of estrogen-treated female rats. *Neuropeptides, 7,* 175–189.
Carmichael, M. S., Humbert, R., Dixen, J., Palmisano, G., Greenleaf, W., & Davidson, J. M. (1987). Plasma oxytocin increases in the human sexual response. *Journal of Clinical Endocrinology & Metabolism, 64,* 27–31.

Carmichael, M., Warburton, V., Dixen, J., & Davidson, J. (1994). Relationships among cardiovascular, muscular, and oxytocin responses during human sexual activity. *Archives of Sexual Behavior, 23,* 59–79.

Carter, C. S. (1998). Neuroendocrine perspectives on social attachment and love. *Psychoneuroendocrinology, 23,* 779–818.

Carter, C. S., & Porges, S. W. (2013). The biochemistry of love: An oxytocin hypothesis. *EMBO Reports, 14,* 12–16.

Carter, C. S., Grippo, A. J., Pournajafi-Nazarloo, H., Ruscio, M. G., & Porges, S. W. (2008). Oxytocin, vasopressin and sociality. In D. N. Inga & L. Rainer (Eds.), *Advances in vasopressin and oxytocin: From genes to behaviour to disease* (pp. 331–336). New York: Elsevier.

Carter, C. S., Harris, J., & Porges, S. W. (2009). Neural and evolutionary perspectives on empathy. In J. Decety & W. Ickes (Eds.), *The social neuroscience of empathy* (pp. 169–182). Cambridge: MIT Press.

Clark, R. D., & Hatfield, E. (1989). Gender Ddifferences in Rreceptivity to Ssexual Ooffers. *Journal of Psychology & Human Sexuality, 2,* 39–55.

Clayton, A. H., Clavet, G. J., McGarvey, E. L., Warnock, J. K., & Weiss, K. (1999). Assessment of sexual functioning during the menstrual cycle. *Journal of Sex & Marital Therapy, 25,* 281–291.

Clifford, R. (1978). Subjective sexual experience in college women. *Archives of Sexual Behavior, 7,* 183–197.

Clutton-Brock, T. (1982). The functions of antlers. *Behaviour, 79,* 108–125.

Cohen, D. L., & Belsky, J. (2008). Avoidant romantic attachment and female orgasm: Testing an emotion-regulation hypothesis. *Attachment & Human Development, 10,,* 1–10.

Colson, M. H. (2010). Female orgasm: Myths, facts and controversies. *Sexologies, 19,* 8–14.

Costa, R. M., & Brody, S. (2007). Women";s relationship quality is associated with specifically penile-vaginal intercourse orgasm and frequency. *Journal of Sex & Marital Therapy, 33,* 319–327.

Costa, R. M., & Brody, S. (2011). Anxious and avoidant attachment, vibrator use, anal sex, and impaired vaginal orgasm. *The Journal of Sexual Medicine, 8,* 2493–2500.

Cross, B., & Wakerley, J. (1977). The neurohypophysis. *International Review of Physiology, 16,* 1–34.

Cushing, B., C., & Carter, C. S. (1999). Prior exposure to oxytocin mimics the effects of social contact and facilitates sexual behaviour in females. *Journal of Neuroendocrinology, 11,* 765–769.

Daly, M., & Wilson, M. I. (1982). Whom are newborn babies said to resemble? *Ethology and Sociobiology, 3,* 69–78.

Daly, M., & Wilson, M. (1983). *Sex, evolution, and behavior* (2nd ed.). Belmont: Wadsworth.

Darling, C., Davidson, J. K., & Jennings, D. (1991). The female sexual response revisited: Understanding the multiorgasmic experience in women. *Archives of Sexual Behavior, 20,* 527–540.

Dawood, K., Kirk, K. M., Bailey, J. M., Andrews, P. W., & Martin, N. G. (2005). Genetic and environmental influences on the frequency of orgasm in women. *Twin Research and Human Genetics, 8,* 27–33.

Ditzen, B., Schaer, M., Gabriel, B., Bodenmann, G., Ehlert, U., & Heinrichs, M. (2009). Intranasal oxytocin increases positive communication and reduces cortisol levels during couple conflict. *Biological Psychiatry, 65,* 728–731.

Dunn, K. M., Cherkas, L. F., & Spector, T. D. (2005). Genetic influences on variation in female orgasmic function: A twin study. *Biology Letters, 1,* 260–263.

Du Vigneaud, V. (1954). Hormones of the posterior pituitary gland: Oxytocin and vasopressin. *Harvey Lectures, 50,* 1–26.

Eibl-Eibesfeldt, I. (1975). *Ethology: The biology of behavior*. New York: Holt, Reinhart, and Winston.

Ellsworth, R., & Bailey, D. (2013). Human female orgasm as evolved signal: A test of two hypotheses. *Archives of Sexual Behavior, 42,* 1545–1554.

Eschler, L. (2004). The physiology of the female orgasm as a proximate mechanism. *Sexualities, Evolution & Gender, 6,* 171–194.

Espmark, Y. (1971). Antler shedding in relation to parturition in female reindeer. *The Journal of Wildlife Management, 35,* 175–177.

Exton, M. S., Bindert, A., Krüger, T., Scheller, F., Hartmann, U., & Schedlowski, M. (1999). Cardiovascular and endocrine alterations after masturbation-induced orgasm in women. *Psychosomatic Medicine, 61,* 280–289.

Fisher, H. E. (1998). Lust, attraction, and attachment in mammalian reproduction. *Human Nature, 9,* 23–52.

Folstad, I., & Karter, A. J. (1992). Parasites, bright males, and the immunocompetence handicap. *American Naturalist, 139,* 603–622.

Fox, C. A., & Fox, B. (1971). A comparative study of coital phsyiologyphysiology, with special reference to the sexual climax. *Journal of Reproduction and Fertility, 24,* 319–336.

Fox, C. A., Wolff, H. S., & Baker, J. A. (1970). Measurement of intra-vaginal and intra-uterine pressures during human coitus by radio-telemetry. *Journal of Reproduction and Fertility, 22,* 243–251.

Fugl-Meyer, K. S., Öberg, K., Lundberg, P. O., Lewin, B., & Fugl-Meyer, A. (2006). On orgasm, sexual techniques, and erotic perceptions in 18- to 74-year-old Swedish women. *The Journal of Sexual Medicine, 3,* 56–68.

Gangestad, S. W., & Thornhill, R. (1997). The evolutionary psychology of extrapair sex: The role of fluctuating asymmetry. *Evolution and Human Behavior, 18,* 69–88.

Gangestad, S. W., & Thornhill, R. (2008). Human oestrus. *Proceedings of the Royal Society B: Biological Sciences, 275,* 991–1000.

Garcia, J. R., Massey, S. G., Merriwether, A. M., & Seibold-Simpson, S. M. (in submission). Orgasm experiences among emerging adult men and women: Gender, relationship context, and attitudes toward casual sex.

Garver-Apgar, C. E., Gangestad, S. W., Thornhill, R., Miller, R. D., & Olp, J. J. (2006). Major histocompatibility complex alleles, sexual responsivity, and unfaithfulness in romantic couples. *Psychological Science, 17,* 830–835.

Gavrilets, S. (2012). Human origins and the transition from promiscuity to pair-bonding. *Proceedings of the National Academy of Sciences, 109,* 9923–9928.

Gebhard, P. H. (1966). Factors in marital orgasm. *Journal of Social Issues, 22,* 88–95.

Gould, S. J. (1987). Freudian slip. *Natural History, 96,* 14–21.

Gould, S. J., & Vrba, E. S. (1982). Exaptation—a missing term in the science of form. *Paleobiology, 8,* 4–15.

Gräfenberg, E. (1950). The role of the urethra in female orgasm. *International Journal of Sexology, 3,* 145–148.

Grammer, K., Fink, B., Møller, A. P., & Thornhill, R. (2003). Darwinian aesthetics: Sexual selection and the biology of beauty. *Biological Reviews, 78,* 385–407.

Graustella, A. J., & MacLeod, C. (2012). A critical review of the influence of oxytocin nasal spray on social cognition in humans: Evidence and future directions. *Hormones and Behavior, 61,* 410–418.

Grewen, K. M., Girdler, S. S., Amico, J., & Light, K. C. (2005). Effects of partner support on resting oxytocin, cortisol, norepinephrine, and blood pressure before and after warm partner contact. *Psychosomatic Medicine, 67,* 531–538.

Hamburg, B. A. (1978). The biosocial basis of sex differences. In S. L. Washburn & E. R. McCown (Eds.), *Human evolution: Biosocial perspectives* (pp. 155–213). Menlo Park: Benjamin/Cummings.

Haning, R. V., O'Keefe, S. L., Randall, E. J., Kommor, M. J., Baker, E., & Wilson, R. (2007). Intimacy, orgasm likelihood, and conflict predict sexual satisfaction in seterosexual male and female respondents. *Journal of Sex & Marital Therapy, 33,* 93–113.

Henshaw, J. (1968). A theory for the occurrence of antlers in females of the genus Rangifer. *Deer, 1,* 222–226.

Hite, S. (1976). *The Hite report: A nationwide study of female sexuality*. New York: Macmillan.

Hoch, Z. (1986). Vaginal erotic sensitivity by sexological examination. *Acta Obstetricia et Gynecologica Scandinavica, 65,* 767–773.

Holmes, B. M., & Johnson, K. R. (2009). Adult attachment and romantic partner preference: A review. *Journal of Social and Personal Relationships, 26,* 833–852.

Holmes, M. M., Putz, O., Crews, D., & Wade, J. (2005). Normally occurring intersexuality and testosterone induced plasticity in the copulatory system of adult leopard geckos. *Hormones and Behavior, 47,* 439–445.

Hosken, D. J. (2008). Clitoral variation says nothing about female orgasm. *Evolution & Development, 10,* 393–395.

Hurlemann, R., Patin, A., Onur, O. A., Cohen, M. X., Baumgartner, T., Metzler, S., et al. (2010). Oxytocin enhances amygdala-dependent, socially reinforced learning and emotional empathy in humans. *The Journal of Neuroscience, 30,* 4999–5007.

Huynh, H. K., Willemsen, A. T. M., & Holstege, G. (2013). Female orgasm but not male ejaculation activates the pituitary. A PET-neuro-imaging study. *NeuroImage, 76,* 178–182.

Ingelman-Sundberg, A. (1997). The anterior vaginal wall as an organ for the transmission of active forces to the urethra and the clitoris. *International Urogynecology Journal, 8,* 50–51.

IsHak, W. W., Bokarius, A., Jeffrey, J. K., Davis, M. C., & Bakhta, Y. (2010). Disorders of orgasm in women: A literature review of etiology and current treatments. *The Journal of Sexual Medicine, 7,* 3254–3268.

Kenrick, D. T., Groth, G. E., Trost, M. R., & Sadalla, E. K. (1993). Integrating evolutionary and social exchange perspectives on relationships: Effects of gender, self-appraisal, and involvement level on mate selection criteria. *Journal of Personality and Social Psychology, 64,* 951–969.

King, R., & Belsky, J. (2012). A typological approach to testing the evolutionary functions of human female orgasm. *Archives of Sexual Behavior, 41,* 1145–1160.

King, M., Holt, V., & Nazareth, I. (2007). Women's views of their sexual difficulties: Agreement and disagreement with clinical diagnoses. *Archives of Sexual Behavior, 36,* 281–288.

King, R., Belsky, J., Mah, K., & Binik, Y. (2011). Are there edifferent types of female orgasm? *Archives of Sexual Behavior, 40,* 865–875.

Knaus, H. (1950). *The physiology of human reproduction* (3rd ed.). Vienna, Austria: Wilhelm Maudlach Verlag.

Komisaruk, B. R., Whipple, B., Crawford, A., Grimes, S., Liu, W.-C., &Kalnin, A., et al. (2004). Brain activation during vaginocervical self-stimulation and orgasm in women with complete spinal cord injury: FMRI evidence of mediation by the Vagus nerves. *Brain Research, 1024,* 77–88.

Kosfeld, M., Heinrichs, M., Zak, P. J., Fischbacher, U., & Fehr, E. (2005). Oxytocin increases trust in humans. *Nature, 435,* 673–676.

Krüger, T. H. C., Haake, P., Hartmann, U., Schedlowski, M., & Exton, M. S. (2002). Orgasm-induced prolactin secretion: Feedback control of sexual drive? *Neuroscience & Biobehavioral Reviews, 26,* 31–44.

Kruger, T. H., Haake, P., Chereath, D., Knapp, W., Janssen, O. E., &Exton, M. S., et al. (2003). Specificity of the neuroendocrine response to orgasm during sexual arousal in men. *Journal of Endocrinology, 177,* 57–64.

Kunz, G., Beil, D., Huppert, P., & Leyendecker, G. (2007). Oxytocin—a stimulator of directed sperm transport in humans. *Reproductive BioMedicine Online, 14,* 32–39.

Levin, R. (2002). The physiology of sexual arousal in the human female: A recreational and procreational synthesis. *Archives of Sexual Behavior, 31,* 405–411.

Levin, R. J. (2011). Can the controversy about the putative role of the human female orgasm in sperm transport be settled with our current physiological knowledge of coitus? *The Journal of Sexual Medicine, 8,* 1566–1578.

Light, K. C., Grewen, K. M., & Amico, J. A. (2005). More frequent partner hugs and higher oxytocin levels are linked to lower blood pressure and heart rate in premenopausal women. *Biological Psychology, 69,* 5–21.

Lightfoot-Klein, H. (1989). The sexual experience and marital adjustment of genitally circumcised and infibulated females in the Sudan. *Journal of Sex Research, 26,* 375–392.

Lloyd, E. A. (2005). *The case of the female orgasm: Bias in the science of evolution.* Harvard: University Press.

Lloyd, E. A. (2006). Response to Puts and Dawood's 'The Evolution of Female Orgasm: Adaptation or Byproduct?'—Been there. *Twin Research and Human Genetics, 9,* 603–608.
Lloyd, J., Crouch, N. S., Minto, C. L., Liao, L. M., & Creighton, S. M. (2005). Female genital appearance: "Normality" unfolds. *BJOG: An International Journal of Obstetrics and Gynaecology, 112,* 643–646.
Lynch, V. J. (2008). Clitoral and penile size variability are not significantly different: Lack of evidence for the byproduct theory of the female orgasm. *Evolution & Development, 10,* 396–397.
MacDonald, K., & Feifel, D. (2012). Dramatic improvement in sexual runction induced by intranasal oxytocin. *The Journal of Sexual Medicine, 9,* 1407–1410.
Mah, K., & Binik, Y. M. (2001). The nature of human orgasm: A critical review of major trends. *Clinical Psychology Review, 21,* 823–856.
Mah, K., & Binik, Y. M. (2002). Do all orgasms feel alike? Evaluating a two-dimensional model of the orgasm experience across gender and sexual context. *Journal of Sex Research, 39,* 104–113.
Mah, K., & Binik, Y. M. (2005). Are orgasms in the mind or the body? Psychosocial versus physiological correlates of orgasmic pleasure and satisfaction. *Journal of Sex & Marital Therapy, 31,* 187–200.
Marazziti, D., Dell'Osso, B., Baroni, S., Mungai, F., Catena, M., & Rucci, P., et al. (2006). A relationship between oxytocin and anxiety of romantic attachment. *Clinical Practice and Epidemiology in Mental Health, 2,* 28.
Masters, W. H., & Johnson, V. E. (1966). *Human sexual response* (p. 185). Boston: Little, Brown. *& Co.*
Matteo, S., & Rissman, E. F. (1984). Increased sexual activity during the midcycle portion of the human menstrual cycle. *Hormones and Behavior, 18,* 249–255.
Mellen, S. L. W. (1982). *The evolution of love*. San Francisco: Freeman.
Messenger, J. C. (1971). Sex and repression in an Irish folk community. In D. S. Marshall & R. C. Suggs (Eds.), *Human sexual behavior: Variations in the ethnographic spectrum* (Vol. 5, pp. 3–37), 5, 3–37. New York: (Basic Books).
Meston, C. M., Levin, R. J., Sipski, M. L., Hull, E. M., & Heiman, J. R. (2004). Women's orgasm. *Annual Review of Sex Research, 15,* 173–257.
Milinski, M. (2006). The major histocompatibility complex, sexual selection, and mate choice. *Annual Review of Ecology, Evolution, and Systematics, 37,* 159–186.
Møller, A. P., & Pomiankowski, A. (1993). Why have birds got multiple sexual ornaments? *Behavioral Ecology and Sociobiology, 32,* 167–176.
Møller, A., & Thornhill, R. (1998). Bilateral symmetry and sexual selection: A meta-analysis. *The American Naturalist, 151,* 174–192.
Morris, D. (1967). *The naked ape*. New York: Dell.
Muehlenhard, C. L., & Shippee, S. K. (2010). Men's and women's reports of pretending orgasm. *Journal of Sex Research, 47,* 552–567.
O'Connell, H. E., Hutson, J. M., Anderson, C. R., & Plenter, R. J. (1998). Anatomical relationship between urethra and clitoris. *The Journal of Urology, 159,* 1892–1897.
O'Connell, H. E., Sanjeevan, K. V., & Hutson, J. M. (2005). Anatomy of the clitoris. *The Journal of Urology, 174,* 1189–1195.
O'Connell, H. E., Eizenberg, N., Rahman, M., & Cleeve, J. (2008). The anatomy of the distal vagina: Towards unity. *The Journal of Sexual Medicine, 5,* 1883–1891.
Ortigue, S., Bianchi-Demicheli, F., de C., Hamilton, A. F. de C., & Grafton, S. T. (2007). The Neural basis of love as a subliminal prime: An event-related functional magnetic resonance imaging study. *Journal of Cognitive Neuroscience, 19,* 1218–1230.
Parsons, P. A. (1990). Fluctuating asymmetry: An epigentic measure of stress. *Biological Reviews, 65,* 131–145.
Parsons, P. A. (1992). Fluctuating asymmetry: A biological monitor of environmental and genomic stress. *Heredity, 68,* 361–364.
Pedersen, C. A., & Boccia, M. L. (2002). Oxytocin maintains as well as initiates female sexual behavior: Effects of a highly selective oxytocin antagonist. *Hormones and Behavior, 41,* 170–177.

Plutchik, R. (1982). A psychoevolutionary theory of emotions. *Social Science Information, 21,* 529–553.

Potts, W. K., & Wakeland, E. K. (1993). Evolution of MHC genetic diversity: A tale of incest, pestilence and sexual preference. *Trends in Genetics, 9,* 408–412.

Prause, N. (2011). The human female orgasm: Critical evaluations of proposed psychological sequelae. *Sexual and Relationship Therapy, 26,* 315–328.

Puts, D. (2007). Of Bugs and boojums: Female orgasm as a facultative adaptation. *Archives of Sexual Behavior, 36,* 337–339.

Puts, D., Dawood, K., & Welling, L. M. (2012a). Why women have orgasms: An evolutionary analysis. *Archives of Sexual Behavior, 41,* 1127–1143.

Puts, D. A., Welling, L. L. M., Burriss, R. P., & Dawood, K. (2012b). Men's masculinity and attractiveness predict their female partners' reported orgasm frequency and timing. *Evolution and Human Behavior, 33,* 1–9.

Regalski, J. M., & Gaulin, S. J. C. (1993). Whom are Mexican infants said to resemble? Monitoring and fostering paternal confidence in the Yucatan. *Ethology and Sociobiology, 14,* 97–113.

Rilling, J. K., DeMarco, A. C., Hackett, P. D., Thompson, R., Ditzen, B., Patel, R., et al. (2012). Effects of intranasal oxytocin and vasopressin on cooperative behavior and associated brain activity in men. *Psychoneuroendocrinology, 37,* 447–461.

Roberts, S. C., & Little, A. C. (2008). Good genes, complementary genes and human mate preferences. *Genetica, 134,* 31–43.

Sanchez, R., Parkin, J. C., Chen, J. Y., & Gray, P. B. (2009). Oxytocin, vasopressin, and human social behavior. In P. T. Ellison & P. B. Gray (Eds.), *Endocrinology of social relationships* (pp. 319–339). Cambridge: Harvard University Press.

Schmitt, D. P. (2005). Sociosexuality from Argentina to Zimbabwe: A 48-nation study of sex, culture, and strategies of human mating. *Behavioral and Brain Sciences, 28,* 247–274.

Schneiderman, I., Zagoory-Sharon, O., Leckman, J. F., & Feldman, R. (2012). Oxytocin during the initial stages of romantic attachment: Relations to couples' interactive reciprocity. *Psychoneuroendocrinology, 37,* 1277–1285.

Setekleiv, J. (1964). Uterine motility of the estrogenized rabbit. *Acta Physiologica Scandinavica, 62,* 79–93.

Shackelford, T., Weekes-Shackelford, V., LeBlanc, G., Bleske, A., Euler, H., & Hoier, S. (2000). Female coital orgasm and male attractiveness. *Human Nature, 11,* 299–306.

Singer, I. (1973). *The goal of human sexuality*. New York: W.W. Norton.

Singh, D., Meyer, W., Zambarano, R., & Hurlbert, D. (1998). Frequency and timing of coital orgasm in women desirous of becoming pregnant. *Archives of Sexual Behavior, 27,* 15–29.

Smith, R. L. (1984). Human sperm competition. In R. L. Smith (Ed.), *Sperm competition and the evolution of animal mating systems* (pp. 601–660). London: Academic Press.

Smith, T. T. (1998). The modulation of sperm function by the oviductal epithelium. *Biology of Reproduction, 58,* 1102–1104.

Spyropoulos, E., Borousas, D., Mavrikos, S., Dellis, A., Bourounis, M., & Athanasiadis, S. (2002). Size of external genital organs and somatometric parameters among physically normal men younger than 40 years old. *Urology, 60,* 485–489;. (dDiscussion 490–481).

Storey, A. E., Walsh, C. J., Quinton, R. L., & Wynne-Edwards, K. E. (2000). Hormonal correlates of paternal responsiveness in new and expectant fathers. *Evolution and Human Behavior, 21,* 79–95.

Suarez, S. S. (1998). The oviductal sperm reservoir in mammals: Mechanisms of formation. *Biology of Reproduction, 58,* 1105–1107.

Suttie, J. M., Fennessy, P. F., Lapwood, K. R., & Corson, I. D. (1995). Role of steroids in antler growth of red deer stags. *Journal of Experimental Zoology, 271,* 120–130.

Symons, D. (1979). *The evolution of human sexuality*. New York: Oxford University Press.

Tabak, B. A., McCullough, M. E., Szeto, A., Mendez, A. J., & McCabe, P. M. (2011). Oxytocin indexes relational distress following interpersonal harms in women. *Psychoneuroendocrinology, 36,* 115–122.

Taylor, S. E., Gonzaga, G. C., Klein, L. C., Hu, P., Greendale, G. A., & Seeman, T. E. (2006). Relation of oxytocin to psychological stress responses and hypothalamic-pituitary-adrenocortical axis activity in older women. *Psychosomatic Medicine, 68,* 238–245.

Taylor, S. E., Saphire-Bernstein, S., & Seeman, T. E. (2010). Are plasma oxytocin in women and plasma vasopressin in men biomarkers of distressed pair-bond relationships? *Psychological Science, 21,* 3–7.

Thornhill, R., Gangestad, S. W., & Comer, R. (1995). Human female orgasm and mate fluctuating asymmetry. *Animal Behaviour, 50,* 1601–1615.

Trivers, R. (1972). Parental investment and sexual selection. In B. Cambell (Ed.), *Sexual selection and the descent of man, 1871–1971* (pp. 136–179). Chicago: Aldine.

Turner, R. A., Altemus, M., Enos, T., Cooper, B., & McGuinness, T. (1999). Preliminary research on plasma oxytocin in normal cycling women: Investigating emotion and interpersonal distress. *Psychiatry, 62,* 97–113.

Udry, J. R., & Morris, N. M. (1968). Distribution of coitus in the menstrual cycle. *Nature, 220,* 593–596.

Unkelbach, C., Guastella, A. J., & Forgas, J. P. (2008). Oxytocin selectively facilitates recognition of positive sex and relationship words. *Psychological Science, 19,* 1092–1094.

van Anders, S. M., & Dunn, E. J. (2009). Are gonadal steroids linked with orgasm perceptions and sexual assertiveness in women and men? *Hormones and Behavior, 56,* 206–213.

van Anders, S., Edelstein, R., Wade, R., & Samples-Steele, C. (2013). Descriptive experiences and sexual vs. nurturant aspects of cuddling between adult romantic partners. *Archives of Sexual Behavior, 42,* 553–560.

Van Valen, L. (1962). A study of fluctuating asymmetry. *Evolution, 16,* 125–142.

Vance, E., & Wagner, N. (1976). Written descriptions of orgasm: A study of sex differences. *Archives of Sexual Behavior, 5,* 87–98.

Walker, R. S., & Bailey, D. H. (2013). Body counts in lowland South American violence. *Evolution and Human Behavior, 34,* 29–34.

Walker, R. S., Hill, K. R., Flinn, M. V., & Ellsworth, R. M. (2011). Evolutionary history of hunter-gatherer marriage practices. *PLOS One, 6,* e19066.

Wallen, K., & Lloyd, E. A. (2008). Clitoral variability compared with penile variability supports nonadaptation of female orgasm. *Evolution & Development, 10,* 1–2.

Welling, L. L. M., Burriss, R. P., & Puts, D. A. (2011). Mate retention behavior modulates men's preferences for self-resemblance in infant faces. *Evolution and Human Behavior, 32,* 118–126.

Whipple, B., Ogden, G., & Komisaruk, B. (1992). Physiological correlates of imagery-induced orgasm in women. *Archives of Sexual Behavior, 21,* 121–133.

Wildt, L., Kissler, S., Licht, P., & Becker, W. (1998). Sperm transport in the human female genital tract and its modulation by oxytocin as assessed by hysterosalpingoscintigraphy, hysterotonography, electrohysterography and Doppler sonography. *Human Reproduction Update, 4,* 655–666.

Williams, G. (1966). *Adaptation and natural selection*. Princeton: Princeton University Press.

Williams, J. R., Catania, K. C., & Carter, C. S. (1992). Development of partner preferences in female prairie voles (Microtus ochrogaster): The role of social and sexual experience. *Hormones and Behavior, 26,* 339–349.

Williams, J. R., Insel, T. R., Harbaugh, C. R., & Carter, C. S. (1994). Oxytocin administered centrally facilitates formation of a partner preference in female prairie voles (Microtus ochrogaster). *Journal of Neuroendocrinology, 6,* 247–250.

Worthman, C. M. (1978). *Psychoendocrine study of human behavior: Some interactions of steroid hormones with affect and behavior in the!Kung San*. Cambridge: Harvard University Press.

Zahavi, A., & Zahavi, A. (1977). *The handicap principle*. New York: Oxford University Press.

Zervomanolakis, I., Ott, H. W., Hadziomerovic, D., Mattle, V., Seeber, B. E., Virgolini, I., et al. (2007). Physiology of upward transport in the human female genital tract. *Annals of the New York Academy of Sciences, 1101,* 1–20.

Zervomanolakis, I., Ott, H. W., Müller, J., Seeber, B. E., Friess, S. C., &Mattle, V.., et al. (2009). Uterine mechanisms of ipsilateral directed spermatozoa transport: Evidence for a contribution of the utero-ovarian countercurrent system. *European Journal of Obstetrics & Gynecology and Reproductive Biology, 144,* (Suppl.ement 1), S45–S49.

Zietsch, B. P., & Santtila, P. (2013). No direct relationship between human female orgasm rate and number of offspring. *Animal Behaviour, 86,* 253–255.

Zumpe, D., & Michael, R. P. (1968). The clutching reaction and orgasm in the female rhesus monkey (*Macaca mulatta*). *Journal of Endocrinology, 40,* 117–123.

Chapter 8
The Functional Design and Phylogeny of Women's Sexuality

Randy Thornhill and Steven W. Gangestad

Introduction

Recent research findings are revolutionizing the science of human sexuality. The major advances in knowledge primarily concern women's sexuality, but men's sexuality, too, is seen in new lights. These advances arise, in part, from the recent discovery of estrus in women, which occurs during the several days prior to ovulation, and may function in good-genes mate choice. This discovery and related findings demand that long-standing conclusions about human sexuality, including that human estrus was lost evolutionarily and that men are totally unknowing of conception probability across women's cycle, be overturned and key established facts reinterpreted. In light of women's estrus, women's sexual motivation outside estrus (i.e., at all times when conception cannot occur) must be viewed as a distinct adaptation, one we propose functions to obtain nongenetic material benefits and services from men—an adaptation we refer to as "extended sexuality" (see also Rodríguez-Gironés and Enquist 2001). Concealed estrus is another distinct adaptation in women. We propose that it functions during estrus to disguise it, especially from pair-bond partners, and to facilitate copulation contingently outside the pair-bond to obtain good genes while retaining the pair-bond partner's nongenetic material benefits.

The traditional assumption of scholars of lost-estrus in the human lineage misled earlier researchers in their interpretations of women's sexuality, including women's sexually attractive features. For example, women's breasts and buttocks were seen

R. Thornhill (✉)
Department of Biology, The University of New Mexico,
Albuquerque, NM 87131-0001, USA
e-mail: rthorn@unm.edu

S. W. Gangestad
Department of Psychology, The University of New Mexico,
Albuquerque, NM 87131-0001, USA

T. K. Shackelford, R. D. Hansen (eds.), *The Evolution of Sexuality*,
Evolutionary Psychology, DOI 10.1007/978-3-319-09384-0_8

as permanent, deceptive signals of nonexistent estrus. Recent investigations indicate that women's estrogen-facilitated sexual ornamentation of both body and face function to honestly signal phenotypic and genetic quality pertaining to residual reproductive value (Thornhill and Gangestad 2008).

There is a heuristic reciprocal relationship between research on human sexuality and research on the sexuality of other vertebrates; research findings from humans inform research on other vertebrates and vice versa. A body of comparative evidence informs an understanding of the function of estrus and extended female sexuality throughout the vertebrates. Comparative evidence also illuminates the phylogeny of female sexuality. Estrus appears to show homology across all vertebrates, whereas extended sexuality evolved independently in the vertebrate lineages in which males deliver nongenetic material benefits to females (Thornhill and Gangestad 2008). Research on human females importantly adds to this comparative evidence.

In our book, *The Evolutionary Biology of Human Female Sexuality* (2008), we explained in detail the aforementioned interpretations of human sexuality and reviewed evidence bearing on them available at that time. This chapter briefly summarizes our interpretations and updates some relevant evidence.

The Lost-Estrus Assumption

With the discovery by 1940 of the hormone estrogen and its causal roles in ovulation in women as well as in ovulation and estrus in nonhuman mammals, researchers turned to the question of estrus in women (Corner 1942). They reasoned that, if estrus exists in women, it will function as estrus was assumed to function in other mammals—to obtain sperm for fertilization. Many research studies searched for a positive association between conception probability across the menstrual cycle and copulation rates with male pair-bond partners in women having ovulatory menstrual cycles. Other studies looked at women's sexual interests in general across the cycle, also anticipating greater female libido during the fertile phase of the cycle if, indeed, women have estrus. Despite some positive findings, a large body of research yielded no consistent evidence of higher sexual motivation toward a pair-bond partner or higher libido in general during the fertile phase than during infertile cycle phases, and that trend continues to this day (reviewed in Thornhill and Gangestad 2008). For example, a study involving over 20,000 normally ovulating women (women not using hormonal contraception) across 13 developing countries detected no variation in rates of mating between women and primary partners, with the exception of a decline during menstruation (Brewis and Meyer 2005). Already by 1960, researchers widely concluded that women had, during their evolutionary history since diverging from common ancestors of close extant primate relatives, lost estrus. This conclusion, in turn, led scholars to the question of why women lost estrus, either as an adaptation itself (e.g., Alexander and Noonan 1979; Burley 1979; Symons 1979) or as a by-product of another adaptation (e.g., Spuhler 1979).

Recent research programs directed at understanding estrus in women are based on a very different assumption about the evolved function of mammalian estrus than that of earlier researchers. Instead of functioning to get sperm in general (hence involving adaptation for conception assurance), estrus is sexual motivation functionally designed to obtain a sire that increases the reproductive value of offspring through enhanced survival and/or sexual attractiveness (hence involving adaptation for obtaining good genes). This hypothesis makes numerous and now well-supported predictions about women's sexuality across the cycle, ranging from women's estrous shift to short-term mate preference for men's traits related to phenotypic quality, increased sexual interest in extra-pair mating when the main partner lacks such traits, and the presence of a differently designed female sexuality outside estrus. Whereas early researchers sought evidence in women for indiscriminate high libido at the fertile cycle phase (or, operationally, sexual interest directed at primary partners), the contemporary approach focuses on a discriminating sexuality at estrus.

The Concept of Estrus

One dictionary defines estrus as "the periodic state of sexual excitement in the female of most mammals, excluding humans, that immediately precedes ovulation and during which the female is most receptive to mating" (*American Heritage Dictionary of the English Language*). In mammalian reproductive biology, the estrous cycle is equivalent to the ovarian cycle. Estrus is synonymous with estrous phase and refers to the period in ovulatory cycles of actual conception probability. Many scholars reserve the term menstrual cycle to refer exclusively to the ovarian cycles of human females, which reflects the widespread assumption—a wrong one, we argue—that, of all mammalian species, humans alone lack estrus (a view reflected in the dictionary definition quoted above). These scholars, then, refer to the cycles of all nonhuman primates as estrous cycles. Typically, *behavioral estrus*, also termed "heat," is defined as a restricted period of proceptivity and receptivity characterized by mammalian females' behavioral readiness to mate, in addition to attractiveness to males, typically coinciding with relatively high probability of conception (e.g., Beach 1976; Nelson 2000).

Women's estrus is comparable in length to estrus of other mammals (and vertebrates in general) (reviewed in Thornhill and Gangestad 2008). Women's "fertile window" is the sequential 5–6-day period of time ending on the day of ovulation. During this window, the probability of conception associated with insemination rises gradually for a couple of days, then more steeply, peaking 1–2 days prior to ovulation, if not on the day of ovulation itself (e.g., Dunson et al. 1999). The day prior to the window and the day following ovulation are associated with very low probabilities of conception. The timing of the window in the menstrual cycle is somewhat unpredictable, even in women with regular cycles (e.g., Wilcox et al. 2000; Dunson et al. 2001; Wilcox et al. 2001). A modal pattern is that, shortly after menses (lasting about 5 days), the risk of conception rises and peaks on about day 12, then sharply declines at ovulation, typically on day 14 in a 28-day cycle (Wilcox et al. 2001).

Women's Fertile-Phase Sexual Preferences

The scientific investigation of changes in women's sexual preferences across the menstrual cycle has been a very active research area, particularly in the last decade. Gildersleeve et al. (2014) have reviewed by meta-analysis many published and unpublished research findings on women's estrous preferences that address the hypothesis that estrus functions to obtain sires of high genetic quality for offspring; some of the studies reviewed by Gildersleeve et al. investigated the related hypothesis that estrous-preference shifts should be particularly strong when women evaluate men's "sexiness" or attractiveness as short-term sex partners, not their attractiveness as long-term, pair-bonded mates. In contrast, according to the hypothesis of women's estrus, women's preferences for traits in long-term partners should be relatively constant across their ovarian cycles. Gildersleeve et al. drew data from 38 published and 12 unpublished studies for a total of 134 empirical findings. Although some studies yielded null results, aggregates of preference changes across the cycle for characteristics thought to reflect male genetic quality ancestrally supported the hypothesis of estrous preferences focused on good-genes markers and especially on short-term male mate attractiveness. There was no consistent evidence that publication bias affected these conclusions. Here we briefly discuss some of the individual published studies across the various types of male traits that have been investigated.

Men's Body Scents

We tested the estrous-preference hypothesis by examining whether estrous women are particularly attracted to the body scents of symmetric men, as symmetry arguably is an (ancestral) indicator of quality (see Thornhill and Gangestad 2008). Low-fluctuating asymmetry reflects a relatively high level of developmental stability which, in turn, may reflect low mutation load, the ability to resist pathogens, and the ability to resist other ontogenetic stresses (Møller and Swaddle 1997; Gangestad and Thornhill 1999, 2003a, b; Møller et al. 2005). With a double-blind protocol, normally ovulating women rated the scent attractiveness of T-shirts worn by men who varied in symmetry. We found a positive correlation between women's probability of conception in their cycle and their preference for the scent of symmetry (see Gangestad and Thornhill 1998). Subsequently, three studies replicated this finding: Thornhill and Gangestad (1999a), Rikowski and Grammer (1999), and Thornhill et al. (2003). Women who were using hormone-based contraceptives (and hence did not experience ovulation and associated estrus) demonstrated no systematic preference for the scent of men's symmetry at any point in the cycle (e.g., Gangestad and Thornhill 1998; Thornhill and Gangestad 1999a).

Havlicek et al. (2005) studied women's preferences for the body scent of men highly dominant socially (as assessed by self-report). Women in estrus, but not women in infertile phases of their cycle, rated the scent of high-dominance men as sexier than that of low-dominance men. Furthermore, preferences for the scent of dominant men were stronger for estrous women in a pair-bond relationship than for estrous women who were single.

Grammer (1993) was the first researcher to test an explicitly stated hypothesis that women, when fertile, particularly prefer a male marker of likely genetic quality. He conjectured that the androgen androstenone may reflect testosterone levels in men which, in turn, may reflect genetic quality. He found that young women not using hormonal contraception rated the scent of androstenone least unpleasant during days of high fertility in the menstrual cycle. Hummel et al. (1991) found the same effect 2 years earlier (though did not cast it in the same theoretical light and involved a very small sample of women).

In our recently published study involving men wearing T-shirts smelled by normally ovulating women of variable cycle times, we found that women prefer the scent of men with high salivary testosterone more strongly when in estrus than when not (Thornhill et al. 2013). An earlier study, however, by Rantala et al. (2006), failed to find this effect, which may have been due to small samples and/or the methods used by Rantala et al. (see Thornhill et al. 2013).

Masculine Faces

Some facial features differ between the sexes because they are influenced during development by reproductive hormones—in particular, testosterone and estrogen, which are sex-specific in quantity (e.g., Johnston and Franklin 1993; Symons 1995; Thornhill and Gangestad 1999b; reviewed in Thornhill and Gangestad 2008). The degree of masculinity/testosteronization in men's facial and other bodily traits arguably reflects genetic quality. A number of researchers have examined women's preferences across their cycle for men's faces that vary in testosterone-effects. For example, Penton-Voak et al. (1999) predicted that normally ovulating women, when fertile in their cycles, prefer a face more masculine than the face they prefer when infertile. They furthermore predicted that this shift would be most pronounced when women evaluate men's attractiveness as a "short-term" mate (i.e., a sex partner) rather than as a long-term, stable mate. In two different studies, Penton-Voak et al. (1999) found the predicted shift toward greater preference for facial masculinity during women's fertile phase. Additional studies revealed precisely the same shift (Penton-Voak and Perrett 2000; Johnston et al. 2001; Gangestad et al. 2010b). The preference shift is not a by-product of a midcycle shift in women's generalized face perception (see Johnston et al. 2001). Some of these studies separately examined changes in women's evaluation of men as short-term and long-term partners. All yielded the effect only when women evaluated, in essence, men's "sexiness," but not when women evaluated men as long-term partners (Penton-Voak et al. 1999; Johnston et al. 2001).

Masculine Male Voices

In general, women prefer lower-pitch male voices (i.e., more masculine) (see review by Puts 2005). Puts (2005) had normally ovulating women rate the sexual attractiveness of men for long-term and short-term relationships based on voice alone. Women in estrus preferred lower-pitched male voices more strongly than did women outside estrus. Moreover, estrous women rated low-pitch male voices as more attractive for short-term sexual relationships than for long-term or committed relationships. But the same was not found in the nonestrous women. In addition, Puts (2006) found that women using hormonal contraception do not have the preference at their midcycle point that was shown by estrous women.

Feinberg et al. (2006) replicated and extended these findings. Women in estrus preferred voices more masculine along two sexually dimorphic dimensions—deeper voices and voices emanating from longer vocal tracts—than did women in infertile phase. Furthermore, Feinberg et al. found that women's preference shifts are specific to *male* vocal traits; they found no evidence that women's preferences for the two vocal traits in *women's* voices change across the cycle.

Dominant and Intrasexually Competitive Behavioral Displays

Simpson et al. (1999) videotaped men's direct intrasexual competitive behavioral displays in a situation in which men competed for a potential lunch date with an attractive woman. Gangestad et al. (2004) had these videotapes coded by women for a variety of specific behaviors (e.g., amount of time spent smiling), impressions (e.g., how confident men appeared), and verbal content (e.g., whether men said they were superior to their competitor). Factor analysis of these ratings revealed two largely independent dimensions. First, men varied in the extent to which they displayed social presence (e.g., appeared composed vs. looked downward). Second, men varied in the extent to which they displayed intrasexual competitiveness (e.g., said they were superior to their competitor vs. reflected the personality of a nice guy). Normally ovulating women then viewed the interviews and rated each man on dimensions of long-term mate attractiveness and attractiveness as a sex or affair partner. As predicted, women were attracted particularly to men who displayed both social presence and intrasexual competitiveness when they were in estrus—but only when they evaluated men as short-term, and not as long-term, mates (see also Gangestad et al. 2007).

In related research, Cantu et al. (2014) investigated women's flirtations with men across the ovulatory cycle. Male actors were portrayed in the video as "sexy cads" (attractive, charismatic, masculine, and socially dominant) and "good dads" (low dominance, agreeable, faithful, and warm) to each woman at estrus and during her infertile midluteal phase. Female participants indicated their attraction to these men as short-term and long-term mates. The study reported that when in estrus women flirted more, but only with "sexy cads." Furthermore, reported short-term mate preferences of estrous women favored sexy cads.

Bodily Masculinity

Testosterone promotes muscle growth, often in the service of intrasexual competition and mating effort (Ellison 2001). Women prefer male bodies that are muscular, particularly through the chest, arms, and back, without being clumsily overbuilt (e.g., Frederick and Haselton 2007). This preference, as we mentioned, may exist because male bodily testosteronization is a marker of superior male condition and, ancestrally, may have been an indicator of good genes.

Using the videos from Simpson et al.'s (1999) study mentioned just above, Gangestad et al. (2007) had women rate the men for their degree of muscularity. Subsequently, other women rated the men for short-term and long-term mate attractiveness. Estrous women, compared to their infertile counterparts, found the men rated as relatively muscular particularly attractive as short-term (relative to long-term) mates. This effect remained when men's behavioral displays (social presence and intrasexual competitiveness) were controlled.

Separate evidence indicates that women, when in estrus, are more sexually attracted to and stimulated by attractive, visually presented male features (Gangestad et al. 2010a). Women (in committed romantic relationships) were asked to twice complete two self-report measures—once when fertile and once when infertile (in the midluteal phase). First, they were asked about their mate preferences for physical attractiveness (with ten items, including, "I place a very high importance on a potential mate's physical attractiveness"). Second, they were asked about their interest in attractive bodily features (with ten items, including, "I find the thought of a very attractive body of the opposite sex very exciting"). The women were asked to think about how they felt *at that moment*, not how they felt in general. As predicted, shifts in both preferences were found. Women reported a greater preference for physical attractiveness and a greater interest in attractive bodily features when in estrus than during the luteal phase. Moreover, Little et al. (2007a) found that estrous women particularly prefer men's bodies that are masculine. Relatedly, Pawlowski and Jasiénska (2005) found that women more strongly preferred a greater degree of sexual dimorphism in stature (i.e., taller men) when they were in the fertile phase of their menstrual cycles and when the men were chosen for short-term relationships.

Some Additional Estrous Preferences

Frost (1994) found that normally ovulating women are more attracted to the faces of men with darker skin pigmentation when they are fertile than when they are infertile in the cycle. This effect is yet another instance in which fertile women particularly prefer masculine male faces. Within all races, skin tone is sexually dimorphic: men have browner skin tone (reflecting greater concentrations of melanin in the skin) and ruddier skin tone (reflecting greater concentrations of hemoglobin in the skin) (Frost 1994; Jones 1996). These sex differences implicate sex hormones. Testosterone tends to enhance melanin production, whereas estrogen suppresses its production in a variety of species, including humans (see review in Manning et al. 2003).

Male facial masculinity may be associated with male symmetry (Gangestad and Thornhill 2003a; Little et al. 2008a, b; but see Van Dongen et al. 2012). Thornhill and Gangestad (2003) found that women's conception risk in the cycle predicts positively their preference for the faces of men whose bodies were measured to be relatively symmetric. Though we did not measure specific facial features in this sample of men to explore which ones were responsible for the preferences, we suspect a role for male masculine facial features (see also Scheib et al. 1999).

Another possibility is that men with symmetric bodies also have more symmetric faces. And, perhaps, fertile women are attracted particularly to male faces that are symmetric. Little et al. (2007b) found that women particularly prefer more symmetric male faces when in the fertile phase of their cycles, especially when pair-bonded and evaluating men as short-term sex partners. However, Cardenas and Harris (2007) found no phase effect. Overall, studies provide mixed evidence that estrous women use symmetry of the face itself as a cue of good genes. Rather, the primary cues that women appear to be attracted to are masculine features (e.g., contained in facial, bodily, vocal, behavioral, and olfactory cues).

In light of Miller's (2000) hypothesis that in humans, creativity is among an array of "mental fitness indicators," Haselton and Miller (2006) asked whether normally ovulating women's preference for men's talent and creativity over their preference for men's material success (wealth) is enhanced when women are fertile in their cycles, and especially when women evaluate men as short-term mates. Women read a pair of vignettes related to artistic talent vs. financial wealth. Estrous women, when evaluating men as short-term partners, favored the talented man over the wealthy man. Their preferences for one sort of man over the other as a long-term, stable mate did not change across the cycle. This pattern may be specific to men's creativity rather than intelligence per se. Consistent with this, Gangestad et al. (2010b) reported that estrous women do not show an increased mate preference for men with high psychometric intelligence.

Not All of Women's Preferences are Exaggerated During Estrus

Women have mate preferences for traits that are not indicators of good genes. For instance, they have evolved preferences for male traits indicative of willingness and ability to provide care (e.g., Symons 1979, Buss 1989). Indeed, Buss (1989) found that the characteristic most valued in a long-term mate by both men and women across 37 cultures was "kindness and understanding." Evidence suggests that men who are kind and understanding to their mates may actually have, on average, lower genetic fitness. For example, symmetric men invest less in their romantic relationships (Gangestad and Thornhill 1997a) and are less faithful to partners (Gangestad and Thornhill 1997b) than asymmetric men. Also, compared to less masculine men, men with masculine faces are viewed as less trustworthy, faithful, and investing in their relationships (e.g., Penton-Voak and Perrett 2001) and as poorer long-term mate choices (e.g.,

Kruger 2006). Overall, men's testosteronization appears to negatively covary with their romantic and paternal investment (Thornhill and Gangestad 2008).

In analyses of the videotapes of men discussed above, Gangestad et al. (2007) had a separate sample of women rate men on 10 different traits women might find attractive in a mate. They then examined, in an analysis of three-way interactions among men's trait level, women's estimated fertility risk in their cycle, and the context of the relationship (short term vs. long term). When in estrus, women were particularly attracted sexually to men perceived as muscular, physically attractive, socially respected, arrogant, and confrontational with other men. By contrast, *no* shifts were detected in women's preference for men perceived as intelligent, kind, likely to be financially successful, or possessing qualities of a good father. Estrous women actually found men perceived to be faithful less sexually attractive when midcycle. (Put otherwise, they found men perceived to be unfaithful—perhaps by virtue of their other features—more sexually attractive during estrus.) Overall, those traits perceived as particularly sexy to women during their estrus were ones that women generally valued more in short-term partners than in long-term partners. The traits valued more in long-term mates than short-term mates—the qualities of good, stable mates—were not particularly sexy to estrous women. As we noted above, Haselton and Miller (2006) found that infertile-phase women prefer uncreative men with wealth over creative, but poorer, men.

Some traits valued in long-term partners appear to be particularly attractive to women with high progesterone levels, characteristic of the infertile luteal phase and pregnancy. Jones et al. (2005) examined normally ovulating women's preferences for femininity in men and women's faces in relation to their estimated progesterone levels based on norms for days across the menstrual cycle. When women's progesterone levels were high, femininity in both men's and women's faces was more attractive than when progesterone levels were low. Both men and women with feminine faces are perceived to be more cooperative and helpful (e.g., Perrett et al. 1998). Women who may be pregnant (during the luteal phase) or who are pregnant (and hence have high progesterone levels) then, prefer investing social partners. During the infertile phase of the cycle, women have other preferences for male traits that may connote the potential for receiving male-provided services and material benefits. Estimated progesterone level in the cycle of normally ovulating women positively predicts their preference for men with less masculine voices (Puts 2006). Similarly, Frost (1994) reported that women in the luteal phase (associated with high levels of progesterone) prefer less masculine facial skin tone.

The hypothesis that women have estrus argues that ancestral females could have benefitted throughout the cycle from having partners who are willing and able to invest in them. Supporting this, Gildersleeve et al.'s (2014) meta-analytic review did not detect significant cycle shifts in women's preferences for men's traits related to partner willingness and ability to invest (warmth, good father, financial success, social respect, and faithfulness).

Moreover, as we have mentioned, women using hormone-based contraception do not show cycle shifts in preferences for putative indicators of genetic quality. Often, hormone-based contraception raises progesterone levels (Gilbert 2000) and softens the midcycle peak in estrogen, hence establishing a hormonal milieu more typical of a woman in the luteal than the fertile phase. That women using hormone-based contraception do not show cycle effects strongly suggests that the hormonal pattern characteristic of the ovulatory cycle plays an important role in sexual-preference shifts across the cycle. As Puts (2006) notes, different estrous preferences may be affected by different hormones or combinations of hormones (see also Garver-Apgar et al. 2008).

In a recent study, Little et al. (2013) studied the change in facial preference for masculinity of women after they initiated use of contraceptive pill. Initiation of pill use significantly decreased women's preferences for male facial masculinity but did not influence their preferences for women's faces. Little et al. (2013) also reported that romantic partners of women who used the pill at the time of partner choice have less masculine faces than partners of women who did not use hormonal contraception at this time. This work indicates that the hormonal condition associated with pill use generates sexual and partner facial preferences in women like those typical of the luteal phase.

Some Women May Strongly Prefer Good Genes Indicators Even During Infertile Cycle Phases

As we explain in a subsequent section, preferences for good-genes indicators when women are in estrus may lead some women to seek genetic benefits from men other than their primary partners (i.e., extra-pair partners). Some women, however, may be able to obtain as primary partners men who possess indicators of good genes. In particular, women with high value on the mating market may be able to attract these men as primary and investing partners. These women, then, may benefit throughout the cycle from a preference for men who possess these features. And, indeed, women who are physically attractive have particularly strong preferences for male facial masculinity and facial symmetry (Little et al. 2001, 2007b). As well, attractive women, compared to unattractive women, show stronger preferences for male facial attractiveness for long-term mateships, whereas unattractive women show stronger preferences for male attractiveness for short-term mateships (Little et al. 2001, 2007b; Penton-Voak et al. 2003; see also Clark 2004, and Rhodes et al. 2005). Consistent with this pattern, the preferences for male vocal masculinity of women who have relatively high estrogen across the cycle (and who thus have higher mate value in terms of general fertility than other women; see, e.g., Jasiénska et al. 2004) shift less (and hence persist at higher levels throughout the cycle) than do the preferences of women with relatively low estrogen (Feinberg et al. 2006). These findings support the hypothesis that attractive women are better able to attract masculine men as long-term, investing mates and hence, tend to prefer masculine traits more consistently across the cycle.

Alternative Hypotheses for Women's Fertile-Phase Choices

One possible alternative explanation for the pattern of preference shifts observed in certain empirical studies is that women, when fertile, prefer masculine and symmetric men to obtain the benefits of physical protection. Although we found that, overall, symmetric men invest less in their relationships, they perceive themselves, and are perceived by their romantic partners, as better able to physically protect those partners (Gangestad and Thornhill 1997a). Protection is a form of male investment in a relationship that perhaps interferes minimally with a man's efforts to attract partners other than his pair-bond mate. Male protective ability may be attractive to women in general and, compared to investment in the form of time, honesty, and sexual exclusivity, men's investment in protection may compete with their pursuit of additional partners to a lesser degree. Quite likely, female preference for male protection occurs in addition to a choice for superior genes. At the same time, it is not obvious that, if females choose men for physical protection alone, they should do so specifically when fertile.

Conceivably, women's preferences for symmetric and masculine men reflect adaptation for obtaining a fertile ejaculate and not good genes for offspring. Manning et al. (1998a, b; also Baker 1997; Firman et al., 2003) found that men's body symmetry positively predicts ejaculate size and sperm quality. Soler et al. (2003) found a positive association between men's facial attractiveness and semen quality. The benefits of obtaining good sperm need not be alternatives to obtaining superior genes. Skau and Folstad (2003) argue that the ability to bear the cost of the immunosuppression necessary for producing high-quality ejaculates is possessed primarily by males of high genetic quality.

It has been proposed that women's fertile-phase sexuality is simply a by-product of stronger sexual desires women express—simply a ramping up of sexual preferences that women possess when infertile (Buller 2005). The studies discussed above, however, indicate that this explanation does not fit the pattern of empirical findings. It is not the case that women, at high-cycle fertility, prefer more of all they prefer when they are infertile. And, contrary to Buller's (2005) claim, fertile women are not simply more "sexual" than infertile women. Women's preferences shift in a more textured way. Some preferences appear to be stronger during the infertile phase relative to the fertile phase. As we discussed, women prefer male features valuable in long-term, stable mates just as much or, in some instances, even more strongly when they are infertile compared to when they are fertile in their cycles. Women's preferences for kind, faithful men who would make good fathers do not shift or are even more pronounced during the luteal phase.

Summary of Cycle Shifts in Sexuality

The body of current evidence, some of which we have briefly reviewed above, is profoundly inconsistent with the traditional view that women have no estrus. It also is inconsistent with the related view that women's sexuality has been liberated from the influences of ovarian-cycle hormones (for discussion, see Wallen 2000). That normally ovulating women, but not users of hormone-based contraception, experience cyclic shifts in sexual preference for putative markers of male genetic quality implicates a role for reproductive hormones in women's estrus.

Furthermore, women, we argue, possess two functionally distinct forms of sexuality and associated adaptations for mate preference, specialized so as to manifest at different times of the menstrual cycle. The evidence we have reviewed above indicates that one set of adaptations—which together comprise what we refer to as extended sexuality—functions at low- or zero-fertility estrus-cycle phases; when lactating, pregnant, or using hormonal contraception, women also manifest extended sexuality. Its function is to increase access to nongenetic material benefits. The other set of adaptations—which together comprise what we refer to as estrus—functions during the fertile window to lead women to prefer particularly male indicators of genetic benefits for offspring.

This is not to say that no adaptations overlap fertile and infertile phases, or that changes across the cycle are not continuously graded (as are hormonal levels, rather than pure step-functions). Indeed, when women evaluate men as long-term partners, whether during fertile or infertile cycle phases, they must concern themselves with characteristics that relate to both a flow of nongenetic material benefits they can receive from a partner as well as genetic benefits for offspring. As expected, then, evidence we have reviewed indicates that adaptations for women's long-term mate choices often are not expressed differently during estrus versus extended sexuality.

The Phylogeny of Estrus

Ichthyologists, herpetologists, and ornithologists rarely describe the reproductive seasonality of female fish, frogs, toads, salamanders, or reptiles (including birds) in terms of estrus (for a review of terminology, see Thornhill and Gangestad 2008). The reproductive cycles of all female vertebrates, however, are regulated by physiological mechanisms, hormonal and neural, that are in part homologous. Female vertebrates share a pattern of hormones that typifies high fertility within female reproductive cycles (reviews in Jones 1978; Liley and Stacey 1983; Crews and Silver 1985; Whittier and Tokarz 1992; Smock et al. 1998; Lombardi 1998; Nelson 2000; Lange et al. 2002). For example, in all nonmammalian, vertebrate species studied, females' estrogen levels are above a basal concentration at the time they mate with males (Crews and Silver 1985)—precisely the pattern observed across taxa of mammals (e.g., Nelson 2000). Additionally, the hormones associated with ovulation appear to promote female attractivity to males in vertebrates in general; typically, the attractivity of fertile females is mediated by effects of estrogen (Nelson 2000). Although studied less intensively than hormonal homology, similar, apparently homologous neurological structures appear to produce heightened female sexual motivation at peak fertility in the reproductive cycle across vertebrate taxa (e.g., Smock et al. 1998; Lombardi 1998).

The convention of using distinct taxon-specific language to describe the sexuality of vertebrate females at peak fertility in their reproductive cycles fails to recognize that important aspects of the physiology underlying vertebrate female fertile-phase sexuality is homologous. Worse yet, it hinders that recognition. For this reason, we use the term "estrus" to refer to the fertile state of all female vertebrates (i.e., gnathostomes = jawed vertebrates) in their reproductive cycles. Estrus first appeared, we propose, in the species of fish-like ancestor of all vertebrates. As estrogen-facilitated female sexual motivation at high fertility in the reproductive cycle apparently characterizes all vertebrates, the logic of the principle of parsimony (applied here to phylogenetic inference) supports this proposal. Two bodies of additional and convergent evidence supporting our proposal about the phylogenetic origin of estrus come from the phylogeny of estrogen receptors and hormonal sexual dimorphism (see Thornhill and Gangestad 2008).

Estrus' Functional Similarity Across Vertebrates

We have proposed that estrus is homologous across vertebrates. We also propose that estrus shares a basic evolved function across all vertebrates: to obtain sires of superior genetic quality. Hence, estrus in fishes and amphibians has been maintained by selection for the same function that has maintained it in reptiles and mammals. Our proposal that estrus is homologous in vertebrates does not imply that estrus is identical in all vertebrate species. Naturally, many specific vertebrate species have evolved specialized, lineage-specific estrous adaptation, which coexists with the homologous features of estrus universally among vertebrates.

In light of modern evolutionary thinking, it makes little sense that estrous females possess indiscriminate sexual desire. We propose that fertile females throughout vertebrates have been shaped to possess adaptations that favor choice of some males—those who possess superior genetic quality—over other males to sire offspring. Separately, we have reviewed much of the literature pertaining to female mate choice at the fertile phase of the reproductive cycle across vertebrates (Thornhill and Gangestad 2008). Across a wide variety of mammals and other vertebrate taxa, estrous females are choosy, not indiscriminate. Furthermore, in a diverse host of vertebrate systems, evidence clearly indicates estrous females are designed to prefer mates of superior genetic quality.

Extended Sexuality

The definition of extended sexuality is straightforward: female receptivity to sex or proceptivity for sex (through which females seek to copulate with males) during periods other than when females are fertile—sex when they cannot conceive (see also Rodríguez-Gironés and Enquist 2001). By definition, then, extended female sexuality involves sex with no direct reproductive benefits *via conception*. By no means do females of all vertebrate species exhibit extended sexuality. In most such species, females copulate only during estrus. In others, however, extended female sexuality exists. Copulation outside estrus occurs in many species of pair-bonding birds, some species of nonprimate mammals, and commonly across species of Old World primates (reviewed in Thornhill and Gangestad 2008). No mammalian female known to biology, however, matches the amplified form of women's extended sexuality.

Why do females engage in extended sexuality? This question is a special case of a broader question that biologists have long pondered: the question of why females in many species have evolved to copulate more than once, up to many times, with the same male or with multiple males within a single reproductive episode (see Parker 1979a, b; Hrdy 1979, 1981; Thornhill and Alcock 1983; Stacey 1982; Baker and Bellis 1995; Jennions and Petrie 2000). Once (or at most a few times) should be optimal if the function of mating, from the female perspective, is merely to obtain enough sperm to fertilize all available eggs. Failure to obtain sufficient sperm through a single copulation is not a problem females likely confront and evolve costly adaptations to solve. After all, sexual selection on males should strongly favor their ability to deliver adequate numbers of sperm per ejaculate to fertilize available eggs.

We can think about this idea within a life history, resource allocations framework. Males in most species must expend considerable costs (e.g., resources on displays, male–male competition, and such) that could be expended on other traits (e.g., those enhancing survival) simply to entice a female to copulate with them. Sperm production itself, however, is in all likelihood relatively cheap. For example, even under conditions of extreme malnutrition, men produce sperm at rates comparable to healthy men (Ellison 2001). (By contrast, women's fertility is highly sensitive to their current energy budget; Ellison 2001.) If sperm production is cheap, it is hard to imagine that males would not benefit often from expending (at some small cost say to viability, or even access to multiple mates), on top of those expended to achieve copulation, whatever costs are required to produce numbers of sperm sufficient to max-out on conception rate, as affected by sperm quantity. Some females (e.g., birds) store sperm. In that instance, selection should operate on males to produce sufficient numbers of sperm that can survive storage (Birkhead and Møller 1998). In light of strong selection on males to deliver adequate sperm numbers, females should not be selected to expend considerable costs to bolster numbers of sperm to which they expose their reproductive tract through an increased rate of mating.

Biologists have advanced several alternative theories for why females copulate multiple times within a single reproductive episode. These theories are discussed and evaluated critically in detail separately (Thornhill and Gangestad 2008). The only one that is consistent with comparative data is the male-material-assistance hypothesis, which proposes that female sexual motivations outside estrus typically function to facilitate females' access to male material benefits and services by exchanging sexual access for them.

The male-assistance hypothesis has been evaluated extensively. On conceptual grounds, the idea is theoretically rigorous. Rodríquez-Gironés and Enquist (2001) explored whether extensive female sexuality as a female strategy and male assistance as a male strategy could stably coevolve. In their model, all males were assumed to be of equal genetic quality and no sperm competition occurred. Hence, though their modeling cannot rule out the possibility that sperm competition and seeking of genetic benefits for offspring could select for female extended sexuality in some cases, it also nicely demonstrates that neither condition is necessary to explain female extended sexuality. Male assistance itself is sufficient to explain extended female sexuality as an evolutionarily stable strategy, when that strategy competes with modeled alternatives. Wakano and Ihara (2005), in game-theoretical and two-locus diploid models involving some different parameters than those modeled by Rodríquez-Gironés and Enquist (2001), similarly demonstrated evolutionary stability of co-occurring female multiple mating and male delivery of material assistance to females. (See also related modeling by Stacey 1982.)

The theory of extended sexuality also has been evaluated empirically. It offers two main predictions. First, comparative studies should show that, in species in which female extended sexuality is found, males should deliver material benefits to females. A second, and perhaps less obvious, prediction also follows: in species with female extended sexuality, males that females favor when in estrus should not be precisely the same males they selectively mate with during extended sexuality. According to our theory, females obtain different forms of reproductive benefits through mating when in estrus than they receive during extended sexuality. Both of these predictions are supported across a range of studies and taxa (reviewed in Thornhill and Gangestad 2008).

The primary alternative explanation for extended sexuality is that it reflects adaptation in males to seek matings in the face of uncertainty of when females are fertile, with an incidental by-product being female copulation during infertile periods (see also Thornhill and Gangestad 2008). In this view, females do not benefit reproductively from copulation during infertile phases. Rather, they are coerced by or simply acquiesce to ardent males. We have argued that, in fact, females only rarely, if ever, evolve signals designed to tell males when they are fertile. In most species, we propose, the most reliable cues of fertility status available to males are incidental by-products of physiological changes associated with fertility (e.g., changes in female scent as a function of changes in concentrations of estrogen), not female signals (e.g., sex swellings or estrous behaviors; Thornhill and Gangestad 2008). According to this scenario, it is not surprising that male detection of females' fertility status is imperfect, with the result that they are not able to discriminate completely matings that could result in conception and those that cannot.

At the same time, there is little denying that, very often, males and females have conflicting interests over the rate of mating (Thornhill and Alcock 1983; Hammerstein and Parker 1987; Clutton-Brock and Parker 1995; Arnqvist and Nilsson 2000; Arnqvist and Rowe 2005). In species with extended female sexuality, females should be expected to strive to control selectively mating frequency, mate identity, and mate number in a way that optimizes their net benefits. For females to effectively garner male-delivered material benefits through a strategy of extended sexuality, however, males must lack perfect knowledge and hence can be expected to evolve to value copulations with females who are infertile. Males are willing to and adaptively accept a high rate of false positives (matings with females that do not lead to conception). Generally, the actual rate of mating can be expected to be some compromise between male and female optima, with each sex imperfectly adapted to the other. Clearly, we do not maintain that all copulations during extended female sexuality occur in females' interest. We do contend that male interests alone generally do not account for the existence of periods of extended female sexuality per se.

No mammalian female known to biology matches the amplified form of extended sexuality exhibited by woman. Although women can possibly conceive on 6 days of their cycles in which ovulation occurs, with pronounced fertility occurring just 2–3 days, women engage in and seek copulation throughout these cycles with little variation in mating frequency across the cycle, aside from a drop at menstruation. Furthermore, women of reproductive age often have nonovulatory cycles and mate frequently within these cycles (reviewed in Thornhill and Gangestad 2008). Also, human females frequently are sexually active during the years of adolescence prior to establishing reliable ovulatory cycles. Indeed, human female adolescents are more sexually active than adolescents of other primates in which adolescent females exhibit sexuality (Anderson and Bielert, 1994; Weisfeld and Woodward, 2004). Finally, women are proceptive and receptive when pregnant or lactating.

We have discussed evidence that, at infertile cycle phases and other infertile times, women prefer male traits indicative of long-term relationship investment. In addition, if the male-assistance hypothesis of extended female sexuality applies specifically to women, women are expected to benefit through male-delivered material assistance by mating during infertile times of their lives. But which male-delivered benefits are paramount?

Men's Material Benefits

Hunting

A traditional view in anthropology is that human pair-bonding derives from the importance of male provisioning for offspring—a form of biparental care (e.g., Lancaster and Lancaster 1983; Lovejoy 1981; Westermarck 1929). In the majority of human foraging populations studied to date, the average adult male generates more calories than he consumes, whereas children and women (until menopause) do not (Kaplan et al. 2000). These food resources yield benefits for reproductive women and juveniles by generating calories and macronutrients such as protein available to them to consume. Marlowe (2001) estimated that, on average, men produce 64 % of the calories in all 95 foraging societies on which sufficient information is available. In Kaplan and colleagues' (2000) analysis of studies that carefully measured produced foods in nine hunter-gatherer societies, men generated on average about 66 % of all calories consumed.

The primary activity through which men generate surplus calories that subsidize women and children's diets is hunting (which is broadly defined to include any activity aimed to harvest animal meat, including fishing). Kaplan et al. (2000) explain male hunting as the outcome of selection for male parental effort. Accordingly, male hunting functions to harvest nutrients not only for self, but also to foster the viability and health of reproductive partners and offspring. Obviously, women (and children) could not have evolved to become dependent on subsidies achieved through men's hunting without men first providing some measure of subsidy. The argument is that male efforts that led to subsidy and the long period of juvenile dependence coevolved over time, in increments, as did human entry into and deepening commitment to an ecological niche requiring capture of high-quality food items. On average, men and women who entered into codependent relationships in which men subsidized the diets of their partners and children, according to this perspective, outreproduced those who did not. An analysis of nearly 100 foraging societies by Marlowe (2001) is consistent with the view that women can and do turn the surplus of calories generated by men into production of offspring and thereby reproductively benefit from this surplus generated through male hunting.

According to Kristen Hawkes (2004; Hawkes et al. 2001), however, men's hunting functions as mating effort—effort to gain access to mates—rather than as parental effort. Men garner prestige through successful hunting exploits, particularly big-game hunting. Ultimately, prestige translates into mating opportunities (including mating with other men's wives) (see also Kaplan and Hill 1985).

We favor a blended model (also see Key and Aiello 1999; Marlowe 1999, 2003). Men's hunting may function as parental effort as well as showing off; historically, men may have benefited reproductively from hunting in currencies of enhanced viability of offspring as well as mating opportunities. Accordingly, the general category of men's hunting may arise from psychological adaptations with two different functions (indeed, at least partly served by distinct adaptations)—parental effort and mating effort. In the mixed model view, different hunting endeavors may benefit men differentially through parental investment and mating effort. Marlowe (2003) presents data on the Hadza people consistent with a blended view.

Male Assistance That Led to Selection for Women's Extended Sexuality

In our model, males must possess imperfect knowledge of their mates' fertility status, as it changes across the cycle. They need not be completely ignorant of females' cycle-related fertility (and, as we discuss below, men are not). Males simply need be unable to completely rule out possibility for conception. When males cannot completely rule out that a female has some risk of conception, they generally will be sexually selected to be motivated to copulate with a female (under appropriate circumstances). Men's interest in copulating across the cycle, even in the absence of female interest (as in sexual coercion), coupled with the female's copulability across the entire cycle and at other infertile times, imply that males do not have unambiguous direct cues of peak fertility. Presumably, it is the lack of unambiguous fertility cues that has selected for men's sexual interests in women throughout their cycles and in adolescent women.

That women's extended sexuality is extreme, in a comparative perspective, is consistent with the view that pair-bonding and male-delivery of associated material services have been highly important to women's fitness in human evolution. As we have emphasized, these benefits include male paternal care (also see Alexander 1979, 1979; Alexander and Noonan 1979; Geary 2000; Geary and Flinn 2001). Another service that male partners purportedly provided to women ancestrally is protection of mates and their female relatives from capture during raids and warfare and from sexual coercion by other men in the same group. (See Smuts 1992; Smuts and Smuts 1993; Thornhill and Palmer 2000 for treatment of the importance cross-culturally to women of protection from sexually coercive males. See also Mesnick 1997, and Wilson and Mesnick 1997, on the bodyguard hypothesis for the evolution of human pair-bonding.)

The benefits that women ancestrally garnered from men and led to their extended sexuality need not have been delivered solely by primary partners. Male mating effort leading them to deliver resources to women in hopes of gaining sexual access also may have selected for women to possess extended sexuality (e.g., Symons 1979; Hill 1982). We do not completely discount these potential benefits. At the same time, we suspect that women's reliance on a continued flow of material benefits delivered by primary partners typically meant that it was not worth the risks of losing those benefits by being unfaithful for the exchange of a single meal. Hence, it would not commonly benefit women to be unfaithful to a partner (at least one she wished to retain) during extended sexuality (particular if she had small children; see Marlowe 2003), unless material benefits gained through infidelity were considerable.

Hrdy's paternity confusion hypothesis is one variant of the male-assistance hypothesis for female extended sexuality. Reduction of maltreatment of offspring by resident males through widespread distribution of paternity confidence (not allowing any male to rule out paternity of one's offspring) has probably been a very important benefit leading to the evolution of female extended sexuality in many nonhuman primates (e.g., Hrdy and Whitten 1987; Wrangham 1993; Palombit 1999; Heistermann et al. 2001; Pazol 2003; van Schaik et al. 2004). Possibly in some circumstances, it has been adaptive for women to distribute paternity confidence widely by mating with multiple men during extended sexuality (see Beckerman et al. 1998; Hrdy 2000).

Luteal-Phase Sexual Proceptivity of Women

In a recent study with colleagues we tested a specific prediction of the extended sexuality hypothesis for women's sexual motivation at times other than estrus (Grebe et al. 2013). If normally ovulating women's sexuality in the infertile luteal phase of their cycle is for securing nongenetic material benefits from men, it follows that luteal-phase women should initiate more sex (show greater sexual proceptivity) with their romantic partner when they have the most to gain from such benefits, and the enhanced proceptivity with the partner should not be seen during estrus. This prediction was supported in a study of romantically involved couples in which the women were normally ovulating. The relationship investment of each man and woman in the relationship was measured and men and women reported their number of sexual initiations during 2 days of the woman's peak estrus and 2 days at the midluteal phase. During the luteal phase, women's sexual proceptivity, but not the male partner's, increased in direct relation to the degree to which the woman's investment exceeded her male partner's investment. This effect was not seen during estrus. Hence, it is in relationships that women value and invest in but the male partners lags behind the female in investment, where luteal sexual proceptivity of women is most frequent. We predict that this same pattern will be seen also in pair-bonded women's sexual motivations at nonconceptive times other than the luteal phase.

Estrus and Extra-Pair Copulation (EPC)

In species in which males and females cooperatively care for offspring, females cannot all be socially paired with males who possess relatively high genetic fitness. Hence, females may be selected to be attracted to and mate with males who do offer such benefits during estrus, particularly if paired with a male who does not. That is, estrous preferences that operate in these species may be maintained by selection, at least partly, through genetic benefits to offspring achieved through extra-pair copulation (EPC). Naturally, in many species countervailing selection pressures do exist, most of which are exerted through male behavioral tendencies to prevent cuckoldry and investment in other males' offspring. In some species, these countervailing pressures may be sufficient to have selected for the loss of estrous preferences in females (see Thornhill and Gangestad 2008). But most generally, female estrous preferences and counteradaptations on the part of males continually coevolve through a sexually antagonistic process. Indeed, in a variety of pair-bonding bird species, females possess adaptations characteristic of estrous sexuality, not operative during extended sexuality, that apparently function to obtain genetic benefits through highly discriminating partner choice and, partly, EPC (Thornhill and Gangestad 2008).

In some of these species, however, there are very low rates of female extra-pair copulation and extra-pair paternity (EPP). These low rates do not imply a lack of sexually antagonistic coevolution. Indeed, once again, female EPC rates are maintained at low levels precisely because of countervailing selection pressures, importantly including male mate-guarding, high rates of within-pair copulation, and withdrawal of investment contingent on nonpaternity. Low but nonzero rates of EPP may often result when male investment in offspring is substantial, such that males are likely to pay relatively high costs for paternity-assurance adaptations and females pay larger costs for male withdrawal of care (see Thornhill and Gangestad 2008).

Estrous sexuality within the human pair-bonded system may also have been maintained and modified through benefits achieved partly through EPC. This perspective on women's estrous preferences implies a number of predictions. First, estrous preference shifts should be observed most strongly in women with primary partners. In a number of cases, researchers have found that women in pair-bond relationships experience preference shifts for putative good genes more strongly than unpaired women. For instance, Penton-Voak et al. (1999) found that women in relationships only or mainly accounted for the effect of cycle fertility status on the shift in preferences for masculine faces. And Havlicek et al. (2005) found that only estrous pair-bonded women particularly preferred the scent of dominant men. Not all studies, we note, have yielded similar effects (In fact, most studies in the literature have not included relationship status as a variable). At the same time, to our knowledge, no study to date has reported a purported good-genes preference shift that occurs selectively in unpaired women.

Second, estrous women with primary partners should report more frequent or stronger attraction to men other than primary partners. On average, estrous women may not report any greater attraction to their own primary partners. Multiple studies support this second prediction: in general, estrous women with a primary partner are more strongly attracted to men other than their partner than are nonestrous women with a primary partner (Gangestad et al. 2002, Gangestad et al. 2005; also Baker and Bellis 1995; cf. Pillsworth et al. 2004).

Third, some women are paired to men who possess features found most attractive by women during their estrous phase. As these women could have lost investment from their main partner through EPC but gained little in genetic benefits, they should not show increased attraction to men other than primary partners when they are in estrus. Indeed, these women may experience increased attraction to their primary partners during estrus. Put otherwise, the expected increase in attraction to men other than primary partners during estrus should be accounted for by women with partners lacking features, notably symmetry and physical attractiveness, particularly favored by women when in estrus. Multiple studies support this prediction (Gangestad et al. 2005, Haselton and Gangestad 2006, Larson et al. 2012). Moreover, estrous women pair-bonded to men with similar (i.e., incompatible) MHC alleles are more strongly attracted to extra-pair men than estrous women with MHC-dissimilar partners, and the former estrous women report more EPCs in their current relationship (Garver-Apgar et al. 2006).

Fourth, compared to the period of extended sexuality, during estrus women should, on average, feel less committed to their partners and more willing to engage in sex with men other than primary partners. Gangestad et al. (2010a), Sheldon et al. (2006), and Jones et al. (2005) provide evidence for this prediction.

Fifth, just as males in bird species are expected to possess counteradaptations that suppress the likelihood of their primary partners' EPC, so, too, men are expected to possess paternity assurance and anti-cuckoldry adaptations. Multiple studies have reported that men can detect estrus and respond to it by increased mate guarding and possibly by copulation patterns that may function in cuckoldry avoidance.

Sixth, in light of male paternity-assurance adaptations that impose costs on female EPC, women's estrous sexuality should be shaped to be responsive not solely to the benefits of potential EPC; it also should be shaped to be responsive to the costs of EPC imposed by males. Hence, women's estrous sexuality should be sensitive to factors that affect the likelihood or size of the cost of male efforts to assure paternity and prevent investment in offspring not their own. Evidence supports this prediction.

Additional findings may reflect estrus' functionality in the context of EPC in humans. Women's high desire for orgasm and sexual satisfaction during short-term mating (Greiling and Buss 2000) may reflect estrous motivation to obtain good genes by extra-pair copulation. Cryptic female choice in the form of orgasmic sperm retention (Thornhill et al. 1995) may play a role in women's good-genes preferences during estrus. Estrous women's nadir in food intake and their increased ambulation and interest in socializing with men may be adaptations that function to promote good-genes mate choice, including by EPC. The feeding nadir may avoid mating with food-providing men of low genetic quality. The increase in ambulatory and socializing activity may be analogous (convergently evolved) or homologous (due to common ancestry) with estrous adaptation to improve good-genes mate choice in certain other female vertebrates (see Thornhill and Gangestad 2008, for review).

Overall, research findings indicate that women's estrous sexuality in the context of pair-bonding, is not merely a by-product of generalized heightened sexual motivation to assure conception, regardless of male quality. Instead, evidence indicates it is adaptation that functions to achieve conception by a male of high genetic quality, including contingently through extra-pair mating.

At the same time, we note, estrus evolved prior to the evolution of pair-bonding in humans. The question of whether estrus has been modified *specifically* for the function of adaptive extra-pair sex (i.e., there exist estrus adaptations specifically designed to promote adaptive EPC) has not yet been fully answered; more evidence is needed.

Concealed Estrus

We have asserted that estrus is not an adaptation to signal cycle-related fertility and that female animals rarely, if at all, advertise their fertility in the reproductive cycle. Instead, males are designed through sexual selection to focus attention on cycle-related fertility cues (not signals) that arise incidentally from female cycle changes in fertility status (for further treatment, see Thornhill and Gangestad 2008).

We propose, too, that women have been designed to suppress these incidental cues, leaving men scant information to glean about their fertility status as it varies across the cycle (although men do not lack cues altogether). If so, women possess adaptation meaningfully referred to as "concealed estrus."

The accuracy with which males can identify females at peak fertility in their cycle varies across species in relation to the conception-related validity of the by-products of reproductive-status changes cycling females emit, *incidentally* resulting in some degree of undisclosure of fertility in some species. As a result, in some vertebrate species, maximum cycle-related fertility in females may not coincide perfectly with peak sexual motivation among males. Some cross-species variation in male ability to detect female fertility status, then, arises for reasons having nothing to do with female adaptation to conceal fertility cues.

Two distinct phenomena, then, must be distinguished. The word "concealment" implies an active "hiding" of information. *Concealed estrus* refers to an evolved outcome due to direct selection on females to suppress information related to cycle fertility, leading to concealed-estrus adaptation. We distinguish concealed estrus from *undisclosed estrus*. The word "undisclosed" does not imply an active process of disguise. Similarly, the term *disclosed estrus* does not imply active advertisement; that is, disclosed estrus does not imply *signaling* of cycle fertility. Undisclosed estrus applies to cases in which the unreliability of information occurs incidentally. Concealed estrus refers to cases in which the unreliability of information occurs by functional design (though, despite selection for concealment, some cues may remain). We propose that disclosed and undisclosed estrus (falling along a continuum, but without active concealment) characterize the vast majority of vertebrate taxa (see Thornhill and Gangestad 2008 for additional discussion).

Note that extended sexuality need not—and probably does not typically—involve concealed estrus. Males need not be "fooled" by females to copulate with them outside of their fertile window. Rather, males will copulate with infertile females whenever information about females' fertility status is imperfect.

Theories and Evidence of Women's Concealed Ovulation

Numerous theories argue that women possess adaptations that function to conceal cycle fertility. Other theories, by contrast, argue that women possess concealed fertility as an incidental by-product. We critically evaluate these theories separately (Thornhill and Gangestad 2008). In short, all of the adaptation hypotheses except one—the cuckoldry hypothesis—are placed in doubt by the presence of estrus in women and men's ability to detect it, because they rely on the absence of estrus and relatedly men's inability to detect it. The by-product hypotheses are cast in doubt by evidence of functional design for estrus' concealment.

The cuckoldry hypothesis for concealed estrus is consistent with the presence of estrus and its apparent design to motivate sex with males of high genetic quality. This hypothesis states that concealed fertility was selected directly in antagonistic coevolution with male traits to mate-guard and monitor female fertility status by pair-bond mates. Concealed estrus functions to reduce effective mate guarding by a pair-bond mate. Ancestral females who concealed estrus hence paid fewer costs imposed by selective mate-guarding, perhaps importantly including greater ability to obtain genetic benefits through EPC (Benshoof and Thornhill 1979; Symons 1979, in part; Schroder 1993, in part).

The cuckoldry hypothesis offers a number of predictions. First, it predicts that women's fertile-phase sexuality is distinct from their extended sexuality, and specifically that, at estrus, women should be attracted particularly to men who exhibit purported indicators of good genes. As discussed, women's estrus is indeed characterized by a variety of enhanced preferences for such indicators. Second, it predicts that, during estrus, women's attraction to men other than primary partners should be enhanced, but selectively so. Specifically, they should be attracted to particular men other than primary partners only if their primary partner lacks indicators of good genes. Again, the evidence to date is consistent with this prediction. Third, the cuckoldry hypothesis for concealed fertility goes beyond these findings and predicts that women's estrus will be concealed in two ways. First, incidental effects associated with estrous adaptations should be reduced or suppressed. Second, some behavioral effects of estrus—selectively greater interest in men other than primary partners—should be concealed from primary partners and, indeed, most anyone other than the men to whom fertile women are attracted. Women need not conceal their estrus from men they seek; there are no benefits to doing so. They do benefit by concealing their estrus from a primary partner, should he not be one of those men. The cuckoldry hypothesis, then, expects that estrus will be concealed selectively.

Before we discuss further evidence that women have been selected to conceal their estrus in various ways, a few additional words about ways women's fertility is not concealed are in order. Cycle-related peak fertility is not hidden from women themselves. As their preferences, patterns of attraction, and experience change across the cycle, women themselves surely can discriminate their fertile periods from their infertile periods. Naturally, there is no presumption that they consciously associate estrous sexuality with "fertility"—but, of course, there is no presumption that females of any other species not exhibiting concealed fertility do so either. Rather, it is presumed generally that females of nonhuman species experience the world differently when in estrus and, hence, can discriminate fertile from nonfertile periods. The same is true of women.

This claim contrasts with opinions frequently expressed in the literature. A number of scholars throughout the history of the study of human sexuality have claimed that women possess virtually no knowledge whatsoever of their peak fertility in the menstrual cycle (e.g., Burley 1979; Strassmann 1981; Turke 1984; Alexander 1990; and many others). As Burley (1979) and Strassmann (1996) point out, even the medical profession did not realize that peak fertility occurs at midcycle until nearly 1930. Similarly, hunter–gatherers do not typically appreciate a link between fertility in the cycle and conception (Marlowe 2004). Clearly, people would not be so ignorant about when the fertile period exists if women were similar to female dogs or cats, which, when confined to a home without a mate, are conspicuously active and exhibit dedicated, intense effort to break out to seek males attracted to their estrous scents. As scholars have noted, there is indeed something different about women's manifestation of fertile-phase sexuality.

As research has documented, however, what is different is not the loss of estrus. Women are different in the extent to which they conceal estrus. The behavioral changes women experience at estrus are much more covert than what is typically observed in other species, as the cuckoldry hypothesis expects.

Colleagues and we examined whether women do engage in attempts to resist men's efforts to track their behavior and whereabouts more frequently when fertile than in the infertile luteal phase. As predicted by the cuckoldry hypothesis, normally ovulating women engaged in efforts to resist their primary partners' mate guarding more frequently during estrus than during the luteal phase. This resistance was particularly manifested in estrous women who reported greater attraction to nonpartner men but not to their primary partner (Gangestad et al. 2014).

Although women are generally more emotionally expressive than men, women also appear to possess a greater ability to control facial and body expressions of emotions (see Bjorklund and Kipp 1996). They also may be better at suppressing unwanted thoughts (e.g., selectively avoiding thoughts about potential mates). Bjorklund and Kipp propose that these emotional features may be adaptation selected in the context of EPC (or, in our terms, concealed estrus). If so, then women's abilities to suppress emotion may vary across the cycle and be particularly keen when women are in estrus.

By-Product Cues of Women's Estrus

Concealed fertility does not imply a complete absence of cues associated with cycle fertility. Selection can operate on women to suppress fertility cues (i.e., conceal estrus) without eliminating cues. Women do exhibit a variety of changes at estrus, which remain not fully suppressed. Studies show that there are discernible cues of estrus in women's behaviors, body scents, voices, and, possibly, aspects of physical beauty (reviewed in part by Haselton and Gildersleeve 2011; also see Gildersleeve et al. 2012, Cantu et al. 2014, Thornhill and Gangestad 2008). We interpret these effects, in large part, as incidental by-products of changes in women's physiology, especially high estrogen, at estrus. In part also they are by-products of women's enhanced, but discriminating, sexual motivation when fertile. We interpret as part of estrous mate choice estrous women's enhanced flirtations specific toward attractive men as short-term mates, which was discovered by Cantu et al. (2014) (discussed above). Similarly, we view the greater willingness of estrous women to dance with an attractive stranger (Gueguen 2009) as a manifestation of estrous good-genes mate choice. None of these estrous effects, including the mate choice behaviors, imply, however, that women seek to compete for insemination or even for insemination by a genetically high-quality male. Indeed, we do not expect that women compete with other women more when fertile than when infertile. The various by-products of women's estrus may well have been acted upon by direct selection for concealment during the evolution of woman's concealed estrus, but their elimination may be constrained by the costs of suppression due to disrupted estrogen regulation. Estrogen-regulated female reproduction is phylogenetically conserved throughout the vertebrates as a result of direct selection for it and hence serves as a constraint that cannot be eliminated in service of concealing estrus (see Thornhill and Gangestad 2008).

Some scholars have argued that the subtle changes in woman's physiology across the cycle that make them more attractive to men in some regards suggest that women do, in fact, signal fertility status (e.g., Scutt and Manning 1996). A variety of considerations argue against these features being signals. First, in general, female animals rarely signal cycle fertility (see Thornhill and Gangestad 2008). Second, no evidence exists that women possess complex design for emitting these signals; indeed, changes in physiology are generally understandable as by-products of changes in women's hormonal status or other estrous adaptations, rather than as outputs of design for signaling. Third, changes in women's attractiveness across the cycle—whether in the form of bodily attractiveness, facial attractiveness, or scent attractiveness—are comparatively subtle. By contrast, fertile females in many other primate, mammalian, and vertebrate species are highly attractive to males—and typically in absence of any signaling system whatsoever. Finally, there is simply no evidence that women benefit through male detection of their fertility status. Indeed, as we show in the next section of the paper, women appear to pay costs as a result of emitting incidental by-products of fertility cues that men detect. Male primary partners pick up on these cues and act on them for their own benefit, sometimes against the interests of female partners.

Concealed estrus adaptation may occur in certain nonhuman species, notably in those with pair-bonding or male–female consortship accompanied by female extra-pair copulation for good genes. In these species, the cuckoldry hypothesis may apply (see discussion in Thornhill and Gangestad 2008).

Men's Counter-Strategies and Women's Responses to Them

Extra-pair copulation leads to intersexual conflicts of interest in pair-bonding species such as *Homo sapiens*. These conflicts of interests arise from the cost of parental care in each sex and cuckolded males investing in other males' offspring. Male counteradaptations to extra-pair mating by a female partner include at least four important categories: mate-guarding, frequent in-pair copulation, detection of female's fertile cycle phase, and detection of paternity of offspring.

Similar to many male pair-bonding birds (Birkhead and Møller 1992; Thornhill and Gangestad 2008), men possess condition-dependent psychological adaptations for mate guarding. According to several studies, men engage in mate guarding more frequently when their mates are in estrus than when their mates are infertile in their cycles (reviewed in Haselton and Pillsworth 2011). Men's interest in and way of copulating are suggestive of anticuckoldry adaptations (Shackelford et al. 2002; Goetz et al. 2005; Goetz and Shackelford 2006; Shackelford et al. 2006). Furthermore, men discriminate paternity based on physical resemblance to offspring and express greater interest in investing in children when cues of resemblance are present than when they are absent (Platek et al. 2004).

Women possess counteradaptations to male adaptations designed to prevent cuckoldry. Women engage in behaviors that undermine their partner's mate-guarding more so when fertile than when infertile in the cycle, and especially when estrous women are motivated to pursue an extra-pair partner. Estrous women perhaps resist copulation with primary partners contingently and adaptively (Garver-Apgar et al. 2006), though more research on this topic is needed. As we have emphasized, women's concealed estrus may be an important counteradaptation. Finally, women may have been designed to show conditional interest in extra-pair men based on cues related to the costs of lost male partner investment and the likelihood of partner detection. This last category of counteradaptations may shed much light on the nature of women's dual sexuality, but has not yet been investigated in depth (see Thornhill and Gangestad 2008).

Current extra-pair paternity and women's extra-pair mating rates are variable across human cultural settings. Whether observed in pair-bonding birds or humans, low rates are fully consistent with a deep-time history of intersexual arms races and associated effective selection for sexually antagonistic adaptations. Researchers should not be misled into inferring from low rates of extra-pair paternity that sexually antagonistic coevolution is unimportant and antagonistic adaptation largely lacking (see review in Thornhill and Gangestad 2008).

The Evolution of Women's Permanent Ornaments

A large body of literature, mostly contributed in the last 15 years, reveals that particular female phenotypic features arouse the sexual attraction of men. Research across the globe shows that men are sexually attracted to hyperfeminine (dimorphic and estrogen-facilitated) facial features and dimensions (e.g., small lower face and eyebrow ridges, large lips). Men are sexually attracted to young women's breasts, and in most places, men prefer women who possess a small waist-to-hip ratio. Smooth feminine skin attracts men, as does feminine voice qualities (studies are reviewed in Thornhill and Gangestad 2008).

A primary question arises about these features: are they ornaments with signal function, or are they by-products to which men are sexually attracted? Women's sexually attractive features depend upon the actions of estrogen (Johnston and Franklin 1993, Thornhill and Grammer 1999; Thornhill and Gangestad 1999b). In turn, women's estrogen levels importantly support their reproductive capacities (Ellison 2001; Jasiénska et al. 2004). These features hence have function independent of their signal value. According to the hypothesis that women's sexually attractive qualities are ornaments, sexual selection has favored exaggeration of these morphological features directly because they signal to men that their bearers possess valued mate qualities, specifically, high estrogenization and hence elevated reproductive capacity, with selection directly operating on men to attend to these signals and benefit women who possess them reproductively. According to the hypothesis that they are merely by-products, men have been selected directly to track cues of women's reproductive capacities, but selection has not operated on woman to display these cues; the fact that men find these features attractive is not because women have benefited directly from male attraction to them (leading to their exaggeration).

A variety of signaling hypotheses for woman's ornaments have been offered. Marlowe (1998) argued that breasts, in particular, signal residual reproductive value, reflecting both age and individual quality, which explains woman's ornaments as honest signals of valued qualities. A variety of alternative theories explain them as deceptive signals, qualities that deceive men by falsely signaling to men valued states. Specifically, breasts and other features have been claimed to deceptively signal youth (Low et al. 1987; Jones 1996) and a pregnant state (Smith 1984; Miller 1995). Because these traits are permanent and change little across the ovulatory cycle, they have also been argued to be deceptive cues of permanent cycle-related fertility status or, relatedly, to function in concealing ovulation (e.g., Szalay and Costello 1991). We have argued that many, if not most, of woman's attractive features of face, body and voice are indeed ornaments. Though many of these features were first selected to support reproduction itself (specialized fat depots of breasts, thighs, and buttocks), subsequently they were exaggerated and shaped by sexual selection for their signal value. Hence, they also function as signals, primarily as condition-dependent ornaments that honestly signal residual reproductive value. The key evidence is revealed in their functional design.

The ratio of gynoid fat to android fat is an honest correlate of women's ability to expend future reproductive effort (see the review in Thornhill and Gangestad 2008). Accordingly, men preferred women possessing depots of gynoid fat, which led to sexual selection on females to display exaggerated fat depots (e.g., in the breasts, buttocks, hips, and thighs) and, in turn, sexual selection on males to increasingly attend to and be sexually motivated by these depots. That is, these ornaments reflect the evolution of signaling systems.

As expected for residual reproductive-value signals, woman's gynoid-fat ornaments reach maximum size in adolescence, prior to woman becoming highly fertile, and are permanently retained throughout their reproductive lives. The other estrogen-facilitated ornaments of women (skin, voice, facial dimensions) also are permanent, most attractive in young women, and may peak in attractiveness at adolescence (Thornhill and Gangestad 2008). Adolescent exaggeration and permanence of woman's ornaments were selected, we propose, in the context of long-term pair-bonding involving discriminative provisioning to females (based on the quality of ornaments) of nongenetic material benefits by males. In the various nonhuman, Old World primates in which female adolescents display ornaments (sometimes in exaggerated form), the ornaments may be adaptations that function to obtain material benefits from males (also, Anderson and Bielert 1994).

As predicted of honest signals of quality, women's attractive features tend to covary positively with each other and evidence indicates relationships between a number of them and fertility, health, developmental stability, and longevity (though exceptions exist; see Thornhill and Gangestad 2008). Thus, women's ornaments appear to be redundant signals of overall fitness, not signals with distinct messages. The chief means by which women's estrogenization of ornamentation acts as an honest signal is likely the trade-off between reproductive effort and somatic effort. In the face of this trade-off, women optimally allocate effort, with those in better condition allocating more effort to reproduction (as seen in the greater display of ornamental estrogenization) compared to women in poorer condition. The maintenance of the very substantial variation in the attractiveness of young women's ornaments probably reflects the evolutionary processes of mutation-selection balance and coevolutionary races, including sexually antagonistic selection (see the review in Thornhill and Gangestad 2008.)

Women's elaborate extensions of sexuality across infertile cycle phases and other infertile periods attests to the importance of male-delivered material benefits and services in enhancing ancestral women's reproductive success. So, too, does women's estrogen-facilitated ornamentation of face and body. The onset of this ornamentation at adolescence, before high conception probability is achieved, and the ornamentation's permanence throughout the female's reproductive life span are design features that function to secure male-delivered material benefits and services, especially in the context of long-term pair-bonds.

Women's gynoid fat depots phylogenetically arose and were elaborated as signals in the hominin lineage. The sexual swellings of female Old World primates are mediated by estrogen, but are water-filled, not fat-filled. One scenario suggested by phylogenetic comparisons is that females of the species ancestral to both the *Pan* lineage (the two species of chimpanzees) and the hominin lineage had slight sex-skin development in the anogenital region (and not exaggerated sexual swellings; Sillén-Tullberg and Møller 1993; see also Pagel and Meade 2006). Hence, the traditional account that woman's evolutionary loss of estrus is equivalent to the loss of sex swellings is inaccurate, because current evidence points to an absence of swellings in ancestral hominins. That traditional account is inaccurate as well, we propose, in misassigning the function of swellings: they are ornaments, not signals to males of need for sperm or cycle-fertility status per se.

Female sexual ornaments are widespread across animal taxa. We review these separately and interpret them as honest signals of female personal quality, not of fertility within female's reproductive cycles—and this interpretation includes female sexual skins and swellings of Old World primates (Pagel 1994; Domb and Pagel 2001; Thornhill and Gangestad 2008). Males are expected to be under strong sexual selection to find fertile females and, hence, costly signals of fertility within the cycle should be rare. Generally, cues that females emit when fertile in their cycles have been interpreted as signals designed to assure insemination. One version of this notion claims that females signal to incite male–male competition, which assures insemination by males with good genes. Typically, females do not have signaling adaptations that function to assure insemination by males in general or males of superior genetic quality, though they do have other adaptations that function to improve female choice (Thornhill and Gangestad 2008).

References

Alexander, R. D. (1990). *How Did Humans Evolve? Reflections on the Uniquely Unique Species.* Special Publication No. 1, Museum of Zoology, The University of Michigan.

Alexander, R. D., & Noonan, K. M. (1979). Concealment of ovulation, parental care, and human social evolution. In N. A. Chagnon & W. G. Irons (Eds.), *Evolutionary biology and human social behavior: An anthropological perspective* (pp. 436–453). North Scituate: Duxbury Press.

Anderson, C. M., & Bielert, C. (1994). Adolescent exaggeration in female catarrhine primates. *Primates, 35,* 283–300.

Arnqvist, G., & Nilsson, T. (2000). The evolution of polyandry: Multiple matings and female fitness in insects. *Animal Behaviour, 60,* 145–164.

Arnqvist, G., & Rowe, L. (2005). *Sexual conflict.* Princeton: Princeton University Press.

Baker, R. R. (1997). Copulation, masturbation and infidelity: State-of-the-art. In A. Schmitt, K. Atzwanger, K. Grammer, & K. Schafer (Eds.) *New aspects of human ethology* (pp. 163–188). New York: Plenum Press.

Baker, R. R., & Bellis, M. A. (1995). *Human sperm competition: Copulation, masturbation and infidelity.* London: Chapman and Hall.

Beach, F. A. (1976). Sexual attractivity, proceptivity and receptivity in female mammals. *Hormones and Behavior, 7,* 105–138.

Beckerman, S., Lizarralde, R., Ballew, C., Schroeder, S., Fingelton, E., Garrison, A., et al. (1998). The Bari partible paternity project: Preliminary results. *Current Anthropology, 39,* 164–167.

Benshoof, L., & Thornhill, R. (1979). The evolution of monogamy and concealed ovulation in humans. *Journal of Social and Biological Structure, 2,* 95–106.

Birkhead, T. R., & Møller, A. P. (1992). *Sperm competition in birds: Evolutionary causes and consequences.* New York: Academic.

Birkhead, T. R., & Møller, A. P. (Eds.), (1998). *Sperm competition and sexual selection.* London: Academic.

Bjorklund, D. F., & Kipp, K. (1996). Parental investment theory and gender differences in the evolution of inhibition mechanisms. *Psychological Bulletin, 120,* 163–188.

Brewis, A., & Meyer, M. (2005). Demographic evidence that human ovulation is undetectable (at least in pair bonds). *Current Anthropology, 46,* 465–471.

Buller, D. J. (2005). *Adapting minds: Evolutionary psychology and the persistent quest for human nature.* Cambridge: MIT Press.

Burley, N. (1979). The evolution of concealed ovulation. *American Naturalist, 114,* 835–858.

Buss, D. M. (1989). Sex differences in human mate preferences: Evolutionary hypotheses tested in 37 cultures. *Behavioral and Brain Sciences, 12,* 1–49.

Cantu, S. M., Simpson, J. A., Griskevicius, V., Weisberg, Y. J., Durante, K. M., & Beal, D. J. (2014). Fertile and selectively flirty: Women's behavior toward men changes across the ovulatory cycle. *Psychological Science, 25,* 431–438.

Cardenas, R. A., & Harris, L. J. (2007). Do women's preferences for symmetry change across the cycle? *Evolution and Human Behavior, 28,* 96–105.

Clark, A. P. (2004). Self-perceived attractiveness and masculinization predict women's sociosexuality. *Evolution and Human Behavior, 25,* 113–124.

Clutton–Brock, T. H., & Parker, G. A. (1995). Sexual coercion in animal societies. *Animal Behaviour, 49,* 1345–1365.

Corner, G. (1942). *The hormones in human reproduction.* Princeton: Princeton University.

Crews, D. & Silver, R. (1985). Reproductive physiology and behavior interactions in non-mammalian vertebrates. In N. T. Adler, D. Pfaff, & R. W. Goy (Eds.), *Handbook of behavioral neurobiology* (pp. 101–182). New York: Plenum Press.

Domb, L. G., & Pagel, M. (2001). Sexual swellings advertise female quality in wild baboons. *Nature, 410,* 204–206.

Dunson, D. B., Baird, D. D., Wilcox, A. J., & Weinberg, C. R. (1999). Day-specific probabilities of clinical pregnancy based on two studies with imperfect measures of ovulation. *Human Reproduction, 14,* 1835–1839.

Dunson, D. B., Weinberg, C. R., Baird, D. D., Kesner, J. S., & Wilcox, A. J. (2001). Assessing human fertility using several markers of ovulation. *Statistics in Medicine, 20,* 965–978.

Ellison, P. T. (2001). *On fertile ground: A natural history of reproduction.* Cambridge: Harvard University Press.

Feinberg, D. R., Jones, B. C., Law Smith, M. J., Moore, F. R., DeBruine, L. M., Cornwell, R. E., et al. (2006). Effects of menstrual cycle, trait estrogen level, and masculinity preferences in the human voice. *Hormones and Behavior, 49,* 215–222.

Firman, R. C., Simmons, L. W., Cummins, J. M., & Matson, P. L. (2003). Are body fluctuating asymmetry and the ratio of 2nd to 4th digit length reliable predictors of semen quality? *Human Reproduction, 18,* 808–812.

Frederick, D. A., & Haselton, M. G. (2007). Why is muscularity sexy? Tests of the fitness indicator hypothesis. *Personality and Social Psychology Bulletin, 33,* 1167–1183.

Frost, P. (1994). Preference for darker faces in photographs at different phases of the menstrual cycle: Preliminary assessment of evidence for a hormonal relationship. *Perception and Motor Skills, 79,* 507–514.

Gangestad, S. W., & Thornhill, R. (1997a). Human sexual selection and developmental stability. In J. A. Simpson & D. T. Kenrick (Eds.), *Evolutionary social psychology* (pp. 169–195). Mahwah: Lawrence Erlbaum.

Gangestad, S. W., & Thornhill, R. (1997b). The evolutionary psychology of extra-pair sex: The role of fluctuating asymmetry. *Evolution and Human Behavior, 18,* 69–88.

Gangestad, S. W., & Thornhill, R. (1998). Menstrual cycle variation in women's women's preference for the scent of symmetrical men. *Proceeding of the Royal Society of London B, 265,* 927–933.

Gangestad, S. W., & Thornhill, R. (1999). Individual differences in developmental precision and fluctuating asymmetry: A model and its implications. *Journal of Evolutionary Biology, 12,* 402–416.

Gangestad, S. W., & Thornhill, R. (2003a). Facial masculinity and fluctuating asymmetry. *Evolution and Human Behavior, 24,* 231–241.

Gangestad, S. W., & Thornhill, R. (2003b). Fluctuating asymmetry, developmental instability, and fitness: Toward model-based interpretation. In M. Polak (Ed.), *Developmental instability: Causes and consequences.* Cambridge: Cambridge University Press.

Gangestad, S. W., Thornhill, R., & Garver, C. E. (2002). Changes in women's sexual interests and their partners' mate retention tactics across the menstrual cycle: Evidence for shifting conflicts of interest. *Proceedings of the Royal Society of London B, 269,* 975–982.

Gangestad, S. W., Simpson, J. A., Cousins, A. J., Garver-Apgar, C. E., & Christensen, P. N. (2004). Women's preferences for male behavioral displays shift across the menstrual cycle. *Psychological Science, 15,* 203–207.

Gangestad, S. W., Thornhill, R., & Garver-Apgar, C. E. (2005). Women's sexual interests across the ovulatory cycle depend on primary partner fluctuating asymmetry. *Proceedings of the Royal Society of London B, 272,* 2023–2027.

Gangestad, S. W., Garver-Apgar, C. E., Cousins, A. J., & Thornhill, R. (2014). Intersexual conflict across women's ovulatory cycle. *Evolution and Human Behavior, 35,* 302–304.

Gangestad, S. W., Garver-Apgar, C. E., Simpson, J. A., & Cousins, A. J. (2007). Changes in women's mate preferences across the ovulatory cycle. *Journal of Personality and Social Psychology, 92,* 151–163.

Gangestad, S. W., Thornhill, R., & Garver-Apgar, C. E. (2010a). Fertility in the cycle predicts women's interest in sexual opportunism. *Evolution and Human Behavior, 31,* 400–411.

Gangestad, S. W., Thornhill, R., & Garver-Apgar, C. E. (2010b). Men's facial masculinity predicts changes in their female partner's sexual interests across the ovulatory cycle, whereas men's intelligence does not. *Evolution and Human Behavior, 31,* 412–424.

Garver-Apgar, C. E., Gangestad, S. W., Thornhill, R., Miller, R. D., & Olp, J. (2006). MHC alleles, sexually responsivity, and unfaithfulness in romantic couples. *Psychological Science, 17,* 830–835.

Garver-Apgar, C. E., Gangestad, S. W., & Thornhill, R. (2008). Hormonal correlates of women's women's mid-cycle preference for scent of symmetry. *Evolution and Human Behavior, 29,* 223–232. doi:10.1016/j.evolhumbehav.2007.12.007.

Geary, D. C. (2000). Evolution and proximate expression of human paternal investment. *Psychological Bulletin, 126,* 55–77.

Geary, D. C., & Flinn, M. V. (2001). Evolution of human parental behavior and the human family. *Parenting, 1,* 5–61.

Gilbert, S. F. (2000). *Developmental biology,* (6th ed). Sunderland: Sinauer Associates.

Gildersleeve, K. A., Haselton, M. G., & Fales, M. R. (2014). Do women's mate preferences change across the ovulatory cycle? A meta-analytic review. *Psychological Bulletin.* http://dx.doi.org/10.1037/a0035438.

Gildersleeve, K. A., Haselton, M. G., Larson, C. M., & Pillsworth, E. G. (2012). Body odor attractiveness as a cue of impending ovulation in women: Evidence from a study using hormone-confirmed ovulation. *Hormones and Behavior, 61,* 157–166.

Goetz, A. T., & Shackelford, T. K. (2006). Sexual coercion and forced in-pair copulation as sperm competition tactics in humans. *Human Nature, 17,* 265–282.

Goetz, A. T., Shackelford, T. K., Weekes-Shackelford, V. A., Euler, H. A., Hoier, S., Schmitt, D. P. et al. (2005). Mate retention, sperm displacement, and human sperm competition: A preliminary investigation of tactics to prevent and correct female infidelity. *Personality and Individual Differences, 38,* 749–763.

Grammer, K. (1993). 5-α-androst-16-en-3α-on: A male pheromone? A brief report. *Ethology and Sociobiology, 14,* 201–214.

Grebe, N. M., Gangestad, S. W., Garver-Apgar, C. E., & Thornhill, R. (2013). Women's luteal-phase sexual proceptivity and the functions of extended sexuality. *Journal in Psychological Science, 24,* 2106–2010.

Greiling, H., & Buss, D. M. (2000). Women's sexual strategies: The hidden dimension of women's short-term mating. *Personality and Individual Differences, 28,* 929–963.

Gueguen, N. (2009). The receptivity of women to courtship solicitation across the menstrual cycle: A field experiment. *Biological Psychology, 80,* 321–324.

Hammerstein, P., & Parker, G. A. (1987). Sexual selection: Games between the sexes. In J. W. Bradbury & M. B. Andersson (Eds.) *Sexual selection: Testing the alternatives* (pp. 119–142). Chichester: Wiley.

Haselton, M. G., & Gangestad, S. W. (2006). Conditional expression of women's desires and men's mate guarding efforts across the ovulatory cycle. *Hormones and Behavior, 49,* 509–518.

Haselton, M. G., & Gildersleeve, K. (2011). Can men detect ovulation? *Psychological Science, 20,* 87–92.

Haselton, M., & Miller, G. F. (2006). Women's fertility across the cycle increases the short-term attractiveness of creative intelligence. *Human Nature, 17,* 40–73.

Havlicek, J., Roberts, S. C., & Flegr, J. (2005). Women's preference for dominant male odour: Effects of menstrual cycle and relationship status. *Biology Letters, 1,* 256–259.

Hawkes, K. (2004). Mating, parenting, and the evolution of human pairbonds. In B. Chapais & C.M. Berman (Eds.), *Kinship and behavior in primates* (pp. 443–473). Oxford: Oxford University Press.

Hawkes, K., O'Connell, J. F., & Jones, N. G. B. (2001). Hunting and nuclear families: Some lessons from the Hadza about men's work. *Current Anthropology, 42,* 681–709.

Heistermann, M., Ziegler, T., van Schaik, C. P., Launhardt, K., Winkler, P., & Hodges, J. K. (2001). Loss of oestrus, concealed ovulation and paternity confusion in free-ranging hanuman langurs. *Proceedings of the Royal Society of London B, 268,* 2445–2451.

Hill, K. (1982). Hunting and human evolution. *Journal of Human Evolution, 11,* 521–544.

Hrdy, S. B. (1979). Infanticide among animals, a review, classification and examination of the implications for the reproductive strategies of females. *Ethology and Sociobiology, 1,* 3–40.

Hrdy, S. B. (1981). *The woman that never evolved.* Cambridge: Harvard University Press.
Hrdy, S. B. (2000). The optimal number of fathers: Evolution, demography, and history in the shaping of female mate preferences. *Annals of the New York Academy of Sciences, 907,* 75–96.
Hrdy, S. B., & Whitten, P. L. (1987). Patterning of sexual activity. In B. B. Smuts, D. L. Cheney, R. M. Seyfarth, R. W. Wrangham, & T. T. Struhsaker (Eds.), *Primate societies* (pp. 370–384). Chicago: University of Chicago Press.
Hummel, T., Gollisch, R., Wildt, G., & Kobal, G. (1991). Changes in olfactory perception during the menstrual cycle. *Experientia, 47,* 712–715.
Jasiénska, G., Ziomkiewicz, A., Ellison, P. T., Lipson, S. F., & Thune, I. (2004). Large breasts and narrow waists indicate high reproductive potential in women. *Proceedings of the Royal Society of London B, 271,* 1213–1217.
Jennions, M. D. & Petrie, M. (2000). Why do females mate multiply? A review of the genetic benefits. *Biological Reviews, 75,* 21–64.
Johnston, V. S. & Franklin, M. (1993). Is beauty in the eye of the beholder? *Ethology and Sociobiology, 14,* 183–199.
Johnston, V. S., Hagel, R., Franklin, M., Fink, B., & Grammer, K. (2001). Male facial attractiveness: Evidence for hormone-mediated adaptive design. *Evolution and Human Behavior, 22,* 251–267.
Jones, D. (1996). *Physical attractiveness and the theory of sexual selection: Results from five populations (with a foreword by D. Symons).* Ann Arbor: Museum of Anthropology, University of Michigan.
Jones, R. E. (1978). *The vertebrate ovary: Comparative biology and evolution.* New York: Plenum Press.
Jones, B. C., Little, A. C., Boothroyd, L., DeBruine, L. M., Feinberg, D. R., Law Smith, M. J., et al. (2005). Commitment to relationships and preferences for femininity and apparent health in faces are strongest on days of the menstrual cycle when progesterone level is high. *Hormones and Behavior, 48,* 283–290.
Kaplan, H., & Hill, K. (1985). Hunting ability and reproductive success among male Ache foragers: Preliminary tests. *Current Anthropology, 26,* 131–133.
Kaplan, H., Hill, K., Lancaster, J., & Hurtado, A. M. (2000). A theory of human life history evolution: Diet, intelligence, and longevity. *Evolutionary Anthropology, 9,* 156–185.
Key, C. A., & Aiello, L. C. (1999). The evolution of social organization. In R. Dunbar, C. Knight, & C. Power (Eds.), *The evolution of culture* (pp. 15–33). New Brunswick: Rutgers University Press.
Kruger, D. J. (2006). Male facial masculinity influences attributions of personality and reproductive strategy. *Personal Relationships, 13,* 451–463.
Lancaster, J. B., & Lancaster, C. S. (1983). Parental investment: The hominid adaptation. In D. Ortner (Ed.), *Parental care in mammals* (pp. 347–387). New York: Plenum.
Lange, I. G., Hartel, A., & Meyer, H. H. D. (2002). Evolution of oestrogen functions in vertebrates. *Journal of Steroid Biochemistry and Molecular Biology, 83,* 219–226.
Larson, C. M., Pillsworth, E. G., & Haselton, M. G. (2012). Ovulatory shifts in women's attractions to primary partners and other men: Further evidence of the importance of primary partner sexual attractiveness. *PLoS One,* 7, e44456. doi:10.1371/journal.pone.0044456.
Liley, N. R., & Stacey, N. E. (1983). Hormones, pheromones, and reproductive behavior in fish. *Fish Physiology, 9,* 1–63.
Little, A. C., Burriss, R. P., Petrie, M., Jones, B. C., & Roberts, S. C. (2013). Oral contraceptive use in women changes preference for male facial masculinity and is associated with partner facial masculinity. *Psychoneuroendocrinology, 38,* 1777–1785. doi:10.1016/j.psyneuen.2013.02.014.
Little, A. C., Burt, D. M., Penton-Voak, I. S., & Perrett, D. I. (2001). Self-perceived attractiveness influences human female preferences for sexual dimorphism and symmetry in male faces. *Proceedings of the Royal Society of London B, 268,* 39–44.
Little, A. C., Jones, B. C., & Burriss, R. P. (2007a). Preferences for masculinity in male bodies change across the menstrual cycle. *Hormones and Behavior, 51,* 633–639.

Little, A. C., Jones, B. C., Burt, D. M., & Perrett, D. I. (2007b). Preferences for symmetry in faces change across the menstrual cycle. *Biological Psychology, 76,* 209–216.
Little, A. C., Debruine, L. M., Jones, B. C., & Corri, W. (2008a). Category contingent aftereffects for faces of different races, ages and species. *Cognition, 106,* 1537–1547.
Little, A. C., Jones, B. C., & DeBruine, L. M. (2008b). Preferences for variation in masculinity in real male faces change when they are more fertile. *Personality and Individual Differences, 45,* 478–482.
Lombardi, J. (1998). *Comparative vertebrate reproduction.* Boston: Kluwer Academic.
Lovejoy, C. O. (1981). The origins of man. *Science, 211,* 341–350.
Low, B. S., Alexander, R. D., & Noonan, K. M. (1987). Human hips, breasts, buttocks: Is fat deceptive? *Ethology and Sociobiology, 4,* 249–257.
Manning, J. T., Scutt, D., Wilson, J., & Lewis-Jones, D. I. (1998a). The ratio of the 2nd to the 4th digit length: A predictor of sperm numbers and concentrations of testosterone, luteinizing hormone and oestrogen. *Human Reproduction, 13,* 3000–3004.
Manning, J. T., Scutt, D., & Lewis-Jones, D. I. (1998b). Developmental stability, ejaculate size, and quality in men. *Evolution and Human Behavior, 19,* 273–282.
Manning, J. T., Bundred, P. E., & Henzi, P. (2003). Melanin and HIV in sub-Saharan Africa. *Journal of Theoretical Biology, 223,* 131–133.
Marlowe, F. (1998). The nubility hypothesis: The human breast as an honest signal of residual reproductive value. *Human Nature, 9,* 263–271.
Marlowe, F. (1999). Male care and mating effort among Hadza foragers. *Behavioral Ecology and Sociobiology, 46,* 57–64.
Marlowe, F. (2001). Male contribution to diet and female reproductive success among foragers. *Current Anthropology, 42,* 755–760.
Marlowe, F. W. (2003). A critical period for provisioning by Hadza men: Implications for pair bonding. *Evolution and Human Behavior, 24,* 217–229.
Marlowe, F. W. (2004). Is human ovulation concealed? Evidence from conception beliefs in a hunter–gatherer society. *Archives of Sexual Behavior, 33,* 427–432.
Mesnick, S. L. (1997). Sexual alliances: Evidence and evolutionary implications. In P. A. Gowaty (Ed.), *Feminism and evolutionary biology: Boundaries, intersections, and frontiers* (pp. 207–260). New York: Chapman and Hall.
Miller, E. M. (1995). Human breasts: Evolutionary origins as a deceptive signal of need for provisioning and temporary infertility. *Mankind Quarterly, 36,* 135–150.
Miller, G. F. (2000). *The mating mind: How sexual choice shaped the evolution of human nature.* New York: Anchor.
Møller, A. P., & Swaddle, J. P. (1997). *Asymmetry, developmental stability, and evolution.* Oxford: Oxford University Press.
Møller, A. P., Thornhill, R., & Gangestad, S. W. (2005). Direct and indirect tests for publication bias: Asymmetry and sexual selection. *Animal Behaviour, 70,* 497–506.
Nelson, R. J. (2000). *An introduction to behavioral endocrinology* (2nd ed.). Sunderland: Sinauer Associates
Pagel, M. (1994). Evolution of conspicuous estrous advertisement in old-world monkeys. *Animal Behaviour, 47,* 1333–1341.
Pagel, M., & Meade, A. (2006). Bayesian analysis of correlated evolution of discrete characters by reversible-jump Markov chain Monte Carlo. *American Naturalist, 167,* 808–825.
Palombit, R. A. (1999). Infanticide and the evolution of pairbonds in nonhuman primates. *Evolutionary Anthropology, 7,* 117–129.
Parker, G. A. (1979a). Sexual selection and sexual conflict. In M. Blum & N. Blum (Eds.), *Sexual selection and reproductive competition in insects* (pp. 123–166). New York: Academic.
Parker, G. A. (1979b). Sperm competition and its evolutionary consequences in the insects. *Biological Reviews, 45,* 525–568.
Pawlowski, B., & Jasiénska, G. (2005). Women's preferences for sexual dimorphism in height depend on menstrual cycle phase and expected duration of relationship. *Biological Psychology, 70,* 38–43.

Pazol, K. (2003). Mating in the Kakamega forest blue monkeys (*Cercopithecus mitis*): Does female sexual behavior function to manipulate paternity assessment? *Behaviour, 140,* 473–499.

Penton–Voak, I. S., & Perrett, D. I. (2000). Female preference for male faces changes cyclically—Further evidence. *Evolution and Human Behavior, 21,* 39–48.

Penton–Voak, I. S., & Perrett, D. I. (2001). Male facial attractiveness: Perceived personality and shifting female preferences for male traits across the menstrual cycle. *Advances in the Study of Behavior, 30,* 219–259.

Penton–Voak, I. S., Perrett, D. I., Castles, D. L., Kobayashi, T., Burt, D. M., Murray, L. K., et al. (1999). Female preference for male faces changes cyclically. *Nature, 399,* 741–742.

Penton–Voak, I. S., Little, A. C., Jones, B. C., Burt, D. M., Tiddeman, B. P., & Perrett, D. I. (2003). Female condition influences preferences for sexual dimorphism in faces of male humans (*Homo sapiens*). *Journal of Comparative Psychology, 117,* 264–271.

Perrett, D. I., Lee, K. J., Penton–Voak, I., Rowland, D., Yoskikawa, S., Burt, D. M., Henzi, S. P., Castles, D. L., & Akamatsu, S. (1998). Effects of sexual dimorphism on facial attractiveness. *Nature, 394,* 884–887.

Pillsworth, E. G., Haselton, M. G., & Buss, D. M. (2004). Ovulatory shifts in female sexual desire. *Journal of Sex Research, 41,* 55–65.

Platek, S. M., Raines, D. M., Gallup, G. G., Mohamed, F. B., Thomson, J. W., Myers, T. E., Panyayin, I. S., Levin, S. L., Davis, J. A., Fonteyn, L. C. M., & Arigo, D. R. (2004). Reactions to children's faces: Males are more affected by resemblance than females are, and so are their brains. *Evolution and Human Behavior, 25,* 394–405.

Puts, D. A. (2005). Mating context and menstrual phase affect women's preferences for male voice pitch. *Evolution and Human Behavior, 26,* 388–397.

Puts, D. A. (2006). Cyclic variation in women's preferences for masculine traits: Potential hormonal causes. *Human Nature, 17,* 114–127.

Rantala, M. J., Eriksson, C. J. P., Vainikka, A., & Kortet, R. (2006). Male steroid hormones and female preference for male body odor. *Evolution and Human Behavior, 27,* 259–260.

Rhodes, G., Simmons, L. W., & Peters, M. (2005). Attractiveness and sexual behavior: Does attractiveness enhance mating success? *Evolution and Human Behavior, 26,* 186–201.

Rikowski, A. & Grammer, K. (1999). Human body odour, symmetry and attractiveness. *Proceedings of the Royal Society of London B, 266,* 869–874.

Rodriguez-Gironés, M. A., & Enquist, M. (2001). The evolution of female sexuality. *Animal Behaviour, 61,* 695–704.

Scheib, J. E., Gangestad, S. W., & Thornhill, R. (1999). Facial attractiveness, symmetry and cues of good genes. *Proceedings of the Royal Society of London B, 266,* 1913–1917.

Schroder, I. (1993). Concealed ovulation and clandestine copulation: A female contribution to human evolution. *Ethology and Sociobiology, 14,* 381–389.

Scutt, D., & Manning, J. T. (1996). Symmetry and ovulation in women. *Human Reproduction, 11,* 2477–2480.

Shackelford, T. K., Goetz, A. T., Guta, F. E., & Schmitt, D. P. (2006). Mate guarding and frequent in-pair copulation in humans: Concurrent or compensatory anti-cuckoldry tactics? *Human Nature, 17,* 239–252.

Shackelford, T. K., LeBlanc, G. J., Weekes-Shackelford, V. A., Bleske-Rechek, A. L., Euler, H. A., & Hoier, S. (2002). Psychological adaptation to human sperm competition. *Evolution of Human Behavior, 23,* 123–138.

Sheldon, M. S., Cooper, M. L., Geary, D. C., Hoard, M., & DeSoto, M. C. (2006). Fertility cycle patterns in motives for sexual behavior. *Personality and Social Psychology Bulletin, 32,* 1659–1673.

Sillén-Tullberg, B., & Møller, A. P. (1993). The relationship between concealed ovulation and mating systems in anthropoid primates—a phylogenetic analysis. *American Naturalist, 141,* 1–25.

Simpson, J. A., Gangestad, S. W., Christensen, P. N., & Leck, K. (1999). Fluctuating asymmetry, sociosexuality, and intrasexual competitive tactics. *Journal of Personality and Social Psychology, 76,* 159–172.

Skau, P. A., & Folstad, I. (2003). Do bacterial infections cause reduced ejaculate quality? A meta-analysis of antibiotic treatment of male infertility. *Behavioral Ecology, 14,* 40–47.
Smith, R. L. (1984). Human sperm competition. In R. L. Smith (Ed.), *Sperm competition and the evolution of animal mating systems* (pp. 601–660). London: Academic.
Smock, T., Albeck, D., & Stark, P. (1998). A peptidergic basis for sexual behavior in mammals. *Progress in Brain Research, 119,* 467–481.
Smuts, B. (1992). Male aggression against women: An evolutionary perspective. *Human Nature, 3,* 1–44.
Smuts, B. B., & Smuts, R. W. (1993). Male aggression and sexual coercion of females in nonhuman primates and other mammals: Evidence and theoretical implications. *Advances in the Study of Behavior, 22,* 1–63.
Soler, C., Núñez, M., Gutiérrez, R., Núñez, J., Medina, P., Sancho, M., et al. (2003). Facial attractiveness in men provides clues to semen quality. *Evolution and Human Behavior, 24,* 199–207.
Spuhler, J. N. (1979). Continuities and discontinuities in anthropoid-hominid behavioral evolution: Bipedal locomotion and sexual receptivity. In N. A. Chagnon & W. Irons (Eds.), *Evolutionary biology and human social behavior: An anthropological perspective* (pp. 454–461). North Scituate: Duxbury Press.
Stacey, P. B. (1982). Female promiscuity and male reproductive success in social birds and mammals. *American Naturalist, 120,* 51–64.
Strassmann, B. I. (1981). Sexual selection, paternal care, and concealed ovulation in humans. *Ethology and Sociobiology, 2,* 31–40.
Strassmann, B. I. (1996). Menstrual cycle visits by Dogon women: A hormonal test distinguishes deceit from honest signaling. *Behavioral Ecology, 7,* 304–315.
Symons, D. (1979). *The Evolution of Human Sexuality*. Oxford: Oxford University Press.
Symons, D. (1995). Beauty is in the adaptations of the beholder: The evolutionary psychology of human female sexual attractiveness. In P. R. Abramson & S. D. Pinkerton (Eds.) *Sexual nature/ sexual culture* (pp. 80–118). Chicago: University of Chicago Press.
Szalay, F. S., & Costello, R. K. (1991). Evolution of permanent estrus displays in hominids. *Journal of Evolution, 20,* 439–464.
Thornhill, R., & Alcock, J. (1983). *The evolution of insect mating systems*. Cambridge: Harvard University Press.
Thornhill, R., & Gangestad, S. W. (1999a). The scent of symmetry: A human sex pheromone that signals fitness? *Evolution and Human Behavior, 20,* 175–201.
Thornhill, R., & Gangestad, S. W. (1999b). Facial attractiveness. *Trends in Cognitive Science, 3,* 452–460.
Thornhill, R., & Gangestad, S. W. (2003). Do women have evolved adaptation for extra-pair copulation? In E. Voland & K. Grammer (Eds.), *Evolutionary aesthetics* (pp. 341–368). Heidelberg: Springer.
Thornhill, R., & Gangestad, S. W. (2008). *The evolutionary biology of human female sexuality*. New York: Oxford University Press.
Thornhill, R., & Grammer, K. (1999). The body and face of woman: One ornament that signals quality? *Evolution and Human Behavior, 20,* 105–120.
Thornhill, R., & Palmer, C. T. (2000). *A natural history of rape: Biological bases of sexual coercion*. Cambridge: MIT Press.
Thornhill, R., Gangestad, S. W., & Comer, R. (1995). Human female orgasm and mate fluctuating asymmetry. *Animal Behaviour, 50,* 1601–1615.
Thornhill, R., Chapman, J. F., & Gangestad, S. W. (2013). Women's preferences or men's scents associated with testosterone and cortisol levels: Patterns across the ovulatory cycle. *Evolution and Human Behavior, 34,* 216–221.
Thornhill, R., Gangestad, S. W., Miller, R., Scheyd, G., Knight, J., & Franklin, M. (2003). MHC, symmetry, and body scent attractiveness in men and women. *Behavioral Ecology, 14,* 668–678.
Turke, P. W. (1984). Effects of ovulatory concealment and synchrony on protohominid mating systems and parental roles. *Ethology and Sociobiology, 5,* 33–44.

Van Dongen, S., Ten Broek, C. M. A., Galis, F., & Wijnaendts, L. C. D. (2012). No association between fluctuating asymmetry in highly stabilized traits and second and fourth digit ratio (2D:4D) in human fetuses. *Early Human Development, 85,* 393–398.

van Schaik, C. P., Prodham, G. R., & Van Noordwijk, M. A. (2004). Mating conflict in primates: Infanticide, sexual harassment and female sexuality. In P. Kappler & C. van Schaik (Eds.), *Sexual selection in primates: New and comparative perspectives* (pp. 131–150). Cambridge: Cambridge University Press.

Wakano, J. Y., & Ihara, Y. (2005). Evolution of male parental care and female multiple mating: Game-theoretical and two-locus diploid models. *The American Naturalist, 166,* E32–E44.

Wallen, K. (2000). Risky business: Social context and hormonal modulation of primate sex drive. In K. Wallen & J. E. Schneider (Eds.) *Reproduction in Context* (pp. 289–323). Cambridge: MIT Press.

Weisfeld, G. E., & Woodward, L. (2004). Current evolutionary perspectives on adolescent romantic relations and sexuality. *Journal of the American Academy of Child and Adolescent Psychiatry, 43,* 11–19.

Westermarck, E. (1929). *Marriage*. New York: Jonathan Cape and Harrison Smith.

Whittier, J. M., & Tokarz, R. R. (1992). Physiological regulation of sexual behavior in female reptiles. In C. Gans & D. Crews (Eds.), *Hormones, brain, and behavior: Biology of the reptilia physiology* (Vol. 18, pp. 24–69), Physiology E. Chicago: University of Chicago Press.

Wilcox, A. J., Dunson, D., & Baird, D. D. (2000). The timing of the "fertile window" in the menstrual cycle: Day-specific estimates from a prospective study. *British Medical Journal, 321,* 1259–1262.

Wilcox, A. J., Dunson, D. B., Weinberg, C. R., Trussell, J., & Baird, D. D. (2001). Likelihood of conception with a single act of intercourse: Providing benchmark rates for assessment of post-coital contraceptives. *Contraception, 63,* 211–215.

Wilson, M., & Mesnick, S. L. (1997). An empirical test of the bodyguard hypothesis. In P.A. Gowaty (Ed.), *Feminism and evolutionary biology* (pp. 505–511). New York: Chapman and Hall.

Wrangham, R. W. (1993). The evolution of sexuality in chimpanzees and bonobos. *Human nature: An interdisciplinary biosocial perspective, 4,* 47–79.

Chapter 9
Transgendered Male Androphilia in the Human Ancestral Environment

Paul L. Vasey and Doug P. VanderLaan

Beyond Gay: Male Sexual Orientation in Cross-Cultural Perspective

Gay men represent an increasingly visible aspect of Western culture. Previous polls have shown that the number of individuals who state that they personally know someone who openly identifies as "gay" grows with every passing year (e.g., Rubin 2000). Consequently, regardless of their sexual orientation, it might come as a surprise to most individuals living in the West that "gay men" do not necessarily exist in other cultures. Indeed, it is not at all uncommon for individuals living in non-Western cultures to claim that "gay men" or "homosexuals" are unknown in their societies. Many of the individuals who are most vehement in making such assertions are biological males who have sex with other biological males.

What are we to make of such claims? Research has demonstrated that the identity categories of "gay" and "homosexual" are culturally and historically "situated" and, as such, do not necessarily translate to other places and times (e.g., Asthana and Oostvogels 2001). Such categories, if they are known at all, might mean something to people in other cultures, but whatever that might be, they are not categories of personhood that individuals draw upon when constructing personal narratives about who they are. As such, the way in which many non-Western, same-sex attracted males think about themselves and pattern their lives (sexual or otherwise) differs radically in many respects from Western gay men. Thus, when individuals from non-Western cultures say that there are no "gays" or "homosexuals" in their societ-

P. L. Vasey (✉)
Department of Psychology, University of Lethbridge, 4401 University Drive,
Lethbridge, Alberta T1K 3M4, Canada
e-mail: paul.vasey@uleth.ca

D. P. VanderLaan
Gender Identity Service, Child, Youth and Family Services,
Centre for Addiction and Mental Health, Beamish Family Wing, Intergenerational
Wellness Centre, 80 Workman Way, 5th Floor, Toronto, ON M6J 1H4, Canada
e-mail: doug.vanderlaan@camh.ca

T. K. Shackelford, R. D. Hansen (eds.), *The Evolution of Sexuality*,
Evolutionary Psychology, DOI 10.1007/978-3-319-09384-0_9

ies, they are not necessarily lying. Based on their understanding of what it means to be "gay" or "homosexual," no one in their societies identifies or behaves as such, and neither do they identify other members of their culture in that way.

Given this, any attempt to undertake comparative cross-cultural research on sexual orientation must focus on the *deep structure* of sexual orientation that transcends differences related to how male same-sex sexual attraction is socially constructed within culturally specific contexts. The deep structure of male same-sex sexuality can be thought of as a set of traits that characterize same-sex attracted males regardless of the cultural context in which they are found. To this end, a focus on cross-culturally universal *sexual feelings* facilitates comparisons in a manner that culturally specific identity categories do not. As such, we employ the terms *androphilia* and *gynephilia* in our discussion of sexual orientation across cultures. Androphilia refers to sexual attraction and arousal to adult males, whereas gynephilia refers to sexual attraction and arousal to adult females. Although same-sex attracted "gay men" in Western societies differ in many dramatic ways from same-sex attracted males in a range of non-Western societies, in terms of the deep structure of sexual orientation, both can be accurately described as androphilic biological males.

An additional advantage of this terminology is that it avoids reference to sexual *behavior,* which may be constrained by cultural circumstances (e.g., taboos against same-sex sexual behavior), or enacted for reasons unrelated to sexual attraction and arousal (e.g., ritual, prostitution, coercion, etc.). As such, the terms "androphilia" and "gynephilia" make no assumptions about whether sexual behavior has been expressed. Consequently, an individual can be androphilic or gynephilic without ever having engaged in sexual behavior.

How Male Androphilia Is Publically Expressed Varies Cross-Culturally

As should be evident from the discussion above, the manner in which male androphilia is publically expressed varies across cultures (Murray 2000). This expression typically takes one of two forms, which are related to gender role enactment and gender identity. These two forms are *sex-gender congruent* and *transgendered* male androphilia. Sex-gender congruent male androphiles occupy the gender role typical of their sex, behave in a relatively masculine manner, and identify as "men." Other authors have referred to sex-gender congruent male androphilia as "egalitarian male homosexuality" (Murray 2000) and "homophilic homosexuality" (Gorer 1966). However, the term "sex-gender congruent" androphilia highlights the critical role of gender-role enactment in distinguishing the two forms of male androphilia under consideration here.

Transgendered androphilic males typically behave in an effeminate manner and often identify as neither "men" nor "women," but as members of some "third" gender category. In some cultures, transgendered male androphilia is linked to particular institutionalized labor practices, which often involve specialized religious activities. Such transgendered male androphilia has been referred to as *role structured homosexuality* (Herdt 1997). For example, on the Indian subcontinent, transgendered male androphiles known as *hijra* bestow blessings from Hindu gods and goddesses for luck and fertility at weddings and at the birth of male babies (Nanda 1999). Similarly, in some cultures, such as the Mohave and the Yorok, all *berdache* (transgendered male androphiles) were shamans (e.g., Devereux 1937; Kroeber 1925).

Both sex-gender congruent and transgendered male androphilia may occur within a given culture, but typically one or the other tends to predominate (Whitam 1983). For example, the sex-gender congruent form tends to be much more common in many Western cultures; in contrast, the transgendered form appears to be more common in many non-Western cultures (Murray 2000). In places where these two forms coexist, albeit with one predominating, the two often consider each other to be members of the same community (Whitam 1983).

In addition to these two forms of male androphilia, a third form, transgenerational homosexuality, has also been reported in the ethnographic literature. Transgenerational homosexuality involves sexual interactions between a sexually immature or younger male and a sexually mature, older male (Murray 2000). It is not clear whether transgenerational homosexuality is motivated by androphilia on the part of either the older or younger partner. For example, in some instances these same-sex interactions might be enacted for primarily ritualistic purposes (e.g., Herdt 1981). Depending on the individual, the older partners in these interactions might be best characterized as either pedophilic (i.e., sexually attracted/aroused to prepubescent individuals), hebephilic (i.e., sexually attracted/aroused to peripubescent individuals), or gynephilic, not androphilic. Similarly, the younger partners might be (pre) gynephilic, not (pre)androphilic. Given these reasons, we do not consider transgenerational homosexuality here. For a discussion of unique properties of transgenerational homosexuality from an ethnological perspective, see Crapo (1995).

Some Correlates of Male Androphilia Reoccur Consistently Across Diverse Cultures

Attempts to draw comparisons between the sex-gender congruent and transgendered forms of male androphilia have been characterized as misguided because, critics argue, these forms are so culturally distinct in terms of what they mean within a particular cultural setting that any comparisons one might make would be largely facile (Johnson et al. 2000). As such, the overall impression one gleans from this social constructionist literature is that a panoply of male "androphilias" exist cross-culturally. Not surprisingly, researchers whose work is informed by evolutionary theory have questioned whether a common biological basis underlies the diverse cultural expressions characterizing this trait. If it were possible to establish that androphilic males from different cultural backgrounds shared traits that are indicators, at least in theory, of common etiology, then this would lend support to the possibility of a common biological basis. Indeed, quantitative research indicates that the sex-gender congruent and transgendered forms of male androphilia share numerous developmental and biodemographic correlates that appear to be cross-culturally invariant.

In terms of biodemographic correlates of male androphilia that exist across cultures, both sex-gender congruent and transgendered male androphiles tend to be later born among their siblings (e.g., Blanchard 2004; VanderLaan and Vasey 2011; Vasey and VanderLaan 2007), have greater numbers of older biological brothers ("fraternal birth order effect,"[1] e.g., Bogaert and Skorska 2011; VanderLaan and Vasey 2011; Vasey and VanderLaan 2007), exhibit larger family sizes (e.g., Blanchard and Lippa 2007; Camperio-Ciani et al. 2004; Iemmola and Camperio Ciani 2009; King et al. 2005; Schwartz et al. 2010; VanderLaan et al. 2012; VanderLaan and Vasey 2011; Vasey and VanderLaan 2007), cluster within families (e.g., Schwartz et al. 2010; VanderLaan et al. 2013a, b), occur at similar prevalence rates across cultures (e.g., Smith et al. 2003; VanderLaan et al. 2013a; Whitam 1983), and exhibit little or no reproductive success (e.g., King et al. 2005; Schwartz et al. 2010; Vasey et al. 2014). In addition, the odds ratios associated with the fraternal birth order effect in various populations of sex-gender congruent and transgendered male androphiles are remarkably consistent, suggesting that the manner in which older brothers influence the development of male androphilia is constant across culturally diverse populations (e.g., Cantor et al. 2002; VanderLaan and Vasey 2011).

Prospective and retrospective cross-cultural research on early psychosocial development among transgendered and sex-gender congruent male androphiles has shown that the childhood behavior of such males is characterized by greater levels of female-typical behavior (e.g., nurturing play with dolls) and lower levels of male-typical behavior (e.g., rough-and-tumble play; Bailey and Zucker 1995; Bartlett and Vasey 2006; Cardoso 2005, 2009; Whitam 1983). In addition, both types of male androphiles express elevated cross-sex beliefs and wishes in childhood (e.g., "I wish I was a girl"; Bailey and Zucker 1995; Vasey and Bartlett 2007; Whitam 1983). Furthermore, both sex-gender congruent and transgendered male androphiles also experience elevated traits of childhood separation anxiety (i.e., anxiety related to separation from major attachment figures such as parents; VanderLaan et al. 2011; Vasey et al. 2011; Zucker et al. 1996), which tends to be more common among girls compared to boys (e.g., Shear et al. 2006; VanderLaan et al. 2011). In adulthood, male androphiles from a range of cultures exhibit preferences for a variety of female-typical occupations and hobbies (e.g., interior design; Lippa 2005; Whitam 1983).

Even though sex-gender congruent androphilic males are relatively feminine as boys compared to their gynephilic counterparts (Bailey and Zucker 1995), they behaviorally defeminize to varying degrees as they develop. It has been suggested that this behavioral defeminization probably occurs in response to culturally-specific gender role expectations, which hold that male-bodied individuals should behave in

[1] The fraternal birth order effect refers to the well-established finding that the number of older biological brothers increases the odds of androphilia in later born males (Blanchard 2004; Bogaert and Skorska 2011).

a masculine manner (Bailey 2003; Berling 2001; Rieger and Savin-Williams 2012). In contrast, in cultures where transgendered male androphilia is the norm, feminine boys develop into feminine adult males. Consequently, adult sex-gender congruent male androphiles are relatively masculine when compared to adult transgendered male androphiles. Conversely, they are, on average, relatively feminine when compared to adult male gynephiles (Bailey 2003; Lippa 2005). Thus, regardless of how it is manifested, male androphilia is associated with gender atypicality in childhood and adulthood. However, the strength of this association varies with the manner in which male androphilia is publically expressed.

Taken together, these numerous, cross-culturally uniform biodemographic and developmental correlates of male androphilia indicate that sex-gender congruent and transgendered male androphilia are cultural variants of what is essentially the same phenomenon with a common biological basis. In this regard, the cross-culturally invariant biodemographic and developmental correlates described above can be thought of as part of the deep structure of male androphilia.

Male Androphilia Is an Evolutionary Paradox

The existence of diverse forms of male androphilia across cultures, which nonetheless share a similar biological etiology, is an evolutionary paradox. There appears to be a genetic influence on male androphilia (e.g., Bailey et al. 2000; Kendler et al. 2000; Långström et al. 2010), yet androphilic males reproduce at significantly lower rates than gynephilic males, if at all (e.g., King et al. 2005; Schwartz et al. 2010; Vasey et al. 2014). Consequently, one would have expected genes for male androphilia to have become extinct given the relative reproductive costs associated with this trait and the reproductive benefits associated with male gynephilia.

Nevertheless, prehistoric rock art and pottery suggests that male-male sexual activity has existed for millennia (e.g., Larco Hoyle 1998; Nash 2001). Further, graves containing male skeletal remains and female-typical artifacts are indicative of transgendered males in the distant past (e.g., Hollimon 1997). Prine (2000) argued that certain architecturally unusual dwellings inhabited by the Hidasta[2] people between 1400 and 1800 AD, were the homes of transgendered males known locally as *miati*. Given what we know about the exclusive androphilic orientation of most transgendered males from comparable populations (e.g., Harrington 1942; Murray 2000; Nanda 1999), archaeological indicators of such individuals are suggestive of the presence of male androphilia in human antiquity.

Furthermore, male androphilia occurs in the vast majority of cultures for which information is available (Murray 2000) and it appears to occur at similar frequencies (e.g., Smith et al. 2003; VanderLaan et al. 2013a; Whitam 1983). Some reports exist of cultures where male–male sexual behavior is unknown (e.g., Hewlett and

[2] The Hidasta are a native North American people that lived in palisaded villages along the modern-day Missouri River in North Dakota.

Hewlett 2010), but it is not clear whether some males in these populations have unexpressed androphilic feelings. Although male–male sexuality may be absent in a minority of cultures, these exceptions do not invalidate the conclusion that male androphlia is a predictably and reliably reoccurring phenomenon in most human cultures. The cross-culturally widespread nature of male androphilia suggests that it is a phylogenetically primitive aspect of the human sexual condition.

In sum, male androphilia has a genetic component, occurs at similar frequencies across many different cultures, and appears to have existed for millennia. Nevertheless, male androphiles reproduce at a fraction of the rate that gynephilic males do, if they reproduce at all. For these reasons, male androphilia is widely considered one of the outstanding paradoxes of evolutionary psychology. A trait that lowers direct reproduction and persists over evolutionary time requires explanation when viewed within the context of natural selection, a process that favors the evolution of reproductively viable traits.

What Was the Human Ancestral Form of Male Androphilia

Given that the manner in which male androphilia is publically expressed varies cross-culturally, the question arises as to which form, sex-gender congruent or transgendered, was the ancestral form? If it were possible to establish that one form of male androphilia was associated, more often than not, with sociocultural conditions thought to characterize ancestral humans, then this would bolster the conclusion that that particular form of male androphilia was ancestral. Implementation of this approach requires establishing at least some of the sociocultural features that characterized ancestral humans. There is widespread consensus that ancestral humans followed a hunter-gatherer pattern of subsistence until the beginning of the Holocene, and archaeological evidence supports this contention (McBrearty and Brooks 2000; Smith 1999). As such, ethnographic data derived from the study of hunter-gatherers has been widely used to model ancestral human sociocultural conditions.

For example, research on hunter-gatherers indicates that the residential groups[3] in which human ancestors lived were likely to have been relatively small (Klein 1999; Ehrlich 2000). Binford (2001) examined group size during the most aggregated phase[4] of subsistence settlement for 219 nonequestrian,[5] hunter-gatherer eth-

[3] "Residential group" refers to the same camp or settlement within which people regularly reside.

[4] Hunter-gatherers exhibit a high degree of residential mobility, which is expressed in terms of a fission-fusion type of group organization whereby the group breaks apart into smaller foraging parties, which then reassemble each day into larger aggregates (Marlowe 2005).

[5] Beginning in the 1700s, after the Spanish introduction of the horse, various North American Plains Indian ethnolinguistic groups subsequently became specialists in hunting bison from horseback (Shimkin 1983). This specialization in foraging pattern influenced the group sizes, home

nolinguistic groups who varied according to primary food source exploited (i.e., terrestrial plants, terrestrial animals, aquatic resources) and mobility (i.e., mobile settlements, semi-sedentary settlements). His analyses indicated that group size for these hunter-gatherers was, on average, 69 individuals. Marlowe's (2005) analysis of warm-climate,[6] non-equestrian hunter-gatherer ethnolinguistic groups ($n = 130$) indicates that residential groups contain a mean (±SD) of 37.46 (±38.28) individuals. Hill et al. (2011) analyzed data from 32 hunter-gatherer societies and found that mean band size was 28.2 individuals. If these results for mean hunter-gatherer residential group size can be taken as representative of the conditions that characterized ancestral humans, then these analyses point to the conclusion that, on average, ancestral humans formed relatively small residential groups of approximately 28–69 individuals.

Research on hunter-gatherers also indicates that ancestral humans were likely egalitarian in terms of their political structure. Contemporary hunting and gathering societies that have economies based on immediate, rather than delayed, return of food resources tend to be egalitarian with respect to power, wealth, prestige, and religious beliefs/practices (Woodburn 1982). In immediate-return systems, all individuals have direct access to food resources, which are owned by no single individual. Food is neither elaborately processed, nor stored. Social groupings are flexible and constantly changing in composition and, as such, there are no fixed dwellings, base camps, storage areas, hunting or fishing apparatuses (i.e., weirs), or ritual sites. Individuals have a choice of whom they associate with in terms of residence, food acquisition, trade, and ritual contexts. Movement between groups does not result in economic penalties. Although sharing and mutuality are stressed, individuals are not dependent on food sharing, nor are they involved in long-term binding commitments and dependencies of the sort that characterize delayed return systems. Moreover, the accumulation of personal possessions is sanctioned. In these societies, there are either no leaders at all, or leaders who are constrained in terms of their ability to exercise authority or influence to acquire wealth and prestige.

With respect to the ancestral form of religion, some scholars have argued that shamanistic[7] activity is depicted in Paleolithic rock art (Clottes and Lewis-Williams 1998; Deacon 1999). Furthermore, shamanism appears to be common in contemporary small-scale hunter-gatherers (Sanderson and Roberts 2008; Winkelman 2010).

---

ranges, hunting success rates, and travel costs of these groups. Because we are interested in reconstructing the sociocultural environment of ancestral humans prior to the domestication of the horse, we do not consider data from equestrian hunter-gatherers here.

[6] It is only during the last 30,000 years that the arctic has been occupied by modern Homo sapiens (Vaughan 1994). Occupation of this biome had concomitant influences on residential group size. Consequently, Marlow (2005) argues that if we are interested in the period prior to 30,000 years ago, it is reasonable to exclude arctic foragers from analysis pertaining to residential group size.

[7] A religion is Shamanic when a shaman is the center of most religious practice, a strong belief in animism is present, there are no calendrical rites, and laypersons rely on a shaman as the sole intermediary between themselves and the supernatural (Sanderson and Roberts 2008).

Taken together, this evidence suggests that shamanism, which is closely associated with animism,[8] represents the form of religion practiced by ancestral humans.

With these insights in mind, VanderLaan et al. (2013c) compared 46 "transgendered societies" (i.e., societies in which transgendered male androphilia predominated) with 146 "non-transgendered societies" using the standard cross-cultural sample (SCCS).[9] Their goal was to ascertain whether human ancestral socio-cultural conditions (i.e., hunting and gathering, smaller group size, egalitarian political structure, and animistic/shamanistic religious beliefs) were more likely to be associated with one of these two types of societies. Their analysis indicated that transgendered societies were characterized by a significantly greater presence of ancestral sociocultural conditions, compared to non-transgendered societies. Given the association between transgendered male androphilia and ancestral human sociocultural conditions, it seems parsimonious to conclude that the ancestral form of male androphilia was the transgendered form.

The existence of two forms of transgendered male androphilia (i.e., *institutionalized role structured* and *non-role structured*) raises the question as to which one preceded the other in evolutionary time. It seems likely that role structure transgendered male androphilia is derived from a more ancestral form of transgendered male androphilia that does not involve role specialization. Once trangendered male androphilia originated in humans, it could then be culturally elaborated upon to serve any number of distinct social roles. This represents the most parsimonious evolutionary sequence in the evolution of transgendered male androphilia because, phylogenetically, less specialized form of traits tend to precede more specialized ones (Dean et al. 2014).

The Fa'afafine of Samoa

To date, tests of evolutionary hypotheses pertaining to male androphilia have been conducted on a single population of transgendered androphilic males—the *fa'afafine* of Samoa. Our research group has conducted this work. Previous discussions pertaining to the evolution of male androphilia have centered almost exclusively sex-gender congruent male androphiles. Consequently, the remainder of this review showcases our Samoan *fa'afafine* research given its unique focus on transgendered male androphiles. For a review of the evolutionary literature that compares both sex-gender congruent and transgendered male androphiles, see Vasey and VanderLaan (2014).

In the Samoan language, *fa'afafine* means: "in the manner of a woman." Within Samoan society, *fa'afafine* are not recognized as "men" or "women," nor do they

---

[8] Animism refers to the belief that spirits inhabit some or all natural objects and phenomena.

[9] The SCCS provides data related to a subset of the world's non-industrial societies and is employed to circumvents Galton's problem (i.e., common cultural derivation and cultural diffusion) when conducting cross-cultural comparisons.

identify as such, and, consequently, they have been described as a type of "third" gender. Like men, *fa'afafine* are biological males. They differ from Samoa men, however, in that they are very feminine with respect to their gender role enactment. From a Western perspective, many *fa'afafine* would be considered transgendered. The majority are not transsexual, however, because they do not experience dysphoria with respect to their genitals (Vasey and Bartlett 2007). Unlike the *hijra* of India, *fa'afafine* have no institutionalized role in Samoa.

Fa'afafine are recognized in childhood by their families and the members of their community based on their tendencies to engage in female-typical activities (e.g., playing with girls) and their aversion toward male-typical activities (e.g., rough-and-tumble play). This process of recognition does not mean that Samoans make boys into *fa'afafine*. Rather, in Samoan culture, boyhood femininity is interpreted to mean that individuals simply are *fa'afafine* and it is understood that such individuals will not grow up to be "men." Some families react negatively to the presence of a *fa'afafine* child with corporal punishment, but many have a *laisser-faire* attitude, some even facilitate the child's feminine behavior—sewing "him" dresses, for example (Bartlett and Vasey 2006; Vasey and Bartlett 2007).

In adulthood, the vast majority of *fa'afafine* are exclusively androphilic and consequently, they do not have children of their own (Vasey et al. 2013). All *fa'afafine* recognize the term "gay" although the precise meaning of this term varies depending on the individual asked. That being said, none of the *fa'afafine* use the term "gay" to describe themselves. "Gays" as one *fa'afafine* told the first author "sleep with each other, but *fa'afafine* don't do that." Indeed, *fa'afafine* express disgust at the thought of engaging in sexual activity with other *fa'afafine* and stress that they do not do so. Instead, they point out, in contrast to "gays," they have sex with "straight men."

In a Samoan cultural context, regardless of sexual orientation, "straight man" means a male who is masculine and who self-identifies as a "man." Some "straight men" in Samoa are gynephilic and only have sex with women. However, other men who are gynephilic may have sex with *fa'afafine* if they are unable to access their preferred sexual partners (i.e., adult females). This may seem perplexing from a Western cultural perspective, however, it is important to note that in cultures where transgendered male androphilia predominates, many male gynephiles may experience relatively little sexual aversion to the idea of engaging in certain types of same-sex sexual interactions because, to a certain extent, transgendered male androphiles represent facsimiles of their preferred sex partners (i.e., adult females). The other men who have sex with *fa'afafine* appear to be a combination of *gynandromorphophilic* (i.e., peak sexual attraction and arousal to she-males), bisexual, or androphilic. In short, the Samoan category of "straight man" is a very heterogeneous one with respect to sexual orientation (Vasey and Petterson, unpublished data).

In Samoa, *fa'afafine* enjoy a high level of social acceptance that, while not absolute, stands in stark contrast to the situation experienced by Western transgendered male androphiles (Namaste 2000; Seil 1996). Indeed, *fa'afafine* are highly visible and active members of Samoa society. They occupy all manner of positions from stay-at-home caregivers to Assistant Chief Executive Officers in the government.

The Prime Minister of Samoa, the Honorabe Tuilaepa Sailele Malielegaoi, is Patron of the National *Fa'afafine* Association and has spoken publically on many occasions about the value of *fa'afafine* for Samoan society.

In the following sections, we describe our research on the Samoan *fa'afafine* that aimed at testing three hypotheses for the evolution of male androphilia, namely, the *Sexually Antagonistic Gene Hypothesis*, the *Over-Dominance Hypothesis* and the *Kin Selection Hypothesis*.

Tests of the Sexually Antagonistic Gene Hypothesis in Samoan Fa'afafine

Balancing selection hypotheses for male androphilia hold that the relatives of male androphiles exhibit increased reproductive success thereby offsetting any reproductive costs associated with male androphilia, itself. For example, the Sexually Antagonistic Gene Hypothesis for male androphilia posits that genes associated with the development of androphilia result in decreased reproductive output in male carriers, but the same genes result in increased reproductive output in female carriers (e.g., Camperio-Ciani et al. 2004). For this reason, this hypothesis is routinely referred to as the Female Fecundity Hypothesis for male androphilia. Given that kin share a disproportionate number of genes in common, the female kin of male androphiles should experience, on average, greater increased reproductive output than females with no androphilic male relatives. In theory, the fitness benefits that accrue to the female relatives of male androphiles balance out the fitness costs associated with male androphilia. Consequently, sexually antagonistic selection occurs for the genes in question owing to their fitness-enhancing properties in female carriers. A by-product of this sexually antagonistic selection is that male androphilia persist in populations over evolutionary time, despite its fitness-reducing consequences. Given all this, the basic prediction that flows from the Sexual Antagonistic Gene Hypothesis is that the female relatives of androphilic males should tend to produce more offspring than those of gynephilic males.

Three studies have been conducted in Samoa by our research group that furnish data pertaining to the Sexually Antagonistic Gene Hypothesis. Vasey and VanderLaan (2007) demonstrated that the mothers of *fa'afafine* produce more offspring than those of gynephilic men. This finding was replicated by VanderLaan and Vasey (2011), who also showed that elevated offspring production among the mothers of *fa'afafine* was not an artifact of the fraternal birth order effect. More recently, VanderLaan et al. (2012) extended these findings by demonstrating that *fa'afafine's* maternal and paternal grandmothers exhibit elevated offspring production, compared to those of gynephilic men. However, elevated reproductive output by the maternal and paternal aunts of *fa'afafine* was not observed (VanderLaan et al. 2012).

Elevated reproductive output by androphilic males' maternal aunts, paternal aunts, or both, would provide the clearest support for the Sexually Antagonistic Gene Hypothesis because androphilic and gynephilic male probands do not share genes with their aunts' male reproductive partners. Nevertheless, the cumulative weight of this evidence suggests that the Sexual Antagonistic Gene Hypothesis is still a tenable explanation for the evolution of male androphilia. Future research is needed, however, to ascertain whether group differences remain nonexistent for maternal and paternal aunts when using a larger sample. In addition, future research will be needed to assess whether VanderLaan et al.'s (2012) finding of group differences for maternal and paternal grandmothers can be replicated.

Apart from the fact that these studies were conducted in a population in which transgendered male androphilia predominates, another major strength of this Samoan-based research is that it examined female reproductive output in a population with a high fertility rate (Central Intelligence Agency 2012). Consequently, anomalous reproductive patterns should be less likely to occur in the Samoan population, compared to lower-fertility Western populations where similar research has been conducted. If the Samoan population is relatively free of susceptibility to anomalous reproductive patterns compared to Western populations, then the study by VanderLaan et al. (2012) indicates that male androphilia is associated with elevated reproductive output in both the maternal and the paternal lines. This is not the case for some of the research that has been presented from certain Western populations (e.g., Camperio-Ciani et al. 2004; Iemmola and Camperio Ciani 2009; Camperio Ciani and Pellizzari 2012; Rahman et al. 2008). On the basis of our Samoan research, it seems reasonable to argue that sexually antagonistic genetic factors are present on the autosomal chromosomes because androphilic males share genetic factors on these chromosomes with both paternal and maternal relatives. Indeed, autosomal linkage of sexually antagonistic genetic factors favoring the evolution of male androphilia is plausible given previously reported mathematical models of sexually antagonistic selection for the evolution of male androphilia (Gavrilets and Rice 2006).

Tests of the Over-Dominance Hypothesis in Samoan Fa'afafine

Another balancing selection hypotheses for male androphilia—the Over-Dominance Hypothesis—takes as its starting point the assumption that male androphilia is not an isolated trait, but rather, is part of a larger package of gender-atypical traits (Miller 2000).[10] Ample empirical evidence exists to support this assumption (Bailey and Zucker 1995; Bartlett and Vasey 2006; Cardoso 2005, 2009; Lippa 2005; VanderLaan et al. 2011; Vasey and Bartlett 2007; Whitam 1983; Zucker et al. 1996). Miller (2000) proposed that multiple genes influence the development of male androphilia and these genes shift male brain development in a female-typical direction. Males who inherit a critical number of these genes become androphilic. Below

[10] This hypothesis is also referred to as the "Balanced Polymorphism Hypothesis" for male androphilia.

this critical threshold, males who inherit some of these genes are gynephilic, but are feminized in terms of certain personality traits, which render them more sensitive, empathetic, tender, and kind. These personality traits, in turn, are thought to render gynephilic males more attractive as mates. Indeed, ample empirical evidence exists to support this assumption (e.g., Barclay 2010; Buss et al. 1990; Buss and Shackelford 2008; Phillips et al. 2008; Tessman 1995). Owing to their increased attractiveness, Miller (2000) argues that these males obtain more female sexual partners and father more children compared to gynephilic males who have no androphilic male relatives. These males are also hypothesized to be better fathers compared to fathers with no androphilic male relatives. The increased reproductive success experienced by the gynephilic male relatives of androphilic males favors selection for the feminizing genes in question. As such, positive selection for these genes occurs despite the reproductive costs associated with male androphilia, itself.

A number of predictions flow from the Over-Dominance Hypothesis. First, androphilic men are more likely to be feminine than masculine. Second, gynephilic males should be more feminine if they have androphilic male relatives, compared to those who do not. Third, gynephilic males should be more attractive if they have androphilic male relatives, compared to those who do not. Fourth, gynephilic males should obtain more female sexual partners if they have androphilic male relatives, compared to those who do not. Fifth, gynephilic males should father more children if they have androphilic male relatives, compared to those who do not. Sixth, gynephilic males should be better fathers if they have androphilic male relatives, compared to those that do not.

To date, only one study has been conducted by our research group that provides a test of the Over-Dominance Hypothesis in Samoa. VanderLaan et al. (2012) found that both the maternal and paternal uncles of Samoan *fa'afafine* did not differ from those of Samoan gynephilic males in terms of offspring production. As such, the research conducted in Samoa provides no support, at present, for the Over-Dominance Hypothesis.

Tests of the Kin Selection in Samoan Fa'afafine

The Kin Selection Hypothesis holds that genes for male androphilia could be maintained in a population if enhancing one's indirect fitness offset the cost of not reproducing directly (Wilson 1975). Indirect fitness is a measure of an individual's impact on the fitness of kin (who share some identical genes by virtue of descent), weighted by the degree of relatedness (Hamilton 1963). Theoretically speaking, androphilic males could increase their indirect fitness by directing altruistic behavior toward kin, which, in principle, would allow those kin to increase their reproductive success. In particular, androphilic males should allocate altruistic behavior toward close kin because they share more genes in common with such individuals.

In formulating this theory, Wilson (1975) stated that "Freed from the special obligations of parental duties, they [androphilic males] could have operated with special efficiency in assisting close relatives" (p. 555). Similarly, Ruse (1982) commented that "...the effect is that in being homosexual, offspring become altruistic towards close relatives in order thereby to increase their own overall inclusive fitness" (p. 20). Given that what is at issue here is a theory that can account for the origin of same-sex sexual attraction, it seems reasonable to interpret these statements as indicating that same-sex sexual attraction, itself, is a prerequisite for the expression of elevated kin-direct altruism, not childlessness. If so, then male androphiles should exhibit elevated kin-directed altruism, whereas male gynephiles (childless or otherwise) should not. Such a pattern would be consistent with the notion that male androphilia is a specially designed adaptation for promoting kin-directed altruism.

Research conducted on transgendered male androphiles in Samoa has repeatedly furnished support for the Kin Selection Hypothesis. Research demonstrates that the avuncular (uncle-like) tendencies of *fa'afafine* are significantly elevated compared to those of Samoan gynephilic males (VanderLaan and Vasey 2012; Vasey et al. 2007; Vasey and VanderLaan 2010a). *Fa'afafine* also exhibited significantly elevated avuncular tendencies compared to the materteral (aunt-like) tendencies of Samoan women (Vasey and VanderLaan 2009). Elevated avuncular tendencies among *fa'afafine* were also documented when comparing them to control groups of childless women and gynephilic men (Vasey and VanderLaan 2009, 2010a). These latter comparisons indicated that the *fa'afafine's* elevated avuncular tendencies cannot be characterized as a simple by-product that is due to a lack of parental care responsibilities and, thus, greater availability of resources for avuncular investment. If this were true, then the avuncular tendencies of *fa'afafine* should be similar to those of childless men and women, but this was not the case. Moreover, these same findings indicate that the elevated avuncular tendencies of *fa'afafine* could not be characterized as a simple by-product that is due to the male members of this "third" gender group adopting feminine gender roles, which included expectations for elevated childcare. If this were true, then the materteral tendencies of Samoan mothers and childless women should be similar to the avuncular tendencies of *fa'afafine*, but again, this was not the case.

We have also demonstrated that *fa'afafine's* avuncular tendencies are significantly higher than their altruistic interest in non-kin children (Vasey and VanderLaan 2010b). As such, *fa'afafine's* elevated avuncular tendencies are not a by-product of general altruistic interest in all children. If this were true, the *fa'afafine's* avuncular tendencies toward nieces and nephews and their altruistic tendencies toward non-kin children would be similar, but this was not the case. This same research also demonstrates that *fa'afafine's* self-reports of elevated avuncular tendencies cannot be explained away as a desire by members of this group to appear more socially virtuous than women or gynephilic men. If this were the case, then one would expect *fa'afafine* to also report that they had elevated altruistic interest in non-kin children, but this was not the case. In fact, the three groups did not differ from each other in this regard.

Additional research indicates that *fa'afafine* exhibit similar levels of sexual/romantic relationships involvement compared to Samoan women and gynephilic men (VanderLaan and Vasey 2012). As such, the *fa'afafine's* relatively elevated avuncular tendencies cannot be characterized as a simple by-product of their failure to form, and invest in, intimate sexual/romantic relationships, which, in turn, leaves them with more time and resources. If that were true, *fa'afafine* should exhibit reduced levels of sexual/romantic relationships involvement compared to men and women, but once again this was not the case.

It should be clear from the research described above that much of our work has focused on falsifying the Kin Selection Hypothesis for male androphilia by examining alternative explanations that might account for the *fa'afafine's* elevated avuncularity. It should be equally clear that none of the alternative explanations we have tested, to date, have been supported. Taken together this body of work is consistent with the conclusion that elevated avuncularity by *fa'afafine* is an adaptation that evolved via kin selection. That being said, establishing that a given trait is an adaptation involves repeatedly satisfying adaptive design criteria empirically while simultaneously ruling out alternatives (Buss et al. 1998). Adaptive design implies complexity, economy, efficiency, reliability, precision, and functionality (Williams 1966).

We have conducted several studies that indicate that compared to Samoan women and gynephilic men, the avuncular cognition of *fa'afafine* appears to be more adaptively designed. First, the avuncular tendencies of the *fa'afafine* are more dissociated from (i.e., co-vary less with) their altruistic interest in non-kin children, compared to Samoan women and gynephilic men (Vasey and VanderLaan 2010b). Such a dissociation would allow *fa'afafine* to channel resources toward nieces/nephews in a more optimal manner (i.e., economical, efficient, reliable, and precise), while minimizing resources directed toward non-kin children. Second, whereas Samoan men and women show a tendency to decrease their willingness to invest in nieces and nephews when they have sexual/romantic relationship partners, the cognition of *fa'afafine* appears to protect against this tendency by maintaining a high level of willingness to invest in nieces and nephews regardless of relationship status (VanderLaan and Vasey 2012). Third, due to the mechanics of human reproduction, individuals can always be certain that their sisters' offspring are their genetic relatives. Yet, due to the possibility of cuckoldry, individuals are necessarily less certain in the case of brothers' offspring. The elevated avuncular tendencies of *fa'afafine* are contingent on the presence of sisters, not brothers, which suggests the avuncular cognition of *fa'afafine* is sensitive to the relative fitness benefits of investing in sisters' versus brothers' offspring (VanderLaan and Vasey 2013). Fourth, compared to gynephlic men and androphilic women, *fa'afafine* are generally better at allocating investment toward indirect fitness-maximizing categories of kin (i.e., sisters' younger daughters) and they do so in a manner that reflects greater sensitivity to non-frivolous versus frivolous investment contexts (VanderLaan and Vasey 2014).

Elevated avuncular tendencies must translate into real-world avuncular behavior if they are to have any impact on the fitness of nieces and nephews and the uncles themselves. Vasey and VanderLaan (2010c) used money given to, and received from, oldest and youngest siblings' sons and daughters as a behavioral assay of expressed kin-directed altruism. In line with the predictions of the Kin Selection Hypothesis, compared to women and gynephilic men, *fa'afafine* gave significantly more money to their youngest siblings' daughters. No other group differences were observed for money given to, or received from, nieces and/or nephews. Moreover, there were no correlations between the number of children parented and monetary exchanges with the niece and nephew categories examined, suggesting, once again, that childlessness cannot account for why *fa'afafine* give more money to their youngest siblings' daughters.

Although quantitative data is lacking, anecdotal evidence suggests that elevated kin-directed altruism characterizes other populations of transgendered male androphiles. For example, Williams (1992, p. 54) quotes a Hupa *berdache* (i.e., a transgendered, androphilic male from the Hupa Valley in Northern California) as saying: "You live your life around your family. My aunt says 'I'm counting on you.' What she means is that someone like me has a special responsibility to help care for the elders." These sorts of statements concerning the focal importance of family for transgendered androphilic males are echoed over and over again in the cross-cultural literature and suggest that elevated attachment and commitment to family is a wide-spread cross-cultural pattern found among such males.

Kin Directed Altruism in an Adaptively Relevant Environment

Adaptively relevant environments (ARE) consist of those features of the environment that must be present in order for an adaptation to be functionally expressed (Irons 1998). Analyses by VanderLaan et al. (2013c) revealed that key aspects of the ARE of transgendered androphilic males likely facilitate elevated kin-directed altruism. For example, relative to non-transgendered societies, transgendered societies are more likely to exhibit bilateral[11] and double descent[12] systems than patrilineal, matrilineal, and ambilineal[13] ones. Ethnologists have argued that bilateral decent systems and bilocal patterns of residence following marriage are maximally inclusive of kin because they do not bias individuals to interact with only one subset of relatives (Alvard 2002; Ember 1975; Kramer and Greaves 2011). Correlational analysis by VanderLaan et al. (2013c) showed that as the presence of ancestral sociocultural conditions increased, so too did the presence of bilateral (and double) descent systems. Consequently, it is reasonable to deduce that, ancestrally, these patterns of descent and post-marital residence would have allowed for more altruistic interactions by transgendered androphlic males with a full range of genetically related kin.

[11] In bilateral descent systems, ego's mother's and father's lineages are equally important for emotional, social, spiritual, and political support, as well as for transfer of property or wealth.

[12] In double descent systems, individuals receive some rights and obligations from the father's side of the family and others from the mother's side.

[13] Some sources treat ambilineal and bilateral descent systems as synonymous, but ambilineal descent systems are defined as existing when individuals have the option of choosing one of their lineages for membership.

The evolution of maximally inclusive kinship systems of descent and residence (i.e., bilateral descent, bilocal residence) would have been contingent on the existence of tribal-level organization (Chapais 2008). Consequently, if kin selection played some role in the evolution of male androphilia within the context of maximally inclusive kinship systems of descent and residence, then tribal-level organization would have been a necessary condition of the sociocultural environment. As Chapais (2008) cogently argued, pair-bonding was a necessary prerequisite for the evolution of tribal-level organization. Some authors have argued that pair-bonding (and presumably tribal-level organization) characterized *Homo erectus* (Wrangham et al. 1999), but others have argued that pair-bonding had not evolved at this stage in the evolution of the genus *Homo* (Hawkes et al. 2003). Given that this debate remains unresolved at present, our conclusions should be taken as representative of *Homo sapiens,* which appear in the fossil record about 195 kya (McDougall et al. 2005), and then only those *Homo sapiens* who exhibited tribal-level organization.

VanderLaan et al. (2013c) also examined the acceptance of homosexuality in 27 transgendered societies for which information could be obtained. The significant majority of these societies expressed no negative reactions to same-sex sexual behavior. Overall then, the same-sex sexual orientation of transgendered males in transgendered societies appears to be socially tolerated. Such tolerance, particularly on the part of the kin of transgendered androphilic males, might be considered essential for kin selection to be deemed as a plausible contributing factor toward the persistence of male androphilia over evolutionary time. Unless transgendered androphilic males are accepted by their families, their opportunity to invest in kin is likely mitigated.

In sum, it is likely that transgendered male androphilia is the ancestral form of this trait, key aspects of the transgendered androphilic male ARE (i.e., bilateral and double descent descent system, social tolerance of same-sex sexuality) would have facilitate elevated kin-directed altruism, and data from contemporary transgendered androphilic males (*fa'afafine*) indicates that they exhibit elevated avuncularity. Given all this, it seems reasonable to suggest that kin selection played some role in the evolution of male androphilia. As such, the elevated kin-directed altruism documented in Samoan *fa'afafine* is more likely to be characteristic of ancestral androphilic males, compared to the lack thereof documented in sex-gender congruent androphilic men from industrialized cultures (e.g., Abild et al., 2014; Bobrow and Bailey, 2001; Rahman and Hull, 2005; Forrester et al. 2011; Vasey and VanderLaan 2012).

Concluding Remarks

In recent years, progress has been made toward understanding how a trait like male androphilia persists over evolutionary time. One of the most important strides in this regard has been the finding that that the ancestral form of male androphilia in humans is likely the transgendered form. In contrast, the sex-gender congruent form of male androphilia is likely to be more derived and may reflect more historically recent, cultural influences. The outcome of evolutionary processes may be obscured when using more derived forms of male androphilia as models. As such, caution needs to be exercised in utilizing sex-gender congruent male androphiles such as “gay men” as models to test hypotheses pertaining to the evolution of male androphilia.

To date, theories pertaining to the evolution of male androphilia have been tested in one population of transgendered male androphiles: the *fa'afafine* of Samoa. In keeping with the predictions of the Sexually Antagonistic Gene Hypothesis, it has been shown that the mothers, maternal grandmothers and paternal grandmothers of *fa'afafine* have more offspring than those of gynephilic males. However, definitive support for this hypothesis, in the form of elevated offspring production among the aunts of *fa'afafine* is lacking at present. In keeping with the predictions of the Kin Selection Hypothesis, it has been repeatedly shown that *fa'afafine* exhibit elevated avuncular tendencies compared to women and gynephilic men. Several studies also suggest that the avuncular cognition of *fa'afafine* exhibits hallmarks of adaptive design.

In light of these results, one potential way that male androphilia could be conceptualized is as a by-product of an adaptation (*sensu* Buss et al. 1998; Gould and Vrba 1982) for increased female fecundity that results from sexually antagonistic selection. By-products of adaptations are characteristics that evolve in association with particular adaptations because they happen to be coupled with those adaptations (Buss et al. 1998). Although they may have some beneficial effect on fitness, they did not originally evolve to solve adaptive problems and, thus, at their point of origin they did not have an evolved fitness-enhancing function, nor were they products of natural selection. In such a situation, increased avuncularity among male androphiles could potentially facilitate reproduction by female kin and thereby have positive “effects” on the genetic factors for both increased fecundity in females and, by extension, its conjectured by-product, male androphilia. Williams (1966) invoked the term “effect” to designate the fortuitous operation of a useful characteristic not built by selection for its current role.

Humans have evolved, via kin selection, to preferentially allocate altruism toward close relatives (e.g., Daly et al. 1997). Consequently, kin nepotism should characterize all individuals, regardless of their sex, sexual orientation, or gender identity. However, markedly elevated avuncularity, such as that observed among *fa'afafine,* might result in distinct fitness advantages that could form a unique basis on which kin selection might act. If so, then the cognitive underpinnings mediating avuncularity in male androphiles may have subsequently undergone *secondary* adaptive modification. Such a conclusion is consistent with our findings that the avuncular cognition of *fa'afafine* exhibits special design features (VanderLaan and Vasey 2012, 2013, 2014; Vasey and VanderLaan 2010b). It is likely that certain features of the ancestral sociocultural environment of transgendered androphilic males, including maximally inclusive descent systems (e.g. double or bilateral descent) and social tolerance of male-male sexuality, would have facilitated this process.

Acknowledgements We thank Resitara Apa, Jean-Baptiste Leca, Nancy Bartlett, Gardenia Elisaia, Vaosa Epa, Vaasatia Poloma Komiti, Anita Latai, Sarah Faletoese Su'a, Vester Fido Collins, Liulauulu Faaleolea Ah Fook, Tyrone Laurenson, Gaualofa Matalavea, Avau Memea, Nella Tavita-Levy, Palanitina Toelupe, Trisha Tuiloma, Avalogo Togi A. Tunupopo, the Kuka family of Savai'i, the Samoan AIDS Foundation, the National University of Samoa, the Samoan Ministry of Health, and the Government of Samoa. We are grateful to all of the individuals who agreed to participate in our studies. We extend special thanks to Alatina Ioelu without whom this research could not have been conducted. Our research on the evolution of male androphilia has taken place over the past decade and has been supported by the University of Lethbridge and a variety of funding agencies. PLV received funding from an Alberta Provincial Government S.T.E.P. Award, an Alberta Innovates Health Solutions (AIHS) Sustainability Fund Grant, a Canadian Institutes of Health Research (CIHR) Catalyst Grant in Methods and Measures for Gender, Sex and Health, three Natural Sciences and Engineering Research Council (NSERC) of Canada Grants, and a Social Sciences and Humanities Research Council of Canada (SSHRC) Grant. DPV received funding from a NSERC of Canada Graduate Scholarship-D3, the Sigma Xi Scientific Research Society Grant-in-Aid-of-Research, a Ralph Steinhauer Award of Distinction, an American Psychological Foundation Henry David Travel Grant, and a Sexual Medicine Society of North America Post-Doctoral Fellowship Stipend.

References

Abild, M., VanderLaan, D.P., & Vasey, P.L. (2014). Does proximity influence the expression of avuncular tendencies in Canadian androphilic males? *Journal of Cognition and Culture, 14,* 40–62.

Alvard, M.S. (2002). Carcass ownership and meat distributions by big game hunters. *Research in Economic Anthropology, 21,* 99–131.

Asthana, S., & Oostvogels, R. (2001). The social construction of male 'homosexuality' in India: Implications for HIV transmission and prevention. *Social Science and Medicine, 52,* 707–721.

Bailey, J.M. (2003). *The man who would be queen: The science of gender-bending and transsexualism*. Washington: Joseph Henry Press.

Bailey, J.M., & Zucker, K.J. (1995). Childhood sex-typed behavior and sexual orientation: A conceptual analysis and quantitative review. *Developmental Psychology, 31,* 43–55.

Bailey, J.M., Dunne, M.P., & Martin, N.G. (2000). Genetic and environmental influences on sexual orientation and its correlates in an Australian twin sample. *Journal of Personality and Social Psychology, 78,* 524–536.

Barclay, P. (2010). Altruism as a courtship display: Some effects of third-party generosity on audience perceptions. *British Journal of Psychology, 101,* 123–135.

Bartlett, N.H., & Vasey, P.L. (2006). A retrospective study of childhood gender-atypical behavior in Samoan fa'afafine. *Archives of Sexual Behavior, 35,* 559–566.

Berling, T. (2001). *Sissyphobia: Gay men and effeminate behavior*. Philadelphia: Harrington Park Press.

Binford, L.R. (2001). *Constructing frames of references: An analytical method for archaeological theory building using hunter-gatherer and environmental data sets*. Berkley: University of California.

Blanchard, R. (2004) Quantitative and theoretical analyses of the relation between older brothers and homosexuality in men. *Journal of Theoretical Biology, 230,* 173–187.

Blanchard, R., & Lippa, R.A. (2007). Birth order, sibling sex ratio, handedness, and sexual orientation of male and female participants in a BBC Internet research project. *Archives of Sexual Behavior, 36,* 163–176.

Bobrow, D., & Bailey, J.M. (2001). Is male homosexuality maintained via kin selection? *Evolution and Human Behavior, 22,* 361–368.

Bogaert, A.F., & Skorska, M. (2011). Sexual orientation, fraternal birth order, and the maternal immune hypothesis: A review. *Frontiers in Neuroendocrinology, 32,* 247–254.

Buss, D.M., & Shakelford, T.K. (2008). Attractive women want it all: Good genes, economic investment, parenting proclivities, and emotional commitment. *Evolutionary Psychology, 6,* 134–146.

Buss, D.M., Abbott, M., Angleitner, A., Asherian, A., Biaggio, A., & Blanco-Villasenor, A. (1990). International preferences in selecting mates: A study of 37 cultures. *Journal of Cross-Cultural Psychology, 21,* 5–47.

Buss, D.M., Haselton, M.G., Shackelford, T.K., Bleske, A.L., & Wakefield, J.C. (1998). Adaptations, exaptations, and spandrels. *American Psychologist, 53,* 533–548.

Camperio Ciani, A., & Pellizzari, E. (2012). Fecundity of paternal and maternal non-parental female relatives of homosexual and heterosexual men. *PLoS ONE, 7*(12), e51088.

Camperio-Ciani, A., Corna, F., & Capiluppi, C. (2004). Evidence for maternally inherited factors favoring male homosexuality and promoting female fecundity. *Proceedings of the Royal Society of London B, 271,* 2217–2221.

Cantor, J.M., Blanchard, R., Paterson, A.D., & Bogaert, A.F. (2002). How many gay men owe their sexual orientation to fraternal birth order? *Archives of Sexual Behavior, 31,* 63–71.

Cardoso, F.L. (2005). Cultural universals and differences in male homosexuality: The case of a Brazilian fishing village. *Archives of Sexual Behavior, 34,* 103–109.

Cardoso, F.L. (2009). Recalled sex-typed behavior in childhood and sports preferences in adulthood of heterosexual, bisexual, and homosexual men from Brazil, Turkey, and Thailand. *Archives of Sexual Behavior, 38,* 726–736.

Central Intelligence Agency. (2012). The World Factbook. https://www.cia.gov/library/publications/the-worldfactbook/rankorder/2127rank. html. Accessed Nov 31, 2012.

Chapais, B. (2008). *Primeval kinship: How pair-bonding gave birth to human society*. Cambridge: Harvard University Press.

Clottes, J., & Lewis-Williams, D. (1998). *The shamans of prehistory: Trance and magic in the painted caves*. New York: Harry N. Abrams.

Crapo, R.H. (1995). Factors in the cross-cultural patterning of male homosexuality: A reappraisal of the literature. *Cross-Cultural Research, 29,* 178–202.

Daly, M., Salmon, C., & Wilson, M. (1997). Kinship: The conceptual hole in psychological studies of social cognition and close relationships. In J. A. Simpson & D. Kenrick (Eds.), *Evolutionary social psychology* (pp. 265–296). Mahwah: Erlbaum.

Deacon, H.J. (1999). *Human beginnings in South Africa: Uncovering the secrets of the stone age*. CA: Altamira Press.

Dean, L.G., Vale, G.L., Laland, K.N., Flynn, E., & Kendal, R.L. (2014). Human cumulative culture: A comparative perspective. *Biology Review, 89,* 284–301.

Devereux, G. (1937). Institutionalized homosexuality of the Mohave Indians. *Human Biology, 9,* 498–527.

Ehrlich, P.R. (2000). *Human natures: Genes, cultures, and the human prospect*. Washington: Island Press.

Ember, C.R. (1975). Residential variation in hunter-gatherers. *Cross-Cultural Research, 10,* 199–227.

Forrester, D.L., VanderLaan, D.P., Parker, J., & Vasey, P.L. (2011). Male sexual orientation and avuncularity in Canada: Implications for the kin selection hypothesis. *Journal of Culture and Cognition, 11,* 339–352.

Gavrilets S., & Rice W.R. (2006) Genetic models of homosexuality: Generating testable predictions. *Proceedings of the Royal Society of London, 273,* 3031–3038.

Gorer, G. (1966). *The danger of equality.* London: Cresset.

Gould, S.J., & Vrba, E.S. (1982). Exaptation—a missing term in the science of form. *Paleobiology, 8,* 4–15.

Hamilton, W.D. (1963). The evolution of altruistic behavior. *American Naturalist, 97,* 354–356.

Harrington, J.P. (1942). Culture element distributions 19: Central California coast. *Anthropological Records, 7*(1), 1–46.

Hawkes, K., O'Connell, J.E., & Blurton Jones, N.G. (2003). Grandmothers and the evolution of human longevity. *American Journal of Human Biology, 15,* 380–400.
Herdt, G. (1981). *Guardians of the flutes*. New York: McGraw-Hill.
Herdt, G. (1997). *Same sex, different cultures.* Colorado: Westview Press.
Hewlett, B. & Hewlett, B.L. (2010). Sex and searching for children among Aka foragers and Ngandu farmers of Central Africa. *African Studies Monographs, 31,* 107–125.
Hill, K.R., Walker, R.S., Bozicevic, M., Eder, J., Headland, T., Hewlett, B., Hurtado, A.M., Marlow, F., Wiessner, P., & Wood, B. (2011). Co-residence patterns in hunter-gatherer societies show unique human social structure. *Science, 331,* 1286–1289.
Hollimon, S.E. (1997). The third gender in California: Two-spirit undertakers among the Chumash, their neighbors. In C. Claassen & R. A. Joyce (Eds.), *Women in prehistory: North America and Mesoamerica,* (pp. 173–188). Philadelphia: University of Pennsylvania Press.
Iemmola, F., & Camperio Ciani, A. (2009). New evidence of genetic factors influencing sexual orientation in men: Female fecundity increase in the maternal line. *Archives of Sexual Behavior, 38,* 393–399.
Irons, W. (1998). Adaptively relevant environments versus the environment of evolutionary adaptiveness. *Evolutionary Anthropology, 6,* 194–204.
Johnson, M., Jackson, P., & Herdt, G. (2000). Critical regionalities and the study of gender and sexual diversity in South East and East Asia. *Culture, Health, and Sexuality, 2,* 361–375.
Kendler, K.S., Thornton, L.M., Gilman, S.E., & Kessler, R.C. (2000). Sexual orientation in a U. S. national sample of twin and nontwin sibling pairs. *American Journal of Psychiatry, 157,* 1843–1846.
King, M.D., Green, J., Osborn, D.P.J., Arkell, J., Hetherton, J., & Pereira, E. (2005). Family size in white gay and heterosexual men. *Archives of Sexual Behavior, 34,* 117–122.
Klein, R. G. (1999). *The human career: Human biological and cultural origins*. Chicago: University of Chicago Press.
Kramer, K.L., & Greaves, R.D. (2011). Postmarital residence and bilateral kin associations among hunter-gatherers: Pumé foragers living in the best of both worlds. *Human Nature, 22,* 41–63.
Kroeber, A.L. (1925). *Handbook of the Indians of California.* Bureau of American Ethnology Bulletin 78.
Långström, N., Rahman, Q., Carlström, E., & Lichtenstein, P. (2010). Genetic and environmental effects on same-sex sexual behavior: A population study of twins in Sweden. *Archives of Sexual Behavior, 39,* 75–80.
Larco Hoyle, R. (1998). *Arte erotico en el antigue Peru.* Lima: Museo Arqueologico Rafael Larco Herrera.
Lippa, R.A. (2005). *Gender, nature, nurture* (2nd ed.). Mahwah: Erlbaum.
Marlowe, F.W. (2005). Hunter-gatherers and human evolution. *Evolutionary Anthropology, 14,* 54–67.
McBrearty, S. & Brooks, A.S. (2000). The revolution that wasn't: A new interpretation of the origins of modern human behavior. *Journal of Human Evolution, 39,* 453–563.
McDougall, I., Brown, F. H., & Fleagle, J. G. (2005). Stratigraphic placement and age of modern humans from Kibish, Ethiopia. *Nature, 433,* 733–736.
Miller, E.M. (2000). Homosexuality, birth order, and evolution: Toward and equilibrium reproductive economics of homosexuality. *Archives of Sexual Behavior, 29,* 1–34.
Murray, S.O. (2000). *Homosexualities.* Chicago: The University of Chicago Press.
Nanda, S. (1999). *Gender diversity: Cross-cultural variations.* Long Grove: Waveland Press.
Namaste, V. (2000). *Invisible lives: The erasure of transsexual and transgendered people*. Chicago: University of Chicago Press.
Nash, G. (2001). The subversive male: homosexual and bestial images on European mesolithic rock art. In L. Bevan (Ed.), *Indecent exposure: Sexuality, society and the archaeological record.* (pp. 43–55). Glasgow: Cruithne Press.
Phillips, T., Barnard, C., Ferguson, E., & Reader, T. (2008). Do humans prefer altruistic mates? Testing a link between sexual selection and altruism toward nonrelatives. *British Journal of Psychology, 99,* 555–572.

Prine, E. (2000). Searching for third genders: Toward a prehistory of domestic space in Middle Missouri villages. In R. A. Schmidt, & B. L. Voss (Eds.), *Archaeologies of sexuality* (pp. 197–219). London: Routledge.

Rahman, Q., Collins, A., Morrison, M., Orrells, J.C., Cadinouche, K., Greenfield, S., et al. (2008). Maternal inheritance and familial fecundity factors in male homosexuality. *Archives of Sexual Behavior, 37,* 962–969.

Rahman, Q., & Hull, M.S. (2005). An empirical test of the kin selection hypothesis for male homosexuality. *Archives of Sexual Behavior 34,* 461–467.

Rieger, G., & Savin-Williams, R.C. (2012). Gender nonconformity, sexual orientation, and psychological well-being. *Archives of Sexual Behavior, 41,* 611–621.

Rubin, A.J. (2000). Public more accepting of gays, poll fines. *Los Angeles Times*, June 19.

Ruse, M. (1982). Are there gay genes? Sociobiology and homosexuality. *Journal of Homosexuality, 6,* 5–34.

Sanderson, S.K., & Roberts, W.W. (2008). The evolutionary forms of the religious life: A cross-cultural, quantitative analysis. *American Anthropologist, 110,* 454–466.

Schwartz, G., Kim, R.M., Kolundziji, A.B., Rieger, G., & Sanders, A.R. (2010). Biodemographic and physical correlates of sexual orientation in men. *Archives of Sexual Behavior, 39,* 93–109.

Seil, D. (1996). Transsexuals: The boundaries of sexual identity and gender. In R.P. Cabaj & T.S. Stein (Eds.), *Textbook of homosexuality and mental health* (pp. 743–62). Washington: American Psychiatric Press.

Shear, K., Jin, R., Ruscio, A.M., Walters, E.E., & Kessler, R.C. (2006). Prevalence and correlates of estimated DSM-IV child and adult separation anxiety disorder in the national comorbidity survey replication. *American Journal of Psychiatry, 163,* 1074–1083.

Shimkin, D.D. (1983). Introduction of the horse. In D'Azevedo W.L. (Ed.), *Great Basin: Vol. 11. Handbook of North American Indians* (pp. 517–524). Washington: Smithsonian Institution Press.

Smith, A.B. (1999). Archaeology and the evolution of hunter-gatherers. In R.B. Lee & R. Daly (Eds.), *The Cambridge encyclopedia of hunters and gatherers* (pp. 384–390). Cambridge: Cambridge University Press.

Smith, A.M., Rissel, C.E., Richters, J., Grulich, A.E., & de Visser, R.O. (2003). Sex in Australia: Sexual identity, sexual attraction, and sexual experience among a representative sample of adults. *Australian and New Zealand Journal of Public Health, 27,* 138–145.

Tessman, I. (1995). Human altruism as a courtship display. *Oikos, 74,* 157–158.

Whitam, F.L. (1983). Culturally invariant properties of male homosexuality: Tentative conclusions from cross-cultural research. *Archives of Sexual Behavior, 12,* 207–226.

Wilson, E.O. (1975). *Sociobiology: The new synthesis*. Cambridge: Belknap Press.

VanderLaan, D.P., & Vasey, P.L. (2011). Male sexual orientation in Independent Samoa: Evidence for fraternal birth order and maternal fecundity effects. *Archives of Sexual Behavior, 40,* 495–503.

VanderLaan, D.P., & Vasey, P.L. (2012). Relationship status and elevated avuncularity in Samoan fa'afafine. *Personal Relationships, 19,* 326–339.

VanderLaan, D.P., & Vasey, P.L. (2013). Birth order and avuncular tendencies in Samoan men and *fa'afafine*. *Archives of Sexual Behavior, 42,* 371–379.

VanderLaan, D.P. & Vasey, P.L. (2014). Evidence of enhanced cognitive biases for maximizing indirect fitness in Samoan fa'afafine. *Archives of Sexual Behavior, 43,* 1009–1022.

VanderLaan, D.P., Gothreau, L., Bartlett, N.H., & Vasey, P.L. (2011). Recalled separation anxiety and gender atypicality in childhood: A study of Canadian heterosexual and homosexual men and women. *Archives of Sexual Behavior, 40,* 1233–1240.

VanderLaan, D.P., Forrester, D.L., Petterson, L.J., & Vasey, P.L. (2012). Offspring production among the extended relatives of Samoan men and *fa'afafine*. *PLoS ONE, 7*(4), e36088.

VanderLaan, D.P., Forrester, D.L., Petterson, L.J., & Vasey, P. L. (2013a). The prevalence of *fa'afafine* relatives among Samoan men and *fa'afafine*. *Archives of Sexual Behavior, 42,* 353–359.

VanderLaan, D.P., Vokey, J.R., & Vasey, P.L. (2013b). Is male androphilia familial in non-Western cultures? The case of a Samoan village. *Archives of Sexual Behavior, 42,* 361–370.
VanderLaan, D.P., Ren, Z., & Vasey, P.L. (2013c). Male androphilia in the ancestral environment: An ethnological analysis. *Human Nature, 24,* 375–401.
Vasey, P.L., & Bartlett, N.H. (2007). What can the Samoan *fa'afafine* teach us about the Western concept of "Gender Identity Disorder in Childhood"? *Perspectives in Biology and Medicine, 50,* 481–490.
Vasey, P.L., & VanderLaan D.P. (2007) Birth order and male androphilia in Samoan *fa'afafine*. *Proceedings of the Royal Society of London B, 274,* 1437–1442.
Vasey, P.L., & VanderLaan, D.P. (2009). Materteral and avuncular tendencies in Samoa: A comparative study of women, men and *fa'afafine*. *Human Nature, 20,* 269–281.
Vasey, P.L., & VanderLaan, D.P. (2010a). Avuncular tendencies in Samoan *fa'afafine* and the evolution of male androphilia. *Archives of Sexual Behavior, 39,* 821–830.
Vasey, P.L., & VanderLaan, D.P. (2010b). An adaptive cognitive dissociation between willingness to help kin and non-kin in Samoan *fa'afafine*. *Psychological Science, 21,* 292–297.
Vasey, P.L., & VanderLaan, D.P. (2010c). Monetary exchanges with nieces and nephews: A comparison of Samoan men, women, and *fa'afafine*. *Evolution and Human Behavior, 31,* 373–380.
Vasey, P.L., & VanderLaan, D.P. (2012). Male sexual orientation and avuncularity in Japan: Implications for the Kin Selection Hypothesis. *Archives of Sexual Behavior, 41,* 209–215.
Vasey, P.L., & VanderLaan, D.P. (2014). Evolutionary perspectives on male androphilia in humans. In V. Weekes-Shackelford & T. Shackelford (Eds.), *Evolutionary perspectives on human sexual psychology and behavior* (pp. 369–392). New York: Springer.
Vasey, P.L., Pocock, D.S., & VanderLaan, D.P. (2007). Kin selection and male androphilia in Samoan *fa'afafine*. *Evolution and Human Behavior*, *28*, 159–167.
Vasey, P.L., VanderLaan, D.P., Gothreau, L., & Bartlett, N.H. (2011). Traits of separation anxiety in childhood: A comparison of Samoan men, women and *fa'afafine*. *Archives of Sexual Behavior, 40,* 511–517.
Vasey, P.L., Parker, J.L., & VanderLaan, D.P. (2014). Comparative reproductive output of Samoan men and fa'afafine. *Archives of Sexual Behavior, 43,* 363–367.
Vaughan, R. (1994). *The Arctic: A history*. Dover: Sutton.
Wilson, E.O. (1975). *Sociobiology: The new synthesis*. Cambridge: Belknap Press.
Williams, G.C. (1966). *Adaptation and natural selection*. Princeton: Princeton University Press.
Williams, W.L. (1992). *The spirit and the flesh: Sexual diversity in American Indian culture*. Boston: Beacon Press.
Winkelman, M. (2010). *Shamanism: A biopsychosocial paradigm of consciousness and healing*. Santa Barbara: Praeger.
Woodburn, J. (1982). Egalitarian societies. *Man, 17,* 431–451.
Wrangham, R. W., Jones, J. H., Laden, G., Pilbeam, D., & Conklin-Brittain, N. (1999). The raw and the stolen: Cooking and the ecology of human origins. *Current Anthropology, 40,* 567–594.
Zucker, K.J., Bradley, S.J., & Sullivan, C.B.L (1996). Traits of separation anxiety in boys with gender identity disorder. *Journal of the American Academy of Child and Adolescent Psychiatry, 35,* 791–798.

Chapter 10
Two Behavioral Hypotheses for the Evolution of Male Homosexuality in Humans

Austin John Jeffery

Two Behavioral Hypotheses for the Evolution of Male Homosexuality in Humans

Homosexuality poses a challenge to evolutionary science. It appears to undermine the logic of reproductive success as the ultimate source of human psychological traits. This challenge rests on four observations: (1) homosexuality is heritable (Mustanski et al. 2005; Pillard and Bailey 1998), (2) homosexuality reduces fecundity (Bell and Weinberg 1978; King et al. 2005), (3) homosexuality is prevalent (Bagley and Tremblay 1998; Johnson et al. 1992; Sell et al. 1995), and (4) homoeroticism is ancient (Crompton 2006). A heritable trait that reduces fecundity is a trait that usually experiences negative selection, eventually being removed from the population. In this chapter, I present two mutually exclusive behavioral hypotheses for the maintenance of male homosexuality in the human population, following an introduction to the psychological study of male homosexuality.

Choosing an appropriate operational definition for homosexuality has challenged researchers for decades (Sell 1997), but three metrics are most often employed: sexual attraction, sexual behavior, and sexual orientation identity (Savin-Williams 2006). Comparing across these metrics and across time, sexual orientation in both sexes is more fluid than the common perception of fixed identities suggests (Diamond 2014; Savin-Williams 2006). In studies that attempt to quantify homosexuality, many respondents report overlapping and inconsistent sexual identities, behaviors, and attractions (Bogaert 2004; Wells et al. 2011; Santtila et al. 2008; Savin Williams 2006). For example, the majority of same-sex attracted individuals in the USA do not identify as homosexual (Laumann et al. 1994). Sixty percent of self-identified homosexual men report experiencing sexual attraction to the opposite sex in the past 12 months, and 25 % of heterosexuals report experiencing same-sex attraction in the past 12 months (Diamond 2014). In a longitudinal interview

---

A. J. Jeffery (✉)
Department of Psychology, Oakland University, Oakland, USA
e-mail: ajjeffer@oakland.edu

T. K. Shackelford, R. D. Hansen (eds.), *The Evolution of Sexuality*,
Evolutionary Psychology, DOI 10.1007/978-3-319-09384-0_10

study of 13,305 high school students, of the boys who reported exclusive same-sex attraction in the first year of the study, only 11 % reported exclusive same-sex attraction in the second year, while 48 % reported exclusive opposite-sex attraction and 35 % reported no attraction to either sex (Udry and Chantala 2004).

The history of homosexuality as we know it in the West as a distinct and visible identity is only about one and a half centuries old, with the coinage of the term "homosexual" and the emergence of a large-scale cultural movement in protest of German anti-sodomy statutes arriving in 1869 (Beachy 2014). If psychological science allows the modern cultural categories of sexual identity to dictate theory, we will be ill-equipped to probe the ancestral functions and contexts that underlie the traits associated with homosexuality. By analogy, it would be unsatisfactory to "explain the Caucasian" without direct reference to the individually heritable, functional, and continuous traits associated with Caucasoids (e.g., reduced pigmentation, straightened hair, lengthened nose, etc.). The homosexual construct (like race) is a cultural heuristic, not a precise diagnostic. Replacing categorical notions of sexual orientation, the term *androphilia* will be used to describe sexual attraction to adult men and *gynephilia* will be used to describe sexual attraction to adult women (VanderLaan et al. 2013). Most sexual orientations can thus be defined as some combination of androphilic and gynephilic interests, the majority of which skew gynephilic in men and androphilic in women. Using this definition, *asexuals* possess neither androphilia nor gynephilia, and *bisexuals* possess some degree of both.

Relative to heterosexual men, self-identifying homosexual men demonstrate reduced gynephilia, in both self-reported and physiological measures of arousal (Cerny and Janssen 2011; Lippa 2013). Most evolutionary perspectives on this fact regard it as the inalienable reproductive cost associated with the genetic benefits of feminization (see Miller 2000). For example, the female relatives of androphilic men show greater fecundity than the female relatives of gynephilic men, in some samples (Camperio-Ciani et al. 2004). This work suggests that the genes that produce androphilia in men facilitate greater fecundity in the women who carry those genes. In this view, called *sexually antagonistic selection*, that men carry and express these genes is a nonadaptive byproduct of selection for genes that build adaptations in women. Alternatively, *overdominance* describes the possibility that homosexuality is reported by men who carry homozygous-recessive alleles, which when expressed in the more common heterozygotic form, confer selective advantages (e.g., more cooperative, more lingual, and less aggressive traits) (Gavrilets and Rice 2006; Zietsch et al. 2008). Here, androphilic men represent the tail of a distribution experiencing selection for feminine traits. Lastly, *kin selection* allows androphilic men to facilitate the transmission of the genes that contribute to homosexuality by providing alloparental care to close relatives who share those genes (Vasey et al. 2007). By improving the reproductive success of kin, the androphilic man may recoup the cost of their reduced participation in reproductive sex. Yet, reduced gynephilia again represents a significant and unexplained sacrifice of reproductive potential attached to an ancestrally adaptive advantage. These views assume a stable linkage between reduced gynephilia and ancestrally advantageous traits, ignoring the possibility that reduced gynephilia itself provides reproductive advantages.

Evolutionary science offers explanations for the presence of increased androphilia in some men (same-sex affiliation and sperm competition are two examples; Baker 1996, 1997; Kirkpatrick 2000; Muscarella 2000), but the relative absence of gynephilia is rarely addressed explicitly. Same-sex behavior has been reported in hundreds of nonhuman species, from birds and mammals to fish and insects (Bagemihl 1999). However, abstinence from heterosexual contact is rare among nonhumans; instead, bisexuality predominates (Bagemihl 1999, p. 50, 53). The reported absence of gynephilia in a majority of adult men who identify as homosexual (Gebhard 1972) does not align with our understanding of male reproductive strategies. In the logic of Trivers (1972) and Buss and Schmitt (1993), the maximum possible male reproductive success is achieved through reproductive frequency, because the minimum necessary investment in each offspring is nearly negligible. Taken in its simplest form, this logic predicts the continual escalation of male gynephilic attraction and intrasexual competition. No obvious reproductive benefits are gained through a relative or total disinterest in reproductive sex. The two arguments outlined below build from an adaptive *de-escalation* of direct intrasexual competition, such that men with reduced gynephilia have reduced motivation to participate in direct competition for sexual access to women. The first strategy follows a slow life-history marked by mate choosiness and committed parenting. The second strategy follows a fast life-history involving sneak copulations and elevated sperm competition. Throughout the discussion, it is essential to keep the ancestral human environment in view—the arguments made herein describe two possible evolutionary histories of homosexuality.

Paternal Investment Hypothesis

Gynephilia motivates heterosexual copulation in men. The more powerful an individual's gynephilic sexual arousal, the more frequently he can be expected to reproduce, all else equal. I speculate that *reduced* gynephilia in men facilitates more deliberate long-term mate-choice, while signaling heterosexual fidelity and long-term mate quality to women. In other words, men who find women less attractive will be less motivated to enter into frequent, hasty sexual relationships with women. Simultaneously, women will find those men who display reduced gynephilic interest to be more attractive as long-term partners, as it suggests low heterosexual promiscuity and the possession of long-term mate qualities (e.g., willingness to invest, parenting ability, low aggression). The primary prediction of this hypothesis is that men will be disinterested in sex with women to the extent that they are predisposed to high levels of paternal care and long-term mate investment. Men who are predisposed to committed investment in offspring cannot afford to be as gynephilic as men who are less disposed to such commitment. Powerful gynephilia motivates men to acquire new sexual relationships, and their ability to invest in any single offspring is reduced with each new reproductive partner. Reduced gynephilia thus complements a long-term, slow life history strategy, marked by reduced fecundity and increased offspring investment. The presence of androphilia in these men and the potential for homosexual infidelity may serve other purposes, but more importantly it does not entail the same dangers as heterosexual infidelity. First, the possibility of extra-pair offspring is removed, freeing the man to enjoy sex without parental consequences. Second, the probability of successfully concealing the affair is improved by the assumption of heterosexuality; the partner is less likely to suspect a homosexual infidelity and perhaps less likely to leave the man if it is discovered (Baker 1996).

To some extent, this strategy undermines the default heterosexual strategy, as the presence of men willing to commit to high paternal care in communities where women have control over their sex lives would draw women's attention away from men pursuing short-term strategies. In this and the following hypothesis, homophobia is interpreted as a cultural adaptation among masculine heterosexuals attempting to identify and degrade men pursuing sexual strategies that would disrupt male-dominance strategies. This will be discussed in greater detail in the final section.

Sex-atypical behavior, personality traits, and physical appearance are closely associated with homosexuality in men (Bailey and Zucker 1995; Lippa and Arad 1997; Udry and Chantala 2006). Consequently, the present hypothesis predicts that women will prefer sex-atypical (feminine) traits in men in long-term contexts. There is evidence that this is the case. Men with feminine facial features are perceived to be warm, kind, honest (Perrett et al. 1998), good long-term partners (Little et al. 2002), and good parents (Kruger 2006). Women have been shown to find men with androgynous personalities (i.e., both highly masculine and feminine) most attractive in all contexts, with feminine characteristics being more attractive than masculine characteristics in long-term mating contexts (Green and Kenrick 1994). There is also evidence that women who choose more feminine men subsequently benefit from improved offspring care. Psychologically feminine men in heterosexual couples spend more time, and a higher percentage of total parenting time, performing day-to-day care and engaging in play with their children than psychologically masculine men (Russell 1978). Gay men anticipate being better fathers than do heterosexual men, and gay men report more rapid gains in parenting skill when they become fathers than do heterosexual men (Goldberg and Smith 2009). Finally, adoptive gay fathers report lower levels of depression and parenting stress, and towards their children they are more interactive, express more warmth, are more responsive, and demonstrate less disciplinary aggression than do heterosexual adoptive parents (Golombok et al. 2014). The present hypothesis implies that freedom from the short-term, high gynephilia mating strategy allows men to be choosey in mate selection and more dedicated, effective fathers.

For this hypothesis to work, the possibility of a sexual relationship between a man with reduced gynephilia and a woman will eventually depend on his interest in heterosexual copulation. The data on sexual fluidity presented in the introduction suggests that this shift is not unlikely. Diamond (2003) offers reason to believe that sexual activities do develop out of platonic friendships of this kind. She points to the chemistry of affection, noting that oxytocin informs the positive feelings of both mutual friendships and sexual bonds, making the introduction of sex into a friendship a matter of escalation rather than renaissance. Once trust is established, copulation can become an open possibility. Diamond (2014) reports that 78 % of adult men who initially come out as homosexual subsequently change their sexual identity (to bisexual, pansexual, unsure, etc.). Accordingly, 20 % of white American men who identify as homosexual report having been married to a woman at some time in their life, and 50 % of white American men who identify as homosexual report having produced at least one child (Bell and Weinberg 1978). The present hypothesis considers the assumption that self-identified homosexual men abstain from reproductive sex to be unwarranted.

One of the most robust correlates of homosexuality in men is the number of older brothers in the family; each older brother, but not older sister, increases the probability of an individual man identifying as homosexual by about 33 % (Blanchard and Bogaert 1996). This is referred to as the fraternal birth order effect. The mechanism of this effect may involve a reduction in prenatal androgen exposure (Hines 2011) or a maternal immunological response (Bogaert and Skorska 2011) producing changes in the developing brain. Regardless of the mechanism, mothers effectively feminize later-born sons. This effect may be explained by the ancestral reproductive futility of later-born boys who practice a highly gynephilic, intrasexually competitive strategy, as they are likely to compete with older, stronger brothers for sexual access to local women. A more effective strategy for a later-born boy may be to exercise a less immediately competitive, long-term strategy, experiencing relatively low gynephilia and thus allowing older brothers to pursue local mating opportunities uncontested. The proposed adaptation for diversified sexual strategies between brothers resides in the mother's prenatal environment—it is ultimately her inclusive fitness that benefits from the reduced gynephilia of later-born sons. In other words, the mother benefits from attenuating the degree to which her sons directly compete with one another for sexual access by predisposing them to differently-timed reproductive careers.

Sneak Copulation Hypothesis

Sperm competition research in humans provides a growing body of evidence for adaptations in human mating psychology and morphology designed to augment the probability of fertilization in multiply-mated females (Baker and Bellis 1995). Here I offer the hypothesis that reduced male gynephilia corresponds with a sexual strategy marked by sneak copulations and enhanced sperm competitive physiology. Reduced gynephilia and elevated androphilia may complement an ability in some males to find high-risk, low-cost reproductive opportunities through their associations with other males, by discretely inseminating females who affiliate with these target males. Reduced gynephilia would allow the male to enter into these relationships without being quickly motivated to compete for sexual access with local females; he is instead perceived by the target male(s) to be sexually nonthreatening, perhaps subordinate. Furthermore, in the effort to parasitize mating opportunities, increased androphilia benefits the male by motivating him to seduce and stimulate the target male, who will be more likely to tolerate his presence and less likely to inseminate his female partner (being sexually satisfied by the male). By sexually preoccupying the target male, he reduces the degree of sperm competition he will need to overcome to fertilize the female. The finding that self-identified homosexuals prefer sex-typical same-sex partners (masculine men and feminine women) may be explained by the greater likelihood that these partners can connect them to opposite-sex partners than more sex-atypical same-sex partners (feminine men and masculine women) (Bailey et al. 1997). More than the paternal investment strategy, this strategy undermines the common, competitive, gynephilic strategy. The sneak copulater has the opportunity to inseminate one or more females and escape, letting the target male invest in offspring that he incorrectly believes to be his genetic offspring. The pattern of feminized male sneak copulaters who engage in homosexual sex and circumnavigate the costs of direct competition is observed in several other species (see Oliveira et al. 2008, for review). To provide perspective on this hypothesis, one example is provided.

Among males of the lekking Ruff shorebird, there are three permanent morphs: independents, satellites, and faeders (female mimics). Independents are large and directly compete to dominate the lek territory. Satellites are moderately sized, orbit the lek, and attempt to copulate with incoming females without directly competing with independents. Faeders are small and resemble females; they enter the lek and allow themselves to be mounted by independents and satellites, while covertly mounting and inseminating females drawn to the lek (Jukema and Piersma 2006). This homosexual activity serves several purposes. On the one hand, it causes independents and satellites to spend reproductive effort on the faeder, directly reducing their reproductive success at the lek; the sexual activity serves to attract more females to the lek, as lek commotion and popularity stimulates female attention; it also allows faeders to remain on the lek, without being ousted by the dominant independents, granting them continued sexual access to arriving females (Jukema and Piersma 2006). The same adaptive logic that builds categorical strategic polymorphisms, as in the Ruff, can build continuous polymorphic blends, as may be present in humans; the major difference is in the number of genes involved (Gavrilets and Rice 2006; MacIntyre and Estep 1993). The purpose of the Ruff example is to affirm the logic of alternative strategies, not to suggest a discrete dimorphism or trimorphism of sexual strategies in humans. In species for which mating strategies, like sneak copulation, are permanent (nonconditional), the population distribution of individuals practicing these alternative strategies exhibits negative frequency-dependent selection (Dominey 1984). This means that an individual pursuing sneak copulations is the most successful when sneak copulaters are uncommon, becoming less successful as the strategy becomes more common. This results in a stable equilibrium point, a balanced polymorphism at which the average reproductive success of sneak copulaters matches the average reproductive success of more common strategies. The persistent, low rate of homosexuality across cultures might qualify as just such an equilibrium. Although self-identified homosexuals represent a minority of individuals who ever engage in sneak copulation, this hypothesis suggests that they are genetically equipped to specialize in this form of reproduction.

Because they do not defend mates and secure long-term access to mating opportunities, faeder Ruffs are under greater pressure to fertilize females in the few mating opportunities they have. Additionally, they can expect to be inseminating females who will be concurrently inseminated by other males in the lek. Because of this, faeders have testes that are 2.5 times the size of the testes of other males, allowing faeders to produce larger, higher-quality ejaculates to outcompete local rivals in the race for fertilization (Jukema and Piersma 2006). Larger testes and higher-quality ejaculates correspond with ancestrally heightened risk of sperm competition, between species with different mating systems and within species containing different sexual strategies (Dixson and Anderson 2004; Simmons et al. 2004). If men who possess reduced gynephilia and increased androphilia have larger testes, it may suggest an evolutionary history of heightened sperm competition associated with these traits. Additionally, there is reason to believe that penis size and shape reflects sperm competitive demands, as the penis can effectively displace and remove rival semen from the vagina (Gallup 2003). Larger penises among self-identified homosexuals have been documented (Bogaert and Hershberger 1999), but only one analysis of testes size has been published, conducted on a small clinical sample of 30 self-identified homosexual men (several of whom were azoospermic; Kolodny et al. 1971). The report found these men to have testes within the normal size range for heterosexuals. Baker (1997) has produced data demonstrating that men with homosexual experience produce smaller ejaculates than men without homosexual experience; however, the sample of men with homosexual experience consisted of just five men. At this time, there is very little data available to directly test the primary prediction that men with low gynephilia have enhanced sperm competitive traits.

It is possible that the two strategies outlined above are not mutually exclusive; in fact, they may represent conditional strategies, activated in the contexts that best facilitated ancestral reproductive success. If both hypotheses generate predictions that find empirical support, the present framework allows for a more nuanced interpretation of the traits associated with homosexuality. For example, it may be that in higher population densities with higher quality of life, the sneak copulatory strategy delivers a greater reproductive payoff, whereas in smaller communities with limited resources, the paternal investment strategy delivers a greater payoff. These context-dependent strategies may be settled during development or in the course of adult life; testes size is a relatively inflexible trait but the activation or deactivation of paternal investment may depend on proximal factors. Fathers of any sexuality with larger testes are, in fact, found to report lower levels of parental care and also show reduced brain activity in regions linked with nurturance and proactive care when viewing images of their own children (Mascaro et al. 2013). The negative correlation between testes size and parental care matches the predictions of life-history theory and suggests that sperm competitive physiology and paternal investment may be distinct developmental pathways. It remains to be seen if the same relationship holds in a nonheterosexual sample.

Identity and Exclusivity

On the seven point Kinsey scale (0 = exclusively heterosexual and 6 = exclusively homosexual) the majority of unmarried men expressing predominant homosexual attraction (scoring 4–6) consider themselves exclusively homosexual (scoring 6) (Gebhard 1972). In other words, men's reported sexual orientations skew towards exclusivity. Without invoking reporting bias or scale inaccuracy, this finding is difficult to integrate into the present framework of homosexuality as a heterosexual strategy; i.e., exclusive homosexuality is never an effective reproductive strategy. My response to this criticism is to make note of the cultural influence of the categorical sexual identity. I hypothesize that the high rate of reported exclusive homosexuality is an artifact of cultural pressures levied against androphilic men to practice and espouse exclusive homosexuality. These pressures may originate from adaptations in heterosexual men targeting men who practice mating strategies that undermine common heterosexual strategies, i.e., strategies not predicated on high gynephilia and direct intrasexual competition (sneak copulation, paternal investment, or otherwise). Almost nowhere in the nonhuman literature is it found that homosexual activity elicits the special attention, condemnation, or segregation demonstrated in humans (Bagemihl 1999, p. 54). The elaborate culture surrounding sexuality in humans has had a unique effect on how sex is perceived and practiced. I will argue that it is responsible, in part, for the reported exclusivity of gay men.

The common public perception of nonexclusive sexuality (bisexuality) is that it represents an immature stage in the progression towards exclusivity, with bisexual men more often perceived to be "really homosexual" and bisexual women more often perceived to be "really heterosexual" (McLean 2007; Yost and Thomas 2012). The process of "coming out" is a well-studied rite of passage among Western homosexuals (Mosher 2001; Vargo 1998). A number of models propose a linear, staged process of overcoming fear, guilt, and doubt, culminating with a public declaration of one's new identity as a homosexual (this is sometimes called the "disclosure imperative"; Cass 1979; Coleman 1982; Dank 1971; McLean 2007). These models include an early recognition of attraction followed by a period of uncertainty and questioning, after which the individual must decide if they are homosexual. The linear progression from questioning to identity confirmation implies a fixed sexuality trajectory, set off in men by the experience of androphilia. Several studies report that bisexuals are perceived more negatively than homosexuals—as more promiscuous, dishonest, and immature (Eliason 1997; Herek 2002; Steffens and Wagner 2004; Yost and Thomas 2012). Heterosexual men show low tolerance for bisexual men and homosexual men, yet bisexual women are more tolerated (in part because they are sexualized). Heterosexual women show low tolerance for bisexual men and women, yet homosexual men are more tolerated (Eliason 1997; Steffens and Wagner 2004). It may be that heterosexual women feel that they can better trust homosexual men than bisexual men, as homosexual men disavow heterosexual intentions. I propose two mechanisms by which these common sentiments influence the sexual self-identification of androphilic men: homophobic bullying and the lesbian gay bisexual transgender queer (LGBTQ) movement.

Bullying is consistently linked with homophobic epithets and violence (Espelage and Swearer 2008; Poteat and Espelage 2005). One study reported that 26 % of bullied boys in a sample of 251 high school students indicated that they were abused by their peers for being perceived as gay. These boys received more severe and frequent bullying and experienced greater psychological distress than peers bullied for other reasons (e.g., getting good grades, wearing certain clothes, etc.; Swearer et al. 2008). A review of 28 school shootings between 1982 and 2001 indicated that the majority of gunmen (always young men) experienced intense, homophobic bullying, yet "from all available evidence, *none* of the school shooters was gay" (Kimmel and Mahler 2003, p. 1449). Espelage et al. (2008) found that students who were sexually questioning received the highest rates of bullying; they were more likely than self-identified heterosexuals *and* homosexuals to be teased and experience suicidal thoughts, abuse drugs and alcohol, and negatively rate their social environment. One interpretation of this form of bullying is that it assigns a homosexual identity to these young men. By labeling his sexually-questioning or bisexual peers as gay, the bully may influence their sexual self-perceptions, pushing them to publicly identify as exclusively homosexual. Supporting this interpretation, adults with more severe childhood histories of homophobic bullying (precipitating PTSD symptoms) "came out" at younger ages and grew up to be *more accepting* of their homosexual identity than less abused individuals (Rivers 2004). Rivers argues that the peer reinforcement of homophobic teasing encourages identity acceptance. I speculate that coming out as homosexual may be protective against the threat of bullying; it may placate the aggressive young men who would benefit from the social castration of their sexual rivals. The psychological and physical violence directed at young androphilic men by bullies may be alleviated with the avowal of homosexuality, the declaration of categorical noncompetition in the heterosexual marketplace.

Another possibility is that the LGBTQ community places a premium on what is anecdotally referred to as a "gold star gay," an individual who has never had sex with the opposite sex and promises that he or she never will (Queen 1999). Self-identified homosexual men who score high on scales measuring LGBTQ community involvement and identification report fewer sexual experiences with women in the past 6 months than self-identified homosexual men who report low community involvement and identification (Vanable et al. 1998, p. 408). This suggests that greater immersion in the LGBTQ community causes men to practice more exclusive homosexuality or bias their reporting to appear more exclusive. Of course, it may be that men who are more exclusively androphilic simply identify more strongly with the LGBTQ community, as they fit less comfortably in the larger heterosexual culture.

I submit that the discrediting of nonexclusive homosexual attraction and the special enmity and distrust targeting bisexual men motivates androphilic men to subdue and conceal their gynephilic tendencies and announce exclusive homosexuality. Men who advertise their homosexuality, who are "out and proud," will be loath to practice the heterosexual mating strategies that would otherwise undermine the predominant heterosexual strategy of direct intrasexual competition. If androphilia aligns with a sneak copulation strategy in men, for example, a culture celebrating conspicuous gay pride and abstinence from heterosexual contact, while cautioning against the dishonesty of bisexuals and unidentified nonheterosexuals, would effectively disarm potential sneak copulators. The homosexual identity may act as a reputational passport, with the relief and psychological benefits of coming out (Juster et al. 2013) coinciding with an escape from the obscurity and intolerance of mere nonheterosexuality, into a welcoming community of sexual refugees who promote identification and solidarity. It would not be controversial to suggest that the LGBTQ community established itself in response to the homophobic forces that perpetrate violence and discrimination against nonheterosexuals (Beachy 2014). What I am suggesting is that the homosexual identity was invented to placate these destructive social forces, that many exclusively homosexual men are the product of an androphilic genetic profile combined with a heterosexual culture that does not tolerate alternative sexual strategies.

A man who considers himself a homosexual today, had he lived in an ancestral environment, may have never applied such a label to himself, nor advertised it to his family and friends. Do we expect that he would have suffered from an inability to mature sexually? Would he be stunted in his sexual expression, stifling a deeper identity? I argue that he would have no need for an intellectually imposed sexual schema, living contentedly with his natural androphilic and gynephilic arousals and repulsions. In many such modern men, low levels of gynephilic interest may be sublimated in the cultural process of homosexual identity formation. Likewise, among heterosexuals, low levels of androphilic interest may be sublimated in the process of heterosexual identity formation (Adams et al. 1996).

References

Adams, H. E., Wright, L. W., & Lohr, B. A. (1996). Is homophobia associated with homosexual arousal? *Journal of Abnormal Psychology, 105,* 440–445.

Bagemihl, B. (1999). *Biological exuberance: Animal homosexuality and natural diversity.* New York: St. Martin's Press.

Bagley, C., & Tremblay, P. (1998). On the prevalence of homosexuality and bisexuality, in a random community survey of 750 men aged 18 to 27. *Journal of Homosexuality, 36,* 1–18.

Bailey, J. M., & Zucker, K. J. (1995). Childhood sex-typed behavior and sexual orientation: A conceptual analysis and quantitative review. *Developmental Psychology, 31,* 43–55.

Bailey, J. M., Kim, P. Y., Hills, A., & Linsenmeier, J. A. (1997). Butch, femme, or straight acting? Partner preferences of gay men and lesbians. *Journal of Personality and Social Psychology, 73,* 960–973.

Baker, R. R. (1996). *Sperm wars: Infidelity, sexual conflict, and other bedroom battles.* London: Fourth Estate.

Baker, R. R. (1997). Copulation, masturbation, and infidelity: State-of-the art. In A. Schmitt, K. A., Atzwanger, K. K., Grammer, & K. Schäfer (Eds.), *New aspects of human ethology* (pp. 163–188). New York: Plenum Press.

Baker, R. R., & Bellis, M. A. (1995). *Human sperm competition: Copulation, masturbation, and infidelity.* London: Chapman & Hall.

Beachy, R. (2014). The German invention of homosexuality. *The Journal of Modern History, 82,* 801–838.

Bell, A. P., & Weinberg, M. S. (1978). *Homosexualities*. New York: Simon and Schuster.

Blanchard, R., & Bogaert, A. F. (1996). Homosexuality in men and number of older brothers. *The American Journal of Psychiatry, 153,* 27–31.

Bobrow, D., & Bailey, J. M. (2001). Is male homosexuality maintained via kin selection? *Evolution and Human Behavior, 22,* 361–368.

Bogaert, A. F. (2004). The prevalence of male homosexuality: The effect of fraternal birth order and variations in family size. *Journal of Theoretical Biology, 230,* 33–37.

Bogaert, A. F., & Hershberger, S. (1999). The relation between sexual orientation and penile size. *Archives of Sexual Behavior, 28,* 213–221.

Bogaert, A. F., & Skorska, M. (2011). Sexual orientation, fraternal birth order, and the maternal immune hypothesis: A review. *Frontiers in Neuroendocrinology, 32,* 247–254.

Buss, D., & Schmitt, D. (1993). Sexual strategies theory: An evolutionary perspective on human mating. *Psychological Review, 100,* 204–232.

Camperio-Ciani, A., Corna, F., & Capiluppi, C. (2004). Evidence for maternally inherited factors favouring male homosexuality and promoting female fecundity. *Proceedings of the Royal Society, Biological Sciences, 271,* 2217–2221.

Cass, V. C. (1979). Homosexual identity formation, a theoretical model. *Journal of Homosexuality, 4,* 219–235.

Cerny, J., & Janssen, E. (2011). Patterns of sexual arousal in homosexual, bisexual, and heterosexual men. *Archives of Sexual Behavior, 40,* 687–697.

Coleman, E. (1982). Developmental stages of the coming-out process. *American Behavioral Scientist, 25,* 469–482.

Crompton, L. (2006). *Homosexuality and civilization*. Boston: Harvard University Press.

Dank, B. (1971). Coming out in the gay world. *Psychiatry, 34,* 180–197.

Diamond, L. M. (2003). What does sexual orientation orient? A biobehavioral model distinguishing romantic love and sexual desire. *Psychological Review, 110,* 173–192.

Diamond, L. M. (2014). *I was wrong! Men are pretty darn sexually fluid, too*. PowerPoint presentation at the annual meeting of the Society for Personality and Social Psychology, Sexuality Preconference, Austin, Texas.

Dixson, A. F., & Anderson, M. J. (2004). Sexual behavior, reproductive physiology and sperm competition in male mammals. *Physiology & Behavior, 83,* 361–371.

Dominey, W. (1984). Alternative mating tactics and evolutionarily stable strategies. *American Zoologist, 24,* 385–396.

Eliason, M. J. (1997). The prevalence and nature of biphobia in heterosexual undergraduate students. *Archives of Sexual Behavior, 26,* 317–326.

Espelage, D. L., & Swearer, S. M. (2008). Addressing research gaps in the intersection between homophobia and bullying. *School Psychology Review, 37,* 155–159.

Espelage, D. L., Aragon, S. R., Birkett, M., & Koeng, W. (2008). Homophobic teasing, psychological outcomes, and sexual orientation among high school students: What influence do parents and schools have? *School Psychology Review, 37,* 202–216.

Gallup, G. (2003). The human penis as a semen displacement device. *Evolution and Human Behavior, 24,* 277–289.

Gavrilets, S., & Rice, W. R. (2006). Genetic models of homosexuality: Generating testable predictions. *Proceedings of the Royal Society, Biological Sciences, 273,* 3031–3038.

Gebhard, P. H. (1972). *Incidence of overt homosexuality in the United States and Western Europe.* National institute of mental health task force on homosexuality: Final report and background papers, pp. 22–29.

Goldberg, A. E., & Smith, J. Z. (2009). Perceived parenting skill across the transition to adoptive parenthood among lesbian, gay, and heterosexual couples. *Journal of Family Psychology Psychology, 23,* 861–870.

Golombok, S., Mellish, L., Jennings, S., Casey, P., Tasker, F., & Lamb, M. E. (2014). Adoptive gay father families: Parent-child relationships and children's psychological adjustment. *Child Development, 85*, 456–468.

Green, B. L., & Kenrick, D. T. (1994). The attractiveness of gender-typed traits at different relationship levels: Androgynous characteristics may be desirable after all. *Personality and Social Psychology Bulletin, 20,* 244–253.

Herek, G. M. (2002). Heterosexuals' attitudes toward bisexual men and women in the United States. *Journal of Sex Research, 39,* 264–274.

Hines, M. (2011). Prenatal endocrine influences on sexual orientation and on sexually differentiated childhood behavior. *Frontiers in Neuroendocrinology, 32,* 170–182.

Johnson, A., Wadsworth, J., Wellings, K., Bradshaw, S., & Field, J. (1992). Sexual lifestyles and HIV risk. *Nature, 360,* 410–412.

Jukema, J., & Piersma, T. (2006). Permanent female mimics in a lekking shorebird. *Biology Letters, 2,* 161–164.

Juster, R. P., Smith, N. G., Ouellet, É., Sindi, S., & Lupien, S. J. (2013). Sexual orientation and disclosure in relation to psychiatric symptoms, diurnal cortisol, and allostatic load. *Psychosomatic Medicine, 75,* 103–116.

Kimmel, M. S., & Mahler, M. (2003). Adolescent masculinity, homophobia, and violence: Random school shootings, 1982–2001. *The American Behavioral Scientist, 46,* 1439–1458.

King, M., Green, J., Osborn, D. P. J., Arkell, J., Hetherton, J., & Pereira, E. (2005). Family size in white gay and heterosexual men. *Archives of Sexual Behavior, 34,* 117–122.

Kirkpatrick, R. C. (2000). The evolution of human homosexual behavior. *Current Anthropology, 41,* 385–413.

Kolodny, R., Masters, W., Hendryx, J., & Toro, G. (1971). Plasma testosterone and semen analysis in male homosexuals. *New England Journal of Medicine, 285,* 1170–1174.

Kruger, D. J. (2006). Male facial masculinity influences attributions of personality and reproductive strategy. *Personal Relationships, 13,* 451–463.

Laumann, E. O., Gagnon, J., Michael, R. T., & Michaels, S. (1994). *The social organization of sexuality: Sexual practices in the United States*. Chicago: University of Chicago Press.

Lippa, R. A. (2013). Men and women with bisexual identities show bisexual patterns of sexual attraction to male and female swimsuit models. *Archives of Sexual Behavior, 42,* 187–96.

Lippa, R. A., & Arad, S. (1997). The structure of sexual orientation and its relation to masculinity, femininity, and gender diagnosticity: Different for men and women. *Sex Roles, 37,* 187–208.

Little, A. C., Jones, B. C., Penton-Voak, I. S., Burt, D. M., & Perrett, D. I. (2002). Partnership status and the temporal context of relationships influence human female preferences for sexual dimorphism in male face shape. *Proceedings of the Royal Society, Biological Sciences, 269,* 1095–1100.

MacIntyre, F., & Estep, K. W. (1993). Sperm competition and the persistence of genes for male homosexuality. *Bio Systems, 31,* 223–233.

Mascaro, J. S., Hackett, P. D., & Rilling, J. K. (2013). Testicular volume is inversely correlated with nurturing-related brain activity in human fathers. *Proceedings of the National Academy of Sciences, 110,* 15746–15751.

McLean, K. (2007). Hiding in the closet? Bisexuals, coming out, and the disclosure imperative. *Journal of Sociology, 43,* 151–166.

Mosher, C. M. (2001). The social implications of sexual identity formation and the coming-out process: A review of the theoretical and empirical literature. *The Family Journal, 9,* 164–173.

Miller, E. M. (2000). Homosexuality, birth order, and evolution: Toward an equilibrium reproductive economics of homosexuality. *Archives of Sexual Behavior, 29,* 1–34.

Muscarella, F. (2000). The evolution of homoerotic behavior in humans. *Journal of Homosexuality, 40,* 51–77.

Mustanski, B. S., Dupree, M. G., Nievergelt, C. M., Bocklandt, S., Schork, N. J., & Hamer, D. H. (2005). A genomewide scan of male sexual orientation. *Human Genetics, 116,* 272–278.

Oliverira, R., Taborsky, M., & Brockmann, H. J. (Eds.). (2008). *Alternative reproductive tactics: An integrative approach*. Cambridge: Cambridge University Press.

Perrett, D., Lee, K., & Penton-Voak, I. (1998). Effects of sexual dimorphism on facial attractiveness. *Nature, 394,* 884–887.
Pillard, R., & Bailey, J. (1998). Human sexual orientation has a heritable component. *Human Biology, 70,* 347–365.
Poteat, V. P., & Espelage, D. L. (2005). Exploring the relation between bullying and homophobic verbal content: The homophobic content agent target (HCAT) scale. *Violence and Victims, 20,* 513–528.
Queen, C. (1999). Strangers at home: Bisexuals in the queer movement. In L. P Gross & J. D. Voods (Eds.), *The Columbia reader on lesbians and gay men in media, society, and politics* (pp. 105–108). New York: Columbia University Press.
Rivers, I. (2004). Recollections of bullying at school and their long-term implications for lesbians, gay men, and bisexuals. *Crisis: The Journal of Crisis Intervention and Suicide Prevention, 25,* 169–175.
Russell, G. (1978). The father role and its relation to masculinity, femininity, and androgyny. *Child Development, 49,* 1174–1181.
Santtila, P., Sandnabba, N. K., Harlaar, N., Varjonen, M., Alanko, K., & von der Pahlen, B. (2008). Potential for homosexual response is prevalent and genetic. *Biological Psychology, 77,* 102–105.
Savin-Williams, R. (2006). Who's gay? Does it matter? *Current Directions in Psychological Science, 15,* 40–44.
Sell, R. L. (1997). Defining and measuring sexual orientation: A review. *Archives of Sexual Behavior, 26,* 643–658.
Sell, R., Wells, J., & Wypij, D. (1995). The prevalence of homosexual behavior and attraction in the United States, the United Kingdom and France: Results of national population-based samples. *Archives of Sexual Behavior, 24,* 235–248.
Simmons, L. W., Firman, R. C., Rhodes, G., & Peters, M. (2004). Human sperm competition: Testis size, sperm production, and rates of extrapair copulations. *Animal Behaviour, 68,* 297–302.
Steffens, M. C., & Wagner, C. (2004). Attitudes toward lesbians, gay men, bisexual women, and bisexual men in Germany. *Journal of Sex Research, 41,* 137–149.
Swearer, S. M., Turner, R. K., Givens, J. E., & Pollack, W. S. (2008). "You're so gay!": Do different forms of bullying matter for adolescent males? *School Psychology Review, 37,* 160–173.
Trivers, R. (1972). Paternal investment and sexual selection. In B. Campbell (Ed.), *Sexual selection and the descent of man* (pp. 136–179). New York: Aldine de Gruyter.
Udry, J. R., & Chantala, K. (2004). Risk factors differ according to same-sex and opposite-sex interest. *Journal of Biosocial Science, 37,* 481–497.
Udry, J. R., & Chantala, K. (2006). Masculinity-femininity predicts sexual orientation in men but not in women. *Journal of Biosocial Science, 38,* 797–809.
Vanable, P. A., McKirnan, D. J., & Stokes, J. P. (1998). Identification and involvement with the gay community scale. In C. M. Davis, W. L. Yarber, R. Bauserman, G. E. Schreer, & S. L. Davis (Eds.), *Handbook of sexuality-related measures* (pp. 407–409). Thousand Oaks: Sage.
VanderLaan, D. P., Ren, Z., & Vasey, P. L. (2013). Male androphilia in the ancestral environment: An ethnological analysis. *Human Nature, 24,* 375–401.
Vargo, M. (1998) *Acts of disclosure: The coming-out process of contemporary gay men.* New York: Harrington Park Press.
Vasey, P. L., Pocock, D. S., & VanderLaan, D. P. (2007). Kin selection and male androphilia in Samoan fa'afafine. *Evolution and Human Behavior, 28,* 159–167.
Wells, J. E., McGee, M., & Beautrais, A. L. (2011). Multiple aspects of sexual orientation: Prevalence and sociodemographic correlates in a New Zealand national survey. *Archives of Sexual Behavior, 40,* 155–168.
Yost, M. R., & Thomas, G. D. (2012). Gender and binegativity: Men's and women's attitudes toward male and female bisexuals. *Archives of Sexual Behavior, 41,* 691–702.
Zietsch, B., Morley, K., Shekar, S., Verweij, K., Keller, M., & Macgregor, S. et al. (2008). Genetic factors predisposing to homosexuality may increase mating success in heterosexuals. *Evolution and Human Behavior, 29,* 424–433.

Chapter 11
The Evolution of Culturally-Variable Sex Differences: Men and Women Are Not Always Different, but When They Are…It Appears *Not* to Result from Patriarchy or Sex Role Socialization

David P. Schmitt

Just like all other sexually reproducing species, male and female humans are more similar than different. Even so, men's and women's psychological traits sometimes differ in important ways, both in terms of typical or average levels (Buss 1989; Del Giudice 2009; Ellis 2011) and in terms of variability (Archer and Mehdikhani 2003; Borkenau et al. 2013; Lippa 2009). Sex differences in numerous traits have been well-established as moderate to large in size[1] and as culturally pervasive. For example, sex differences in negative emotion-related traits have been documented across several meta-analyses (Feingold 1994; Miettunen et al. 2007; Whissell 1996), integrative neuroscientific reviews (Hyde et al. 2008; Stevens and Hamann 2012), and large cross-cultural surveys (Costa et al. 2001; Hopcroft and McLaughlin 2012; Lippa 2009; McCrae and Terracciano 2005; Schmitt et al. 2008; Van de Velde et al. 2010). Using a multivariate approach, Del Giudice et al. (2012) documented across 16 personality traits—ranging from dominance and liveliness to perfectionism and tension—that sex differences in personality are astonishingly large, with only 10% overlap in men's and women's overall distributions.

Beyond sex differences in personality traits, psychologists have uncovered dozens of ways that men and women differ in affect, behavior, and cognition across most cultures (Archer 2014; Browne 1998; Mealey 2000). In one comprehensive review, Ellis (2011) identified 63 psychological sex differences that have been replicated across multiple cultures and at least 10 studies, with not a single replication failure (probably an overly strict exclusionary criterion; see Schmitt et al. 2014). In another wide-ranging review, Archer (2014) reported culturally-pervasive sex differences are reliably found in the assessment of negative emotions (e.g., fear, anxiety, depression), anti-social behaviors (e.g., aggression, violence, criminality), cognitive abilities (e.g., mental rotation, object location,

[1] According to Cohen (1988), effect sizes expressed in terms of the *d* statistic are considered small if 0.20, medium if 0.50, and large above 0.80.

D. P. Schmitt (✉)
Department of Psychology, Bradley University, 75 Bradley Hall, Peoria, IL 61625, USA
e-mail: dps@bradley.edu

T. K. Shackelford, R. D. Hansen (eds.), *The Evolution of Sexuality*,
Evolutionary Psychology, DOI 10.1007/978-3-319-09384-0_11

verbal fluency), personality traits (e.g., agreeableness, neuroticism, sociability), motor activities (e.g., strength, throwing ability, activity level), sexual attitudes and behaviors (e.g., mate preferences, sociosexuality, sexual coercion), and numerous other characteristics such as interest in infants and occupations (see also, Lippa 2005).

The pervasive nature of so many psychological sex differences necessitates serious questions concerning the origin of men's and women's panculturally distinctive psychologies (Brown 1991; Norenzayan and Heine 2005). Among the most likely forces behind culturally ubiquitous sex differences include the specialized design of men's and women's evolved psychological adaptations (Buss 1995; Mealey 2000), universal features of human sex role socialization (which may be generated, in part, by evolved features of gendered sexual psychology; see Maccoby 2000; Pirlott and Schmitt 2014; Williams and Best 1990), and a wide range of other developmental and biocultural factors that produce the profoundly pervasive sex differences exhibited by our species (Geary 1998; Low 2000; Miller and Halpern 2014).

Despite growing evidence that many sex differences are at least partially the result of specially-designed differences in the evolved psychology of men and women (Archer 2014; Buss and Schmitt 2011; Ellis 2011), many contemporary social scientists still assume—in accordance with the Standard Social Science Model (SSSM; see Tooby and Cosmides 1992)—that men's and women's psychological differences, if they exist at all (Hyde 2005), are solely the result of extensive sex role socialization processes and sociopolitical power differentials (Eagly and Wood 1999). As Wood and Eagly (2002) assert, "it is likely that extensive socialization is required to orient boys and girls to function differently" (p. 705). Given this blank slate approach to psychological sex differences, it is unsurprising that social role theorists further assume that sex differences should be conspicuously smaller in cultures with more egalitarian sex role socialization or greater sociopolitical gender equity (e.g., greater representation of women in parliament; see Kasser and Sharma 1999; Wood and Eagly 2002; Zentner and Mitura 2012). Indeed, there can be no more ironclad prediction than the "demise of many sex differences with increasing gender equality is a prediction of social role theory" (Eagly et al. 2004, p. 289).

In this chapter, evidence is marshalled across 21 data sources that directly challenge this foundational assumption of social role theory and, more generally, the SSSM. In fact, most psychological sex differences—in personality, sexuality, attitudes, and cognitive abilities—are conspicuously *larger* in cultures with more egalitarian sex role socialization and greater sociopolitical gender equity. Even sex differences in physical traits such as height, body mass index, obesity, and blood pressure are larger in cultures with more egalitarian sex role socialization and greater sociopolitical gender equity (Schmitt et al. 2014). In order to explain these counter-intuitive patterns, it is helpful to understand four basic evolutionary perspectives on the generation of psychological sex differences.

Obligate Sex Differences: Culturally Insensitive Sex-Specific Adaptations

There are four basic ways that evolved psychological adaptations generating sex differences across cultures (see Fig. 11.1). The first is for men and women to be adaptively designed to manifest relatively uniform sex differences across cultures. Evolutionary theories such as Sexual Selection Theory (Darwin 1871) and Parental Investment Theory (Trivers 1972) suggest that, as mammals, men and women likely pursue somewhat different mating strategies and at least some of the adaptive design of men's and women's evolved mating psychologies will be robustly manifested across cultures (Lippa 2005; Schmitt 2005a). According to Lippa (2009), sex differences in sex drive demonstrate a particularly persistent and uniform magnitude across all cultures and are entirely unrelated to factors such as degree of sex role socialization and sociopolitical gender equity. Much like human sex differences in physical strength and stature (Deaner et al. 2012; Gaulin and Boster 1985; Puts 2010; Van Damme et al. 2008), it is not the case that all men must have higher sex drives than all women for the sex difference to have resulted from evolved psychological adaptations. Nor must evolved sex differences be present at birth, a common misconception (Voyer et al. 2007; Wood and Eagly 2012). Instead, sex differences in traits such as sex drive and physical strength likely result from obligate sex-specific adaptations consistently generating, on average and to about the same degree, observable sex differences across all cultural forms (or at least across all cultures with similar ecologies; see also Baumeister et al. 2001).

According to the neuroandrogenic theory of psychological sex differences (Ellis 2001), a key mechanism that generates panculturally uniform sex differences is the organizational effect of prenatal sex hormone exposure on the developing human brain (see also Baron-Cohen 2004). In a sweeping review of the literature, Ellis (2011) identified 63 psychological sex differences that are culturally universal and likely result from human males (but not females) experiencing testosterone-related brain masculinization between 12 and 22 weeks of gestation (Archer 2014; Baron-Cohen 2002; Ellis 2011). It is not that all 63 sex differences—ranging from personality traits to work preferences to consuming behavior—are immediately present at birth. Instead, the prenatal masculinization (or not) of the human brain adaptively biases the development of future psychological traits, particularly in terms of risk-taking (Byrnes et al. 1999), sensation-seeking (Cross et al. 2011), and systematizing-versus-empathizing (Baron-Cohen 2004).

Supportive findings of this perspective on sex differences come from studies of girls exposed to male-typical levels of testosterone in utero (compared to their unaffected sisters) developing more male-typical personalities and play behaviors (Alexander et al. 2009; Auyeung et al. 2009). Indeed, prenatal testosterone exposure within normal levels also predicts sexually differentiated childhood behavior in girls and boys (Hines 2006; Udry et al. 1995), including dose-dependent relationships between degree of testosterone exposure and male-typical behavior (Nordenström et al. 2002), results that are not explainable from measured degree of parental sex

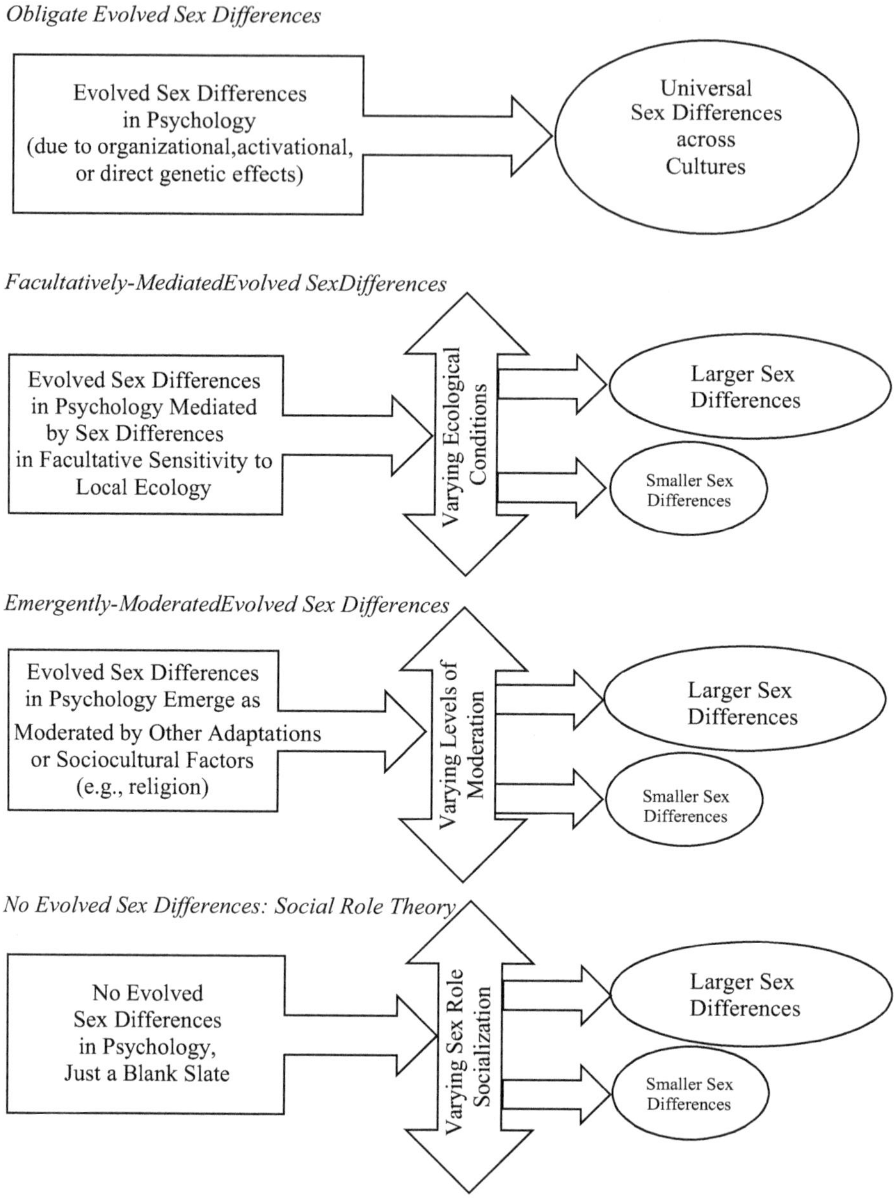

Fig. 11.1 Four approaches to evolved sex difference variation across cultures

role socialization (Pasterski et al. 2005). Further support of neuroandrogenic theory includes findings that girls with gene variants that make them more sensitive to testosterone develop more masculine personalities (with especially strong effects among those girls whose mothers tried the hardest to socialize their daughters to be highly feminine; Udry 2000) and studies that show gene variants and biomarkers linked to brain masculinization influence transsexualism in predictably ways in adults (Hare et al. 2009; Schneider et al. 2006). In short, the seeds of many psychological sex differences appear to be sewn before birth, with future developmental experiences being shaped by the degree of prenatal neuroandrogenic brain masculinization (Alexander and Wilcox 2012).

Some psychological sex differences are not established before birth, of course, and instead result from activational effects that emerge in early childhood (Del Giudice and Belsky 2010; Ellis 2004) or at puberty (Galambos et al. 2009; Hyde et al. 2008; Ruigrok et al. 2013). From a life history perspective (Kaplan and Gangestad 2005), just as sex differences in body hair, muscle mass, and voice pitch emerge most strongly at puberty (Puts 2010), many psychological sex differences may be designed to emerge most strongly with the onset of such factors as ambulatory exploration (Silverman et al. 2007) or mating effort (Burke et al. 2014). Other psychological sex differences may result from direct genetic effects in which specific genes outside the sex chromosomes function differently in the brains of men and women (Becker et al. 2007; McCarthy and Arnold 2011; Ngun et al. 2011; Trabzuni et al. 2013).

Regardless, compared to the "ethnographic hyperspace" of all possible sexual cultures one can imagine (Cronk 1999), it appears some psychological adaptations in men and women are obligate enough to generate persistent and relatively uniform psychological sex differences across all cultures (Brown 1991; Gaulin 1997; Karremans et al. 2010; Lippa 2010). Of course, even obligate sex differences are not immutable, as there always exists a continuous interplay of biological and environmental factors that can alter the degree of human sexual differentiation. Sometimes the size of psychological sex differences is variable as a direct result of specially-designed psychological adaptations. That is, sometimes evolution generates culturally-variable sex differences by design.

Facultatively-Mediated Sex Differences: Differential Sensitivity to Local Conditions

A second way evolution generating sex differences across cultures is for men and women to be specially designed with differentially-sensitive facultative adaptations. For instance, men's and women's adaptations may be specially designed to be differentially sensitive to local ecological information. Sometimes men are adaptively designed to be more sensitive to ecological conditions, at other times women are designed to be more sensitive (see Baumeister 2000; Ellis 2004; Schmitt 2011). Critically, it is the differential sensitivity of men's and women's adaptations, combined with varying local ecologies, that facultatively mediates variation in size or degree of psychological sex differences across cultures (see Fig. 11.1).

The notion that evolved biology may be specially designed to anticipate that ecological variability is a foundational component of phenotypic plasticity

(Gomulkiewicz et al. 1995; Pigliucci 2001; Stearns and Koella 1986). Across species, phenotypic plasticity has been implicated in strategies of predator avoidance, physical polymorphisms, the timing of developmental experiences, and alternative reproductive tactics (Kelly et al. 2012; Via et al. 1995). For example, depending on the type of plant a certain caterpillar feeds on, its physical structure changes to match its surroundings (Whitman and Agrawal 2009). Similarly, some species of grasshopper facultatively alter their color depending on local ecology. Grasshoppers in darker ecologies grow a dark body, whereas grasshoppers in green ecologies grow a green body (Burtt 1951). Any differences between green and dark grasshoppers are not genetic differences; they represent the same genome facultatively responding in a mediated way to ecological variation (Nettle 2009).

In humans, phenotypic plasticity (Hughes et al. 2003), Life History Theory (Geary 2002; Kaplan and Gangestad 2005), and the related concept of "evoked culture" (Tooby and Cosmides 1992) have generated a wealth of findings on the adaptive effects of ecological variability on human psychology (Ellis et al. 2009; Del Giudice and Belsky 2010; Griskevicius et al. 2011; Nettle 2009). For example, Gangestad et al. (2006) has shown that both men's and women's long-term mate preference adaptations for health, attractiveness, and intelligence in potential mates are facultatively evoked in cultures with high pathogens. In the realm of mating adaptations, local pathogen levels appear to facultatively mediating a culture's level of polygyny (Low 1990), fertility rate (Guégan et al. 2001), sociosexuality levels (Schaller and Murray 2008), expressed mate preferences (DeBruine et al. 2010; Moore et al. 2013), and degree of parental care (Quinlan 2007).

In this view, human psychological adaptations may be built in a way that generates predictable patterns of cultural variability, but not in an agnostic blank slate way. Instead, human psychological adaptations are designed to pay specific attention to particular sources of ecological information and generate specially designed, highly functional, sex-specific forms of behavior (Gangestad and Simpson 2000; Hartung 1985; Hill et al. 2014; Lancaster 1994; Marcinkowska et al. 2014; Schmitt and Rohde 2013).

Schmitt et al. (2003) found that men are generally higher in dismissing attachment and this sex difference is nearly universal across cultures, but not always (Del Giudice 2011). Sex differences in dismissing attachment become negligible in cultures with high ecological stress (such as having high pathogen levels), in part, due to women's psychology being specially-designed to be more sensitive to stressful ecological contexts (see also Belsky 2012; Schmitt 2011). In other words, both men's and women's attachment psychology reacts to stressful ecologies with greater dismissing attachment, but it is more adaptive for women's "dismissing reaction" psychology to respond more strongly. This greater sensitivity in women's psychology results in women's dismissing attachment levels becoming nearly as high as men's in stressful ecological contexts. From a facultatively-mediated perspective, human cultures—including the degree of sex differences in human cultures—can be both evolved and variable.

Emergently-Moderated Sex Differences: On the Freedom to Express Sex-Specific Psychology

A third way evolution can generate psychological sex differences across cultures for men and women to be adaptively designed to manifest psychological sex differences across cultures (in either obligate or facultative forms), but the full expression of those sex-specific adaptations is moderated by other factors (Massimini and Delle Fave 2000). One example of emergently-moderated sex differences comes in the form of biological suppression. Biological suppression involves the nondevelopment of an adaptation that is normally present in the species.

A prime instigator of adaptation suppression in humans may be the potent cultural factor of religion. For example, the eighteenth century religious sect called the Shakers abolished the practice marriage, insisted upon complete sexual celibacy, and eradicated nearly all physical contact between men and women (Foster 1981). As a consequence, the adaptive expression of most, if not all, evolved sex differences in mating psychology that was greatly suppressed in Shaker culture. Another potent factor in the emergently-moderated suppression of sex differences is ecological stress. For instance, although sex differences in height are largely obligate across cultures, they can be emergently suppressed in cultures with especially poor nutrition and extremely stressful ecological conditions (Guégan et al. 2000; Gustafsson and Lindenfors 2008; Katzmarzyk and Leonard 1998; Nettle 2002). As Gaulin and Boster (1992) noted in their review of sex differences in stature across 155 human societies, "substandard nutrition could cause individuals to fall short of their genetically set growth potential, and, importantly, males seem to be more sensitive to such developmental perturbations than females" (p. 474). Hence, in high stress ecologies sex differences in height can be attenuated.

Another instance of emergently-moderated or suppressed sex differences comes from the "mismatch" perspective in evolutionary psychology (Crawford 1998; Nesse and Williams 1994). Evolutionary mismatch perspectives explain psychological variation across cultures by the degree of mismatch between contemporary environmental conditions and those within which early humans evolved—namely, hunter-gatherer environments (Brown 1991; Tooby and Cosmides 1990). When contemporary environments are different from hunter–gatherer environments in critical ways, the adaptive development of innate psychological sex differences can be suppressed (Schmitt et al. 2008). It is also possible, though, for contemporary environments to accentuate evolved sex differences in an emergently-moderated way.

An illustrative example of the emergently-moderated sex differences approach can be found in cross-cultural studies of human values. Schwartz and Rubel (2005) documented that most sex differences in expressed personal values emerge more strongly in nations with egalitarian sex role socialization and greater sociopolitical gender equity. They note that men and women may have evolved sex differences in certain values (e.g., men's mating psychology may have evolved to value power, achievement, and hedonism more than women; whereas women's tendency toward child-rearing may have evolved to value benevolence and universalism more than men). They speculate that "Increased independence and equality of women in the labor force may encourage them to express distinctive values rather than to accommodate their values to those of their husbands" (p. 1023). In their view, it is possible that in cultures with egalitarian sex role socialization and greater sociopolitical gender equity, men and women are freer to express their evolved, sex-specific psychological adaptations. As they noted in 2009, egalitarian sex role socialization and greater sociopolitical gender equity could "permit both sexes to pursue more freely the values they inherently care about more." (Schwartz and Rubel-Lifshitz 2009, p. 171). It is not that humans are adapted to anticipate this freedom and have special design features that facultatively respond to it, instead it is the case that sex differences in values may be obligate (or perhaps facultative) and the degree of psychological sex difference is an emergently-moderated response this to freedom (see also, Barber 2014).

Noting differences between mediation and moderation effects has led to conceptual advances in social psychology (Baron and Kenny 1986). A key conceptual difference between the facultatively-mediated adaptation approach and the emergently-moderated approach outlined here is that the facultatively-mediated approach views cultural variation in psychological sex differences as adaptive and design-specific—facultative adaptations are designed to interact with only certain ecological factors and to generate only certain functional outcomes. In contrast, the emergently-moderated approach views cultural variation in psychological sex differences as a domain-general byproduct of extraneous factors (Barber 2014). It is possible these moderating extraneous factors involve evolved psychological adaptations, such as adaptations for the generation of religion (Kirkpatrick 2011; Weeden and Kurzban 2013). Moreover, there may be some factors that biologically intensify, rather than suppress, psychological adaptations, such as religions intensifying adaptions involving violence and sexuality (Keller 1990; Sela et al. 2014) and bifurcated sex roles intensifying adaptations that normally generate merely small to moderate psychological sex differences (see Pirlott and Schmitt 2014).

The key conceptual point is that the facultatively-mediated approach, but not the emergently-moderated approach, views cross-cultural variation in sex differences as resulting from adaptations specially-designed for properly generating those very sex difference variabilities. In the emergently-moderated approach, cross-cultural variation in sex differences is merely a functionally-disruptive side effect of some other factor. Finally, there can exist combinations of these first three evolutionary explanations of sex differences, such as obligate sex differences that are emergently-moderated in certain contexts (e.g., height differences suppressed by poor nutrition), facultatively-mediated adaptations that are emergently-moderated in certain contexts (e.g., mate preference differences suppressed by religion), and so forth. For instance, although the variation appearing in local sex ratios is strongly associated with shifts in men's and women's sexuality (Guttentag and Secord 1983), it remains unclear which sex differences in sexuality are facultatively versus emergently responding to local sex ratios (Hudson and Den Boer 2004; Lazarus 2002; Marlowe and Berbesque 2012; Pedersen 1991). Only by clearly specifying the precise mechanisms—obligate, facultative, and emergent—will researchers be able to fully explain the connections between human sex ratio variation and psychological sex differences.

Social Role Theory: Sex Differences Drawn on a Blank Slate

A fourth way psychological sex differences might be generated across cultures is through domain-general learning combined with sex role socialization. According to social role theory, most psychological sex differences result from exposure to sex role socialization, a process whereby a culture defines and enforces the appropriate ways of thinking, feeling, and behaving for men and women (Eagly 1987; Ruble and Martin 1998). For some social role theorists, the motivated origin of these sex roles primarily involves men's patriarchal attempts to subjugate and control women (Dworkin 1987; MacKinnon 1982; Rudman et al. 2013) and the privileging of men via androcentrism (Bem 1993). For others, evolved physical differences between the sexes create pancultural divisions of labor which, in turn, generate emergent sex roles (Alesina et al. 2013; Wood and Eagly 2002). Regardless, it is assumed by social role theorists that sex roles are the most direct cause amongst all observed *psychological* differences between men and women—with the only innate psychology of men and women presumed, by default, to consist of a single blank-slated domain-general learning mechanism (Katz 1995; Kitzinger 1994).

As a consequence, assuming that sex roles are the sole cause of sex differences, social role theorists expect that when men and women occupy more similar roles, sex differences will erode (Eagly and Wood 1999; Wood and Eagly 2002). Without extensive sex role socialization, in this view, there would be no observable psychological sex differences. Indeed, Wood and Eagly (2002) have specifically argued that "it is likely that extensive socialization is required to orient boys and girls to function differently" (p. 705) and Eagly et al. (2004) have asserted the "demise of many sex differences with increasing gender equality is a prediction of social role theory" (p. 289). Thus, the social role approach unambiguously predicts that psychological sex differences will be attenuated or even eliminated in cultures with more egalitarian sex role socialization and greater sociopolitical gender equity (Bem 1993).

Measuring Psychological Sex Differences, Sex Role Socialization, and Sociopolitical Gender Equity Across Cultures

Evaluating these four perspectives—obligate, facultative-mediation, emergent-moderation, and social role—on how psychological sex differences vary (or not) across cultures is the focus of this chapter. Utilizing findings on sex differences across dozens of large cross-cultural studies, 21 sources of evidence are presented suggesting psychological sex differences that can be simultaneously evolved and variable across cultures. Moreover in most cases, evidence suggests counter-intuitively so that sex differences emerge more strongly in cultures with egalitarian sex role socialization and greater sociopolitical gender equity.

National indicators of sex role socialization and sociopolitical gender equity come in two basic forms. First, there are many international organizations that rate nations along dimensions related to sex role socialization and sociopolitical gender equity (see Else-Quest and Grabe 2012; Hyde 2012). Perhaps the most commonly used metric among this form estimating sociopolitical gender equality is the Gender Empowerment Measure (GEM) of the United Nations (Wood and Eagly 2002). The GEM is a composite index measuring gender inequality in three basic dimensions of empowerment—economic participation and decision-making, political participation and decision-making, and power over economic resources. Also in this category of national gender equity indicators are the Gender Gap Index of the World Economic Forum, the Gender Gap Index of Social Watch, and the Standardized Index of Gender Equality (SIGE; Dijkstra 2002).

The second basic type of indicator of sex role socialization and sociopolitical gender equity comes from cross-cultural studies that provide direct assessments of individual attitudes toward gender equality, women's freedom, and women's place in family and work. Based on these individual responses, researchers generate overall national averages along gender equity-related dimensions. Examples of this form of national gender equity include attitudes toward gender equality from nationally-representative attitude surveys (Inglehart and Norris 2003), surveys of hostile and benevolent sexism (Napier et al. 2010), and surveys of sex role ideology (SRI; Schmitt et al. 2014). Schmitt et al. (2014) reported in a 58-nation study that there are very high correlations among both type of measures, with the progressive SRI correlating positively with both the United Nations' GEM, $r(56)=.65$, $p<.001$, and the nationally-representative indexes of gender equality attitudes (Inglehart and Norris 2003), $r(30)=.76$, $p<.001$. In this chapter, reviewed findings will generally refer to how national levels of psychological sex differences are related to both types of indicators.

Evaluating Social Role Theory's Ability to Explain the Size of Sex Differences Across Cultures

According to the SSSM and various social role theories of sex differences, men's and women's psychological differences are solely the result of extensive sex role socialization processes and sociopolitical power differentials (Eagly et al. 2004; Eagly and Wood 1999). As a result, social role theorists assume that sex differences should be conspicuously smaller in cultures with more egalitarian sex role socialization or greater sociopolitical gender equity (Kasser and Sharma 1999; Wood and Eagly 2002; Zentner and Mitura 2012). Several data sources are relevant for evaluating this empirical claim, including sex differences in self-reported Big Five personality traits.

Big Five Personality Traits Several studies have found pervasive sex differences in Big Five personality traits, with women typically scoring higher in agreeableness and neuroticism (Feingold 1994). Social role theory predicts that sex differences in all Big Five personality traits will be smaller in nations with more egalitarian sex role socialization and greater sociopolitical gender equity. Several studies have evaluated this prediction. In almost every instance, the observed pattern of sex differences across cultures strongly disconfirms social role theory's predictions.

For instance, in one of the earliest and largest studies of sex differences in personality across cultures, Costa, Terracciano, and McCrae (2001) found in most countries women were higher on the extraversion facet of warmth, agreeableness, neuroticism, and the openness facet of feelings, whereas men scored higher on scales measuring the extraversion facet of assertiveness and the openness facet of ideas (see also Lynn and Martin 1997; McCrae 2002; Weisberg et al. 2011). In direct disconfirmation of social role theory, Costa et al. (2001) reported that sex differences in most Big Five personality traits were larger in egalitarian cultures where women have more equal opportunities with men (McCrae 2002). Both in self-report and other-report data, Asian and African cultures generally showed the smallest sex differences, whereas European and American cultures—where the egalitarian sex role socialization and sociopolitical gender equity indexes were generally higher—showed the largest differences (McCrae and Terracciano 2005).

In 2008, Schmitt and his colleagues reported findings from a cross-cultural study of sex differences across 55 nations of the International Sexuality Description Project (n=17,637). Based on responses to the Big Five Inventory (Benet-Martínez and John 1998), women reported higher levels of extraversion, agreeableness, conscientiousness, and neuroticism than men across most nations. Overall, more egalitarian sex role socialization and greater sociopolitical gender equity were associated with larger sex differences in personality. The same pattern of findings also has been replicated in a large cross-cultural study by Lippa (2009), further disconfirming sex role theory.

In a study of 58 nations called the International Sexuality Description Project-2 (ISDP-2; Schmitt et al. 2014), data were collected from a more diverse set of cultures than previous studies, including samples from several Northern European nations with relatively high levels of gender egalitarianism (e.g., Denmark, Finland, Iceland, and Norway) and several new samples from less egalitarian nations (e.g., Colombia, Ecuador, Nigeria, and Swaziland). Men's and women's nation-level personality traits were related to Sex Role Ideology (SRI as directly measured in the ISDP-2), an index of gender equality attitudes from a nationally-representative study (Inglehart and Norris 2003), the Standardized Index of Gender Equality (SIGE), the Gender Empowerment Measure (GEM), and other indicators of sex role socialization and sociopolitical gender equity across this more diverse set of nations. Schmitt et al. (2014) reported across nearly all Big Five traits that egalitarian sex role socialization and greater sociopolitical gender equity were associated with larger sex differences in personality.

Overall, women generally score higher than men on measures of extraversion (Schmitt et al. 2014). As noted in Table 11.1, increasing levels of egalitarian sex role socialization and greater sociopolitical gender equity are generally associated with

Table 11.1 Social role theory and predictions about the effects of sociopolitical gender equity on sex differences across cultures

Trait	Effects of increasing sociopolitical gender equity on sex differences across cultures	Social role prediction
Traits typically higher in women		
Extraversion	Gender equity increases extraversion, more so in women—sex differences widen	Disconfirmed
Agreeableness	Gender equity increases agreeableness, more so in women—sex differences widen	Disconfirmed
Conscientiousness	Gender equity increases conscientiousness, more so in women—sex differences widen	Disconfirmed
Neuroticism	Gender equity decreases neuroticism, more so in men—sex differences widen	Disconfirmed
Love	Gender equity increases love, more so in women—sex differences widen	Disconfirmed
Resources mate preference	Gender equity decreases preferences for resources, more so in women—sex differences narrow	Confirmed
Intimate partner violence	Gender equity decreases intimate partner violence, more so in men—sex differences widen	Disconfirmed
Spatial location ability	Gender equity increases spatial location ability, more so in women—sex differences widen	Disconfirmed
Crying	Gender equity increases crying, more so in women—sex differences widen	Disconfirmed
Depression	Gender equity decreases depression, more so in men—sex differences widen	Disconfirmed
Benevolence values	Gender equity increases benevolence values, more so in women—sex differences widen	Disconfirmed
Empathetic occupation preference	Gender equity unrelated to empathetic occupation preference, sex differences relatively stable	Disconfirmed
Traits typically higher in men		
Openness	Gender equity unrelated to openness, sex differences relatively stable	Disconfirmed
Machiavellianism	Gender equity decreases Machiavellianism, more so in women—sex differences widen	Disconfirmed
Narcissism	Gender equity decreases Narcissism, more so in women—sex differences widen	Disconfirmed

Table 11.1 (continued)

Trait	Effects of increasing sociopolitical gender equity on sex differences across cultures	Social role prediction
Psychopathy	Gender equity decreases psychopathy, more so in women—sex differences widen	Disconfirmed
Social dominance orientation	Gender equity unrelated to social dominance, sex differences relatively stable	Disconfirmed
Dismissing attachment	Gender equity decreases dismissing attachment, more so in women—sex differences widen	Disconfirmed
Sociosexuality (SOI) overall	Gender equity increases sociosexuality, more so in women—sex differences narrow	Confirmed
SOI "enjoy casual sex"	Gender equity increases sociosexual "enjoy casual sex," more so in men—sex differences widen	Disconfirmed
Attractiveness mate preference	Gender equity decreases attractiveness preference, more so in women—sex differences widen	Disconfirmed
Self-esteem	Gender equity unrelated to self-esteem, sex differences relatively stable	Disconfirmed
Subjective well-being	Gender equity unrelated to subjective well-being, sex differences relatively stable	Disconfirmed
Height	Gender equity increases height, more so in men—sex differences widen	Disconfirmed
Body mass index/obesity	Gender equity increases body mass index and obesity, more so in men—sex differences widen	Disconfirmed
Blood pressure	Gender equity increases blood pressure, only in men—sex differences widen	Disconfirmed
Spatial rotation ability	Gender equity increases spatial rotation ability, more so in men—sex differences widen	Disconfirmed
Occupation preference	Gender equity unrelated to systematic occupation preference, sex differences relatively stable	Disconfirmed

increases in extraversion among both men and women, but the increase is greater among women, leading to wider sex differences in nations with higher gender equity. This is not always the profile of sex difference variation across cultures. However, in most cases, higher levels of egalitarian sex role socialization and greater sociopolitical gender equity are linked with larger sex differences in extraversion across cultures, in direct contradiction to social role theory.

Women also score higher in agreeableness and conscientiousness than men. Increasing levels of egalitarian sex role socialization and greater sociopolitical gender equity are generally associated with increases in agreeableness and conscientiousness among both men and women, but increases are greater among women, leading to wider sex differences in nations with higher gender equity. Neuroticism is also higher in women than men across cultures. Increasing levels of egalitarian sex role socialization and greater sociopolitical gender equity are generally associated with lower neuroticism among both men and women, but the decrease is greater among men, leading to wider neuroticism sex differences in nations with higher gender equity. Men are slightly higher in openness to experience, but this sex difference is unrelated to gender equity. Nonetheless, because social role theory predicts sex difference to be smaller in nations with more egalitarian sex role socialization and greater sociopolitical gender equity, social role theory is disconfirmed by this finding, as well (see Table 11.1).

Dark Triad Personality Traits Several studies have found sex differences in Dark Triad personality traits, with men typically scoring higher in Machiavellianism, Narcissism, and psychopathy (Foster et al. 2003; Jonason et al. 2009; McHoskey 2001). Social role theory predicts that sex differences in Dark Triad personality traits will be smaller in nations with more egalitarian sex role socialization and greater sociopolitical gender equity. The ISDP-2 empirically evaluated these predictions of social role theory (Schmitt et al. 2014). In almost every instance, the observed pattern of sex differences across cultures strongly disconfirmed social role theory's predictions (see Table 11.1).

For example, Schmitt et al. (2014) found that both men and women are lower in Machiavellianism in nations with more egalitarian sex role socialization and greater sociopolitical gender equity. However, the effect on women, $r(46)=-0.48$, $p<.05$, was much stronger than the effect on men, $r(46)=-0.19$. As result, the degree of sex difference in Machiavellianism was wider or larger in magnitude in nations with more egalitarian sex role socialization and greater sociopolitical gender equity (e.g., the effect size of national sex differences in Machiavellianism correlated positively with national GEM scores, $r(46)=0.57$, $p<.001$. As an illustration, larger sex differences in Machiavellianism were found in high egalitarian cultures such as Netherlands ($d=0.63$), Iceland ($d=0.61$), New Zealand ($d=0.60$), and Denmark ($d=0.55$). Smaller sex differences in Machiavellianism were found in less egalitarian cultures such as Ethiopia ($d=-0.09$), Malaysia ($d=-0.10$), Bangladesh ($d=-0.17$), and Swaziland ($d=-0.19$). Very similar cross-cultural results were observed for sex differences in the Dark Triad traits of Narcissism and psychopathy. As noted in Table 11.1, cross-cultural patterns of sexual differentiation in Dark Triad traits strongly disconfirmed sex role theory.

Social Dominance Orientation Sex differences in social dominance orientation have been documented such that men report significantly higher social dominance orientation than women (Sidanius et al. 2000; Sidanius and Pratto 1999). Generally, these sex differences have been found to be invariant across cultures. Schmitt et al. (2014) assessed social dominance orientation across 54 nations of the ISDP–2 and found neither men's nor women's reported social dominance levels were related to egalitarian sex role socialization or greater sociopolitical gender equity. As noted in Table 11.1, cross-cultural patterns of sexual differentiation in social dominance orientation strongly disconfirmed sex role theory.

Romantic Attachment Styles Sex differences in adult romantic attachment have been documented such that men report significantly more dismissing attachment levels than women (Bartholomew and Horowitz 1991; Del Giudice 2011). The finding of sex differences in dismissing romantic attachment seems to fit with social role theory, in that men are often socialized to be less emotional, less nurturing, and less willing to connect with others (Bem 1993). However, Schmitt et al. (2003) found sex differences in dismissing attachment were larger in nations with more egalitarian sex role socialization and greater sociopolitical gender equity. As noted in Table 11.1, cross-cultural patterns of sexual differentiation in dismissing attachment strongly disconfirmed sex role theory.

Love Sex differences in love have been documented such that women report significantly higher levels of many love styles and emotional investment tendencies than men (Bailey et al. 1987; Hendrick and Hendrick 1995; Schmitt and Buss 2000). Schmitt et al. (2009) found both men and women report higher levels of emotional investment in nations with more egalitarian sex role socialization and greater sociopolitical gender equity. However, the effects on women were more profound, leading to larger sex differences in emotional investment in nations with more egalitarian sex role socialization and greater sociopolitical gender equity. As noted in Table 11.1, cross-cultural patterns of sexual differentiation in emotional investment strongly disconfirmed sex role theory.

Sociosexuality Sex differences in sociosexuality (i.e., attitudes toward having sex without commitment) have been documented such that men report significantly more unrestricted or permissive sociosexuality than women (Lippa 2009; Schmitt 2005b; Simpson and Gangestad 1991; see also Petersen and Hyde 2010). Schmitt (2005b) found both men and women reported higher sociosexuality in nations with more egalitarian sex role socialization and greater sociopolitical gender equity. However, the effects on women were more profound, leading to more moderate sex differences in sociosexuality in nations with more egalitarian sex role socialization and greater sociopolitical gender equity. Lippa (2009) replicated these results in a 53-nation study, concluding "although culture moderated the magnitude of sex differences, it was never the case that culture eliminated these sex differences, which remained quite powerful overall, despite the presence of significant cultural main effects and interactions" (p. 644). Even so, as noted in Table 11.1, cross-cultural patterns of sexual differentiation in sociosexuality strongly confirmed sex role theory.

Sociosexual—Enjoy Casual Sex with Different Partners Sex differences in sociosexuality item "I can imagine myself being comfortable and enjoying 'casual' sex with different partners" have been documented such that men report significantly more enjoyment of casual sex with different partners than women (Schmitt et al. 2014). Schmitt et al. (2014) found both men and women report higher enjoyment of casual sex with different partners in nations with more egalitarian sex role socialization and greater sociopolitical gender equity. However, the effects on men were more profound, leading to *larger* sex differences in enjoyment of casual sex with different partners in nations with more egalitarian sex role socialization and greater sociopolitical gender equity. As noted in Table 11.1, cross-cultural patterns of sexual differentiation in enjoyment of casual sex with different partners strongly disconfirmed sex role theory.

Why the difference between the cross-cultural patterns of overall sociosexuality and the cross-cultural patterns with the item "enjoy casual sex" with different partners? According to Sexual Strategies Theory (Buss and Schmitt 1993), among those men and women who pursue a short-term sexual strategy, it is expected that men will seek larger numbers of partners than women (Schmitt et al. 2004). When women engage in short-term mating, they are expected to be more selective than men, particularly over genetic quality of short-term mates (Thornhill and Gangestad 2008). As a result, when more egalitarian sex role socialization and greater sociopolitical gender equity "set free" or release men's and women's short-term mating psychology, it can be expected that the specific item "enjoy casual sex with different partners" taps the release of men's evolved short-term mating psychology more than women's evolved short-term mating psychology.

Mate Preferences for Resources Sex differences in long-term mate preferences for cues to resource provisioning ability have been documented such that women report significantly more desire for long-term mates with status and resources than men do (Buss and Schmitt 1993; Ellis 1992; Feingold 1992; Li et al. 2002; Sprecher et al. 1994). Eagly and Wood (1999) found both men and women report less desire for long-term mates with resources in nations with more egalitarian sex role socialization and greater sociopolitical gender equity. However, the effects on women were more profound, leading to smaller sex differences in the desire for long-term mates with resources in nations with more egalitarian sex role socialization and greater sociopolitical gender equity (see also Kasser and Sharma 1999; Zentner and Mitura 2012).

Specifically, Eagly and Wood (1999) found in one of four statistical tests that nations with greater sociopolitical gender equity had smaller sex differences in "Good Financial Prospects" preferences. In a recent replication, Zentner and Mitura (2012) found sex differences in "Ambition" shrink from a moderate effect size ($d=-.65$) in lowest gender equity nations to a still moderate effect size ($d=-0.48$) in highest gender parity nations (see Schmitt 2012). Similar results were found for sex differences in desires for Social Status ($d=-0.31$) and Good Financial Prospects ($d=-0.55$). Across only the highest gender equity nations, the average sex difference $|d|$ was 0.42 which places sex differences in long-term mate preferences for resources in the 81st percentile of all meta-analytically documented psychological sex differences (Hyde 2005). Thus, although these results are consistent with social role theory, the results do not suggest sex differences in resource preferences eliminated in high equity cultures. Additionally, Gangestad et al. (2006) demonstrated the ability of social role theory to explain sex differences in resource preferences is negated after controlling for ecological factors. Even so, as noted in Table 11.1, cross-cultural patterns of sexual differentiation in the desire for long-term mates with resources at least partially confirmed sex role theory.

Mate Preferences for Attractiveness Sex differences in long-term mate preferences for cues to beauty and youth have been documented such that men report significantly more desire for physically attractive long-term mates than women do (Buss and Schmitt 1993; Buss 1989; Feingold 1990; Li et al. 2002; Lippa 2007; Wheatley et al. 2014). These sex differentiated mate preferences have been documented in studies of real-life personal ads, online dating choices, and actual marital choice; in studies of older adults and nationally representative samples; and in studies of the outcomes of these preferences on patterns of jealousy, mate retention, and fertility (for a review, see Schmitt 2014). These sex differences also have been shown to vary in adaptive ways across cultures (Gangestad et al. 2006).

Zentner and Mitura (2012) found sex differences in preferences for physical attractiveness increase from a small effect size ($d=0.24$) in lowest gender equity nations to a moderate effect size ($d=0.51$) in highest gender parity nations (see Schmitt 2012). Schmitt et al. (2014) also found sex differences in long-term mate preferences for physical attractiveness are largest in nations with more egalitarian sex role socialization and greater sociopolitical gender equity, particularly because women, but not men, reduce their desire for physical attractiveness in long-term mates within egalitarian sex role socialization and greater sociopolitical gender equity nations. As noted in Table 11.1, cross-cultural patterns of sexual differentiation in long-term mate preferences for attractiveness strongly disconfirmed sex role theory.

Self-Esteem Sex differences in self-esteem have been documented such that men report significantly higher self-esteem than women (Kearney-Cooke 1998; Kling et al. 1999). In the ISDP-2, Schmitt et al. (2014) found both men and women report higher levels of self-esteem in nations with more egalitarian sex role socialization and greater sociopolitical gender equity. However, the effects on men were more profound, leading to larger sex differences in self-esteem in nations with more egalitarian sex role socialization and greater sociopolitical gender equity. As noted in Table 11.1, cross-cultural patterns of sexual differentiation in self-esteem strongly disconfirmed sex role theory.

Subjective Well-Being Sex differences in subjective well-being have been documented such that men report somewhat higher subjective well-being than women, though this is mainly due to women's heightened negative affect responsivity (Fujita et al. 1991) and differences are often negligible after controlling for other demographic factors (Diener et al. 1999; Lucas and Gohm 2000). Even so, in the ISDP-2 Schmitt et al. (2014) found both men and women report higher levels of subjective well-being in nations with more egalitarian sex role socialization and greater sociopolitical gender equity. However, the effects of egalitarian sex role socialization on men were more profound, leading to larger sex differences in subjective well-being in nations with more egalitarian sex role socialization. The effects of greater sociopolitical gender equity on men and women were similar, leading to no association between sociopolitical gender equity and sex differences in subjective. As noted in Table 11.1, these cross-cultural patterns of sexual differentiation in subjective well-being strongly disconfirmed sex role theory.

Intimate Partner Violence (IPV) Sex differences in IPV have been documented such that women report significantly higher perpetration of IPV than men (Archer 2000; Magdol et al. 1997). For example, Archer (2000, 2006) analyzed reports of IPV across more than 70 nations, finding that women self-report perpetrating IPV more than men do. Examining sex differences in the Conflict Tactics Scale (Straus 1979, 2008), Archer found more women than men perpetrated most acts of IPV. In the ISDP-2, Schmitt et al. (2014) also examined sex differences in perpetration of IPV using a self-report measure (Dobash et al. 1998), finding both men and women report lower levels of IPV in nations with more egalitarian sex role socialization and greater sociopolitical gender equity. However, the effects of egalitarian sex role socialization on men were more profound, leading to larger sex differences in IPV in nations with greater sociopolitical gender equity and more egalitarian sex role socialization. As noted in Table 11.1, these cross-cultural patterns of sexual differentiation in IPV strongly disconfirmed sex role theory.

Height Sex differences in height have been documented such that men are taller than women across all cultures (Gaulin and Boster 1985; Lippa 2009), a difference that likely has been consistent since 150,000 years ago or even earlier (Ruff 2002). In the ISDP-2, Schmitt et al. (2014) found both men and women report taller height in nations with more egalitarian sex role socialization and greater sociopolitical gender equity. However, the effects on men were more profound, leading to larger sex differences in height in nations with more egalitarian sex role socialization and greater sociopolitical gender equity (see also, Lippa 2009). As noted in Table 11.1, cross-cultural patterns of sexual differentiation in height strongly disconfirmed sex role theory (cf. Touraille 2013).

Body Mass Index (BMI) and Obesity Sex differences in body mass index and obesity have been documented such that men are often slightly higher on these characteristics than women (Eveleth and Tanner 1990; Pasco et al. 2012). In the ISDP-2, Schmitt et al. (2014) found both men and women report larger BMI and obesity rates in nations with more egalitarian sex role socialization and greater sociopolitical gender equity. However, the effects on men were more profound, leading to larger sex differences in BMI and obesity in nations with more egalitarian sex role socialization and greater sociopolitical gender equity. As noted in Table 11.1, cross-cultural patterns of sexual differentiation in BMI and obesity strongly disconfirmed sex role theory.

Blood Pressure Sex differences in blood pressure have been documented such that men have higher blood pressures than women (Hottenga et al. 2005). Schmitt et al. (2014) found men report higher blood pressure in nations with more egalitarian sex role socialization and greater sociopolitical gender equity, whereas women's blood pressure is unrelated, leading to larger sex differences in blood pressure in nations with more egalitarian sex role socialization and greater sociopolitical gender equity. As noted in Table 11.1, cross-cultural patterns of sexual differentiation in blood pressure disconfirmed sex role theory.

Spatial Rotation Ability Sex differences in spatial rotation ability have been documented such that men have better spatial rotation ability than women (Silverman et al. 2007; Silverman et al. 1996; Voyer et al. 1995). In a large cross-cultural study, Lippa et al. (2010) found both men and women report better spatial rotation ability in nations with more egalitarian sex role socialization and greater sociopolitical gender equity. However, the effects on men were more profound, leading to larger sex differences in spatial rotation ability in nations with more egalitarian sex role socialization and greater sociopolitical gender equity. As noted in Table 11.1, cross-cultural patterns of sexual differentiation in spatial rotation ability strongly disconfirmed sex role theory.

Spatial Location Ability Sex differences in spatial location ability have been documented such that women have better spatial location ability than men (Silverman et al. 2007; Silverman et al. 1996; Voyer et al. 2007). In a large cross-cultural study, Silverman et al. (2007) found men's and women's spatial location abilities were unrelated to egalitarian sex role socialization and sociopolitical gender equity. As noted in Table 11.1, cross-cultural patterns of sexual differentiation in spatial location ability disconfirmed sex role theory.

Crying Sex differences in crying behavior have been documented such that women cry more than men (Becht et al. 2001; Lombardo et al. 2001; Santiago-Menendez and Campbell 2013; van Hemert et al. 2011; Vingerhoets and Scheirs 2000). Evaluating national data from van Hemert et al. (2011), Schmitt et al. (2014) found both men and women report crying more in nations with more egalitarian sex role socialization and greater sociopolitical gender equity. However, the effects on women were more profound, leading to larger sex differences in crying behavior in nations with more egalitarian sex role socialization and greater sociopolitical gender equity. As noted in Table 11.1, cross-cultural patterns of sexual differentiation in crying behavior strongly disconfirmed sex role theory.

Depression Sex differences in depression have been documented such that women have nearly twice the rate of depression as men (Fischer and Manstead 2000; Hopcroft and McLaughlin 2012; Nolen-Hoeksema 2001). Hopcroft and McLaughlin (2012) found both men and women report lower rates of depression in nations with more egalitarian sex role socialization and greater sociopolitical gender equity. However, the effects on men were more profound, leading to larger sex differences in depression in nations with more egalitarian sex role socialization and greater sociopolitical gender equity. As noted in Table 11.1, cross-cultural patterns of sexual differentiation in depression strongly disconfirmed sex role theory.

Values Sex differences in values have been documented such that women report higher levels of benevolence and universalism values, whereas men report higher levels of power, achievement, and hedonism values (Schwartz and Rubel 2005; Schwartz and Rubel-Lifshitz 2009). Schwartz and Rubel (2005) and Schwartz and Rubel-Lifshitz (2009) found both men and women report higher benevolence in nations with more egalitarian sex role socialization and greater sociopolitical gender equity. However, the effects on women were more profound, leading to larger sex differences in benevolence in nations with more egalitarian sex role socialization and greater sociopolitical gender equity. Similar findings were observed for sex differences in universalism, power, achievement, and hedonism. As noted in Table 11.1, cross-cultural patterns of sexual differentiation in values strongly disconfirmed sex role theory.

Occupational Preferences and Interests Sex differences in occupational preferences and interests have been documented such that men report significantly more interest in systematic, thing-oriented professions and women report significantly more interest in empathetic, people-oriented professions (Konrad et al. 2000; see also Nettle 2007; Su et al. 2009). Lippa (2010) found across 53 nations that men's and women's occupational preferences are entirely unrelated to egalitarian sex role socialization and greater sociopolitical gender equity. As noted in Table 11.1, cross-cultural patterns of sexual differentiation in occupational preferences disconfirmed sex role theory.

Nation-Level Covariates as Alternative Explanations The highly consistent pattern of sex differences getting larger in nations with more egalitarian sex role socialization and greater sociopolitical gender equity could be explained by forces that statistically overwhelm the power of social role theory to explain sex differences. For instance, sex role socialization and sociopolitical gender equity are highly correlated with wealth, prevalence of education, and overall human development of nations. Schmitt et al. (2014) noted that after controlling for these and other factors, virtually all of the above findings that disconfirmed social role theory still emerge. Indeed, many of the counter-intuitive associations between sex differences and sociopolitical gender equity *increased* after controlling for potential covariates. For instance, the finding that sex differences in neuroticism were larger in nations with more progressive sex role ideology was intensified after controlling for gross domestic product of nation. The same was true for links between sex differences in neuroticism and the gender gap index, the gender equality index, and the SIGE. Future research should examine multiple control variables in ways that allow for a more complete evaluation of these alternative explanations (see Nettle 2009).

Reference-Group Effects as Alternative Explanations Guimond et al. (2007, 2008) have suggested that sex differences in self-reported personality traits are suppressed in less progressive nations because of reference-group effects. That is, men and women may compare themselves only to their own gender when completing surveys in less progressive cultures, but in more progressive nations men and women compare themselves to everyone, resulting in more accurate sex differences in nations with egalitarian sex role socialization and greater sociopolitical gender equity (Biernat et al. 1991). If true, this has dire implications for gender similarities theory (Hyde 2005).

It would suggest that the moderate to large sex differences commonly observed in more progressive Northern European nations are "truer" estimates of psychological sex differences (after all, men and women are comparing themselves to everyone, not just their own gender), whereas in more traditional cultures researchers are merely observing masked versions of what, according to social role theory, must be incredibly larger sex differences in personality (Lukaszewski et al. 2013). Finding that so many sex differences in psychology range from large in egalitarian nations to extremely large in more traditional cultures would provide strong refutation of gender similarities theory (Hyde 2005). According to Fischer (2010), Lippa (2009), and Schmitt et al. (2014), such a perspective is unlikely for several reasons.

First, if reference-group effects were masking sex differences in traditional cultures, researchers should observe smaller standard deviations in men's and women's distributions in traditional cultures compared to more progressive cultures (where men and women compare themselves to everyone). Empirically, this is typically not the case (Fischer 2010; Lippa 2009; Schmitt et al. 2014). Second, if reference-group effects were the driving force behind counter-intuitively larger sex differences in egalitarian nations, researchers should observe all survey items are equally effected by reference-group effects. Again, this is not the case empirically (Fischer 2010; Lippa 2009; Schmitt et al. 2014). Third, if reference-group effects were the driving force behind counter-intuitively larger sex differences in egalitarian nations, researchers should observe both men and women are equally biased in their responses. Again, this is not the case (Fischer 2010; Lippa 2009; Schmitt et al. 2014).

An additional limitation of reference-group effects for discounting the robust disconfirmation of social role theory in this review is that many of the findings reviewed earlier transcend of the reference-group limitations of self-report methods. Reference-group effects cannot explain the finding that sex differences in height, BMI, obesity, and blood pressure are larger in cultures with more egalitarian sex role socialization and greater sociopolitical gender equity. Reference-group effects cannot explain the finding that sex differences in tested spatial rotation and spatial location abilities are larger in cultures with more egalitarian sex role socialization and greater sociopolitical gender equity. Such measures represent fair "common rule" tests of social role theory (Biernat et al. 1991), and in nearly every case social role theory was strongly disconfirmed.

Additional Psychological Traits to Consider Not all psychological sex differences have been measured across large numbers of nations in a way that allows researchers to relate the size of nation-level sex differences to indicators of sex role socialization and sociopolitical gender equity. It is possible that additional psychological traits will be discovered, measured, and documented as sexually differentiated across cultures that either support or further refute social role theory. Given the reviewed findings here, researchers should probably expect the vast majority of future research findings that will refute social role theory.

In addition, there are some psychological traits that do not typically show sex differences, on average, but still relate to sex role socialization and sociopolitical gender equity in interesting ways. For instance, although sex differences in the high-end distribution of math scores can be substantial (Halpern et al. 2007), mean-level sex differences in math ability have been shown to be minimal (Hyde et al. 1990; Hyde and Mertz 2009). In one of two cross-cultural datasets, Else-Quest et al. (2010) found no significant links between the size of a nation's mean-level sex difference in math and indicators of global sociopolitical gender equity. However, in another dataset they did find some links. Importantly, the size of a nation's sex difference in math was most closely associated with national indicators of women being given greater access to education and research jobs, with such indicators having positive associations with women's (more so than men's) math scores. The precision with which Else-Quest et al. (2010) identified this potential sociological source of psychological sex difference is laudable and informative, but their overall findings did not provide broad sweeping confirmation of social role theory.

Evaluating Life History Theory as a Facultative Mediator of Sex Differences Across Cultures

According to evolutionary psychologists who take life history approaches (Ellis et al. 2009; Geary 2002; Kaplan and Gangestad 2005), many human psychological adaptations are facultatively designed to attend to particular sources of ecological information and generate specially-designed, highly functional forms of behavior. As a result, many aspects of human culture—including the magnitude or degree of sex differences across human cultures—can be both evolved and variable. In particular, ecological stress is thought to have a profound effect on men's and women's psychology, and does so in a slightly different way in men and women.

According to Psychosocial Acceleration Theory (Belsky 2012; Belsky et al. 1991), in stressful ecological contexts, both men and women are thought to facultatively pursue a faster life history strategy composed of psychological traits including dismissing attachment, anti-sociality, short-term temporal orientations, and prolific short-term mating. Many of these traits display a sex difference such that men are higher in the trait than women. Because the effects of ecological stress are thought to be more profound among women on many of these traits (Ellis 2004), researchers often expect smaller sex differences in high stress cultures (Schmitt 2011). As shown in Table 11.2, based on measures of national pathogen stress (Fincher and Thornhill 2012) and responses to personality scales across multiple studies, eight psychological traits relevant to Psychosocial Acceleration Theory followed the predicted pattern of sexual differentiation across cultures, including sex differences in love, resource mate preferences, Narcissism, psychopathy, social dominance orientation, dismissing attachment, attractiveness mate preferences, and self-esteem. Four traits disconfirmed this theory, including sex differences in agreeableness, Machiavellianism, sociosexuality, and sociosexual enjoy casual sex (see also, Schmitt 2005b).

Table 11.2 Life history theory and predictions about the effects of local ecology on sex differences across cultures

Trait	Effects of ecological stress on sex differences across cultures	Life history prediction
Traits typically higher in women		
Agreeableness	Ecological stress increases agreeableness in men and women—sex differences stable	Disconfirmed
Love	Ecological stress decreases love, more so in women—sex differences narrow	Confirmed
Resources mate preference	Ecological stress increases resources preferences, more so in women—sex differences widen	Confirmed
Traits typically higher in men		
Machiavellianism	Ecological stress decreases Machiavellianism, more so in men—sex differences narrow	Disconfirmed
Narcissism	Ecological stress increases Narcissism, more so in women—sex differences narrow	Confirmed
Psychopathy	Ecological stress increases psychopathy, only in women—sex differences narrow	Confirmed
Social dominance orientation	Ecological stress increases social dominance orientation, only in women—sex differences narrow	Confirmed
Dismissing attachment	Ecological stress increases dismissing attachment, more so in women—sex differences narrow	Confirmed
Sociosexuality (SOI) overall	Ecological stress decreases sociosexuality, more so in women—sex differences widen	Disconfirmed
SOI "enjoy casual sex"	Ecological stress decreases "enjoy casual sex" in men and women—sex differences stable	Disconfirmed
Attractiveness mate preference	Ecological stress increases attractiveness preferences, more in women—sex differences narrow	Confirmed
Self-esteem	Ecological stress increases self-esteem, only in women—sex differences narrow	Confirmed

Evaluating Religiosity as an Emergent Moderator of Sex Differences Across Cultures

As noted earlier, religion may be an especially potent "emergent moderator" of psychological adaptations in men and women across cultures (e.g., Kirkpatrick 2011; McCullough et al. 2012; Mealey 1985). Some researchers expect most religions will suppress psychological adaptations involving short-term mating, and do so more

Table 11.3 Religion's emergently-moderated sex differences across cultures

Trait	Effects of religiosity on sex differences across cultures	Suppression prediction
Traits typically higher in women		
Agreeableness	Religiosity increases agreeableness, more so in men—sex differences narrow	Confirmed
Love	Religiosity decreases love, more so in women—sex differences narrow	Confirmed
Traits typically higher in men		
Machiavellianism	Religiosity decreases Machiavellianism, only in men—sex differences narrow	Confirmed
Narcissism	Religiosity decreases Narcissism, only in men—sex differences narrow	Confirmed
Psychopathy	Religiosity decreases psychopathy, only in men—sex differences narrow	Confirmed
Social dominance orientation	Religiosity decreases social dominance orientation, more so in men—sex differences narrow	Confirmed
Dismissing attachment	Religiosity increases dismissing attachment, more so in women—sex differences narrow	Confirmed
Sociosexuality (SOI) overall	Religiosity decreases sociosexuality, more so in women—sex differences widen	Disconfirmed
SOI "enjoy casual sex"	Religiosity decreases "enjoy casual sex" in men and women—sex differences relatively stable	Disconfirmed

to women's than men's short-term mating psychology (Baumeister and Twenge 2002). In addition, religion might enhance psychological adaptations having to do with prosociality and reduce antisociality (Fincher and Thornhill 2012; Weeden and Kurzban 2013), and do so more in one sex than the other. As shown in Table 11.3, based on measures of religiosity and personality in the ISDP-2, seven psychological traits followed this pattern, including sex differences in agreeableness, love, Machiavellianism, Narcissism, psychopathy, social dominance orientation, and dismissing attachment. Two traits disconfirmed these predictions, including sex differences in sociosexuality and sociosexual enjoy casual sex.

Conclusion

Only a few decades ago, parental socialization of children was thought to be a primary force in the constructivist shaping of children's psychological traits (Bruner 1986; Fosnot 1996). Today, psychologists know from genetically-informative designs that children's genes often evoke parental behaviors that only appear to be socializing factors (Kagan 1999), and shared genes among biological parents and their children account for much of the ostensible association between parental behavior and childhood personality (Krueger and Johnson 2008; Plomin 2008). Psychologists no longer place most of the blame for children's personality traits and mental health outcomes on parental socialization (Harris 1998; Maccoby 2000). In a similar manner, new knowledge from cross-cultural research suggests that sex role socialization may no longer be to blame for most patterns of psychological sex differences across cultures (see also, Udry 2000). Instead, evolved combinations of men's and women's obligate, facultatively-mediated, and emergently-moderated psychological adaptations may better account for sex differences across cultures.

In this chapter, evidence was marshaled across 21 data sources that evaluated the foundational assumption of social role theory that psychological sex differences directly result from sex role socialization and that, as consequences, psychological sex differences will be smaller in cultures with more egalitarian sex role socialization and greater sociopolitical gender equity. Empirically, sex differences in most psychological traits—in personality, sexuality, attitudes, and cognitive abilities—are conspicuously *larger* in cultures with more egalitarian sex role socialization and greater sociopolitical gender equity. Even sex differences in many physical traits such as height, obesity, and blood pressure were shown to be larger in cultures with more egalitarian sex role socialization and greater sociopolitical gender equity. Three alternative evolutionary perspectives on sex differences—obligate, facultative-mediation, and emergent-moderation—appeared to better explain the universal and culturally-variable sex differences reliably observed across cultures.

From an evolutionary perspective, it would be remarkable if men and women have not evolved at least some differences in psychological design. Indeed, as Vandermassen (2011) has noted, "that human males and females should have evolved to be psychologically identical, for example, is a theoretical impossibility, and, indeed, turns out to be untrue" (p. 733). At this point, it is perhaps unscientific to assert absolutely no sex-specific psychological adaptations exist in humans (Buss and Schmitt 2011; Kenrick et al. 2010), yet many social scientists continue to assert this is so (Winegard et al. 2014). One reason behind the extreme popularity of the SSSM's sex difference denialism is that magnitude of many sex differences noticeably varies across cultures. Because this variation is in a few cases linked to sex role variation in ways consistent with social role theory across cultures (e.g., mate preferences for resources), some researchers have spotlighted these few peculiar cases and advanced all-encompassing theories about the primary causal force of sex roles in generating psychological sex differences across cultures (Eagly and Wood 1998; Wood and Eagly 2002). In this chapter, the legitimacy of social role explanations as the sole source of psychological sex differences has been called into serious question (see also Udry 2000). In most cases, the cross-cultural evidence has directly refuted patriarchal social role explanations of sex differences. Social role theory, as a theory that purportedly explains the degree of psychological sex differences across cultures, should probably be considered a scientific dead end. At the very least, psychological science needs other perspectives to explain the sweeping patterns of culturally variable sex differences documented in this chapter.

Evolutionary perspectives take a different and ultimately more profitable approach to understanding the relationship between culture and sex differences in psychology (Campbell 2002). First, evolutionary psychology perspectives offer mechanisms for why cultural universals *and* cultural variations exist in sex differences (see Pirlott and Schmitt 2014). For example, Gangestad et al. (2006) explained that universal sex differences in mate preferences for physical attractiveness are universally generated by evolved sex-specific psychological adaptations, but these mechanisms are also facultatively mediated by local ecology—if environments have high pathogen prevalence, then mate preferences for physical attractiveness are adaptively enhanced. If the local environment has low pathogen prevalence, then adaptive desires de-emphasize physical attractiveness in potential mates.

Second, evolutionary psychology perspectives offer explanations of cultural *change* (Mesoudi et al. 2006). When levels of pathogen prevalence shift within a culture, the corresponding emphasis on physical attractiveness should shift in that culture, as well. Moreover, the degree to which pathogens affect the adaptations one sex more than the other would explain why the size of sex differences varies across cultures. Thus, evolutionary perspectives on phenotypic plasticity and evoked culture such as Gangestad et al. (2006) possesses the ability to explain pancultural universals and facultative variations, something very much missing from SSSM accounts of culture and sexuality (Maccoby 2000). It is not the case that all aspects of culture will be subject to evolutionary explanations, at least in terms of genetic evolution (see Brown et al. 2011). Still, by combining genetic and cultural levels of evolution in sophisticated ways—utilizing obligate, facultatively-mediated, emergently-moderated, and social role approaches—more complete and scientifically fruitful understandings of psychological sex differences will prosper.

References

Alesina, A., Giuliano, P., & Nunn, N. (2013). On the origins of gender roles: Women and the plough. *The Quartrly Journal of Economics, 128,* 469–530.

Alexander, G. M., & Wilcox, T. (2012). Sex differences in early infancy. *Child Development Perspectives, 6,* 400–406.

Alexander, G. M., Wilcox, T., & Farmer, M. B. (2009). Hormone behavior associations in early infancy. *Hormones and Behavior, 56,* 498–502.

Archer, J. (2000). Sex differences in aggression between heterosexual partners: A meta-analytic review. *Psychological Bulletin, 126,* 651–680.

Archer, J. (2006). Cross-cultural differences in physical aggression between partners: A social role analysis. *Personality and Social Psychology Review, 10,* 133–153.

Archer, J. (2014). The reality and evolutionary significance of psychological sex differences. Manuscript under review.

Archer, J., & Mehdikhani, M. (2003). Variability among males in sexually selected attributes. *Review of General Psychology, 7,* 219–236.

Auyeung, B., Baron-Cohen, S., Ashwin, E., Knickmeyer, R., Taylor, K., Hackett, G., & Hines, M. (2009). Fetal testosterone predicts sexually differentiated childhood behavior in girls and in boys. *Psychological Science, 20,* 144–148.

Bailey, W. C., Hendrick, C., & Hendrick, S. S. (1987). Relation of sex and gender role to love, sexual attitudes, and self-esteem. *Sex Roles, 16,* 637–648.

Barber, N. (2014). Why behavior matches ecology: Adaptive variation as a novel solution. *Cross-Cultural Research.* E-print ahead of publication.

Baron, R. M., & Kenny, D. A. (1986). The moderator-mediator variable distinction in social psychological research: Conceptual, strategic, and statistical considerations. *Journal of Personality and Social Psychology, 51,* 1173–1182.

Baron-Cohen, S. (2002). The extreme male brain theory of autism. *Trends in Cognitive Sciences, 6,* 248–254.

Baron-Cohen, S. (2004). *Essential difference: Male and female brains and the truth about autism.* New York: Basic Books.

Bartholomew, K., & Horowitz, L. M. (1991). Attachment styles among young adults: A test of a four-category model. *Journal of Personality and Social Psychology, 61,* 226–244.

Baumeister, R. F. (2000). Gender differences in erotic plasticity: The female sex drive as socially flexible and responsive. *Psychological Bulletin, 126,* 347.

Baumeister, R. F., & Twenge, J. M. (2002). Cultural suppression of female sexuality. *Review of General Psychology, 6,* 166–203.

Baumeister, R. F., Catanese, K. R., & Vohs, K. D. (2001). Is there a gender difference in strength of sex drive? Theoretical views, conceptual distinctions, and a review of relevant evidence. *Personality and Social Psychology Review, 5,* 242–273.

Becht, M. C., Poortinga, Y. H., & Vingerhoets, A. J. J. M. (2001). Crying across countries. In A. J. J. M. Vingerhoets & R. R. Cornelius (Eds.), *Adult crying: A biopsychosocial approach* (pp. 135–158). East Sussex: Brunner-Routledge.

Becker, J. B., Berkley, K. J., Geary, N., & Hampson, E. (2007). *Sex differences in the brain: From genes to behavior.* New York: Oxford University Press.

Belsky, J. (2012). The development of human reproductive strategies: Progress and prospects. *Current Directions in Psychological Science, 21,* 310–316.

Belsky, J., Steinberg, L., & Draper, P. (1991). Childhood experience, interpersonal development, and reproductive strategy: An evolutionary theory of socialization. *Child Development, 62,* 647–670.

Bem, S. L. (1993). *The lenses of gender: Transforming the debate on sexual inequality.* New Haven: Yale University Press.

Benet-Martínez, V., & John, O. P. (1998). Los cinco grandes across cultures and ethnic groups: Multitrait multimethod analysis of the Big Five in Spanish and English. *Journal of Personality and Social Psychology, 75,* 729–750.

Biernat, M., Manis, M., & Nelson, T. F. (1991). Comparison and expectancy processes in human judgment. *Journal of Personality and Social Psychology, 61,* 203–211.

Borkenau, P., Hřebíčková, M., Kuppens, P., Realo, A., & Allik, J. (2013). Sex differences in variability in personality: A study in four samples. *Journal of Personality, 81,* 49–60.

Brown, D. E. (1991). *Human universals.* New York: McGraw-Hill.

Brown, G. R., Dickens, T. E., Sear, R., & Laland, K. N. (2011). Evolutionary accounts of human behavioural diversity. *Philosophical Transactions of the Royal Society B, 366,* 313–324. doi:10.1098/rstb.2010.0267.

Browne, K. (1998). *Divided labours.* New Haven: Yale University Press.

Bruner, J. (1986). *Actual minds, possible worlds.* Cambridge: Harvard University Press.

Burke, S. M., Cohen-Kettenis, P. T., Veltman, D., Klink, D. T., & Bakker, J. (2014). Hypothalamic response to the chemo-signal androstadienone in gender dysphoric children and adolescents. *Neuroendocrine Science, 5,* 60. doi:10.3389/fendo.2014.00060.

Burtt, E. (1951). The ability of adult grasshoppers to change colour on burnt ground. *Proceedings of the Royal Entomological Society of London. Series A, General Entomology, 26,* 45–48.

Buss, D. M. (1989). Sex differences in human mate preferences: Evolutionary hypotheses tested in 37 cultures. *Behavioral and Brain Sciences, 12,* 1–49.

Buss, D. M. (1995). Psychological sex differences: Origins through sexual selection. *American Psychologist, 50,* 164–171.

Buss, D. M. & Schmitt, D. P. (1993). Sexual strategies theory: An evolutionary perspective on human mating. *Psychological Review, 100,* 204–232.
Buss, D. M. & Schmitt, D. P. (2011). Evolutionary psychology and feminism. *Sex Roles, 64,* 768–787.
Byrnes, J. P., Miller, D. C., & Schafer, W. D. (1999). Gender differences in risk taking: A meta-analysis. *Psychological Bulletin, 125,* 367–383.
Campbell, A. (2002). *A mind of her own: The evolutionary psychology of women.* New York: Oxford University Press.
Cohen, J. (1988). *Statistical power analysis for the behavioral sciences* (2nd ed.). Hillsdale: Lawrence Erlbaum.
Costa, P. T., Terracciano, A., & McCrae, R. R. (2001). Gender differences in personality traits across cultures: Robust and surprising findings. *Journal of Personality and Social Psychology, 81,* 322–331.
Crawford, C. (1998). Environments and adaptations: Then and now. In C. Crawford & D. L. Krebs (Eds.), *Handbook of evolutionary psychology: ideas, issues, and applications* (pp. 275–302). Mahwah: Erlbaum.
Cronk, L. (1999). *That complex whole: Culture and the evolution of human behavior.* Boulder: Westview Press.
Cross, C. P., Copping, L. T., & Campbell, A. (2011). Sex differences in impulsivity: A meta-analysis. *Psychological Bulletin, 137,* 97–101.
Darwin, C. R. (1871). *The descent of man and selection in relation to sex.* London: Murray.
Deaner, R. O., Geary, D. C., Puts, D. A., Ham, S. A., Kruger, J., Fles, E., et al. (2012). A sex difference in the predisposition for physical competition: Males play sports much more than females even in the contemporary US. *PloS One, 7,* e49168.
DeBruine, L. M., Jones, B. C., Crawford, J.R., Welling, L. L., & Little, A. C. (2010). The health of a nation predicts their mate preferences: Cross-cultural variation in women's preferences for masculinized male faces. *Proceedings of the Royal Society B: Biological Sciences, 277,* 2405–2410.
Del Giudice, M. (2009). On the real magnitude of psychological sex differences. *Evolutionary Psychology, 7,* 264–279.
Del Giudice, M. (2011). Sex differences in romantic attachment: A meta-analysis. *Personality and Social Psychology Bulletin, 37,* 193–214.
Del Giudice, M., & Belsky, J. (2010). Sex differences in attachment emerge in middle childhood: An evolutionary hypothesis. *Child Development Perspectives, 4,* 97–105.
Del Giudice, M., Booth, T., & Irwing, P. (2012). The distance between Mars and Venus: Measuring global sex differences in personality. *PloS One, 7,* e29265.
Diener, E., Suh, E. M., Lucas, R. E., & Smith, H. L. (1999). Subjective well-being: Three decades of progress. *Psychological Bulletin, 125,* 276–304.
Dijkstra, A. G. (2002). Revisiting UNDP's GDI and GEM: Towards an alternative. *Social Indicators Research, 57,* 301–338.
Dobash, R. P., Dobash, R. E., Cavanaugh, K. & Lewis, R. (1998). Separate and intersecting realities: A comparison of men's and women's account of violence against women. *Violence Against Women, 4,* 382–414.
Dworkin, A. (1987). *Intercourse.* New York: Basic Books.
Eagly, A. H. (1987). *Sex differences in social behavior: A social-role interpretation.* Hillsdale: Erlbaum.
Eagly, A. H., & Wood, W. (1999). The origins of sex differences in human behavior: Evolved dispositions versus social roles. *American Psychologist, 54,* 408–423.
Eagly, A. H., Wood, W., & Johannesen-Schmidt, M. C. (2004). Social role theory of sex differences and similarities: Implications for the partner preferences of women and men. In A. H. Eagly, A. E. Beall, & R. Sternberg (Eds.), *The psychology of gender* (2nd ed., pp. 269–295). New York: Guilford.

Ellis, B. J. (1992). The evolution of sexual attraction: Evaluative mechanisms in women. In J. H. Barkow, L. Cosmides, & J. Tooby (Eds.), *The adapted mind* (pp. 267–288). New York: Oxford University Press.
Ellis L. (2001). The biosocial female choice theory of social stratification. *Biodemography and Social Biology, 48,* 297–319.
Ellis, B. J. (2004). Timing of pubertal maturation in girls: An integrated life history approach. *Psychology Bulletin, 130,* 920–928.
Ellis, L. (2011). Identifying and explaining apparent universal sex differences in cognition and behavior. *Personality and Individual Differences, 51,* 552–561.
Ellis, B. J., Figueredo, A. J., Brumbach, B. H., & Schlomer, G. L. (2009). Fundamental dimensions of environmental risk. *Human Nature, 20,* 204–268.
Else-Quest, N. M., & Grabe, S. (2012). The political is personal: Measurement and application of nation-level indicators of gender equity in psychological research. *Psychology of Women Quarterly, 36,* 131–144.
Else-Quest, N. M., Hyde, J. S., & Linn, M. C. (2010). Cross-national patterns of gender differences in mathematics: A meta-analysis. *Psychological Bulletin, 136,* 103–127.
Eveleth, P. B., & Tanner, J. M. (1990). *Worldwide variation in human growth* (2nd ed.). Cambridge: Cambridge University Press.
Feingold, A. (1990). Gender differences in effects of physical attractiveness on romantic attraction: A comparison across five research paradigms. *Journal of Personality and Social Psychology, 59,* 981–993.
Feingold, A. (1992). Gender differences in mate selection preferences: A test of the parental investment model. *Psychological Bulletin, 112,* 125–139.
Feingold, A. (1994). Gender differences in personality: A meta-analysis. *Psychological Bulletin, 116,* 429–456.
Fincher, C. L., & Thornhill, R. (2012). Parasite-stress promotes in-group assortative sociality: The cases of strong family ties and heightened religiosity. *Behavioral and Brain Sciences, 35,* 61–79.
Fischer, A., & Manstead, A. (2000). The relation between gender and emotion in different cultures. In A. Fischer (Ed.), *Gender and emotion: Social psychological perspectives* (pp. 71–94). London: Cambridge University Press.
Fischer, R., Vauclair, C. M., Fontaine, J. R., & Schwartz, S. H. (2010). Are individual-level and country-level value structures different? Testing Hofstede's legacy with the Schwartz Value Survey. *Journal of Cross-Cultural Psychology, 41,* 135–151.
Fosnot, C. T. (1996). (Ed.) *Constructivism: Theory, perspectives, and practice*. New York: Teachers College Press.
Foster, L. (1981). *Religion and sexuality: The Shakers, the Mormons, and the Oneida community.* University of Illinois Press.
Foster, J. D., Keith Campbell, W., & Twenge, J. M. (2003). Individual differences in narcissism: Inflated self-views across the lifespan and around the world. *Journal of Research in Personality, 37,* 469–486.
Fujita, F., Diener, E., & Sandvik, E. (1991). Gender differences in negative affect and well-being: The case for emotional intensity. *Journal of Personality and Social Psychology, 61,* 427–434.
Galambos, N. L., Berenbaum, S. A., & McHale, S. M. (2009). Gender development in adolescence. In R. M. Lerner & L. Steinberg (Eds.), *Handbook of adolescent psychology: Vol. 1: Individual bases of adolescent development* (3rd ed., pp. 305–357). Hoboken: Wiley.
Gangestad, S. W., & Simpson, J. A. (2000). The evolution of human mating: Trade-offs and strategic pluralism. *Behavioral and Brain Sciences, 23,* 573–644.
Gangestad, S. W., Haselton, M. G., & Buss, D. M. (2006). Evolutionary foundations of cultural variation: Evoked culture and mate preferences. *Psychological Inquiry, 17,* 75–95.
Gaulin, S. J. (1997). Cross-cultural patterns and the search for evolved psychological mechanisms. In G. R. Bock & G. Cardew (Eds.), *Characterizing human psychological adaptations* (pp. 195–207). Chichester: Wiley.

Gaulin, S. J., & Boster, J. S. (1985). Cross-cultural differences in sexual dimorphism: Is there any variance to be explained? *Ethology and Sociobiology, 6,* 219–225.

Gaulin, S. J., & Boster, J. S. (1992). Human marriage systems and sexual dimorphism in stature. *American Journal of Physical Anthropology, 89,* 467–475.

Geary, D. C. (1998). *Male, female: The evolution of human sex differences.* Washington, DC: American Psychological Association.

Geary, D. C. (2002). Sexual selection and human life history. *Advances in Child Development and Behavior, 30,* 41–104.

Griskevicius, V., Delton, A. W., Robertson, T. E., & Tybur, J. M. (2011). The environmental contingency of life history strategies: Influences of mortality and socioeconomic status on reproductive timing. *Journal of Personality and Social Psychology, 100,* 241–254.

Guégan, J. F., Teriokhin, A. T., & Thomas, F. (2000). Human fertility variation, size-related obstetrical performance and the evolution of sexual stature dimorphism. *Proceedings of the Royal Society of London Series B-Biological Sciences, 267,* 2529–2535.

Guégan, J. F., Thomas, F., Hochberg, M. E., Meeus, T. D., & Renaud, F. (2001). Disease diversity and human fertility. *Evolution, 55,* 1308–1314.

Guimond, S. (2008). Psychological similarities and differences between women and men across cultures. *Social and Personality Psychology Compass, 2,* 494–510.

Guimond, S., Branscombe, N. R., Brunot, S., Buunk, A. P., Chatard, A., Desert, M., Garcia, D. M., Haque, S., Martinot, D., & Yzerbyt, V. (2007). Culture, gender, and the self: Variations and impact of social comparison processes. *Journal of Personality and Social Psychology, 92,* 1118–1134.

Gustafsson, A., & Lindenfors, P. (2008). Latitudinal patterns in human stature and sexual stature dimorphism. *Annals of Human Biology, 36,* 74–87.

Guttentag, M., & Secord, P. F. (1983). *Too many women? The sex ratio question.* Beverly Hills: Sage.

Halpern, D. F., Benbow, C. P., Geary, D. C., Gur, R. C., Hyde, J. S., & Gernsbacher, M. A. (2007). The science of sex differences in science and mathematics. *Psychological Science in the Public Interest, 8,* 1–51.

Hare, L., Bernard, P., Sánchez, F. J., Baird, P. N., Vilain, E., Kennedy, T., & Harley, V. R. (2009). Androgen receptor repeat length polymorphism associated with male-to-female transsexualism. *Biological Psychiatry, 65,* 93–96.

Harris, J. R. (1998). *The nurture assumption: Why children turn out the way they do.* New York: Simon and Schuster.

Hartung, J. (1985). Matrilineal inheritance: New theory and analysis. *Behavioral and Brain Sciences, 8,* 661–668.

Hendrick, S. S., & Hendrick, C. (1995). Gender differences and similarities in sex and love. *Personal Relationships, 2,* 55–65.

Hill, S. E., Delpriore, D. J., Rodeheffer, C. D., & Butterfield, M. E. (2014). The effect of ecological harshness on perceptions of the ideal female body size: An experimental life history approach. *Evolution and Human Behavior, 35,* 148–154.

Hines, M. (2006). Prenatal testosterone and gender-related behaviour. *European Journal of Endocrinology, 155*(suppl 1), S115–S121.

Hopcroft, R. L., & McLaughlin, J. (2012). Why is the sex gap in feelings of depression wider in high gender equity countries? The effect of children on the psychological well-being of men and women. *Social Science Research, 41,* 501–513.

Hottenga, J. J., Boomsma, D. I., Kupper, N., Posthuma, D., Snieder, H., Willemsen, G., & De Geus, E. J. C. (2005). Heritability and stability of resting blood pressure. *Twin Research and Human Genetics, 8,* 499–508.

Hudson, V. M., & Den Boer, A. M. (2004). *Bare branches: The security implications of Asia's surplus male population.* Cambridge: MIT Press.

Hughes, K. A., Burleson, M. H., & Rodd, F. H. (2003). Is phenotypic plasticity adaptive? In *The Biodemography of human reproduction and fertility* (pp. 23–42). New York: Springer.

Hyde, J. S. (2005). The gender similarities hypothesis. *American Psychologist, 60,* 581–592.

Hyde, J. S. (2012). Nation-level indicators of gender equity in psychological research: Theoretical and methodological issues. *Psychology of Women Quarterly, 36,* 145–148.

Hyde, J. S., & Mertz, J. E. (2009). Gender, culture, and mathematics performance. *Proceedings of the National Academy of Sciences, 106,* 8801–8807.

Hyde, J. S., Fennema, E., & Lamon, S. J. (1990). Gender differences in mathematics performance: A meta-analysis. *Psychological Bulletin, 107,* 139–155.

Hyde, J. S., Mezulis, A. H., & Abramson, L. Y. (2008). The ABCs of depression: Integrating affective, biological, and cognitive models to explain the emergence of the gender difference in depression. *Psychological Review, 115,* 291–313.

Inglehart, R., & Norris, P. (2003). *Rising tide: Gender equality and cultural change around the world.* New York: Cambridge University Press.

Jonason, P. K., Li, N. P., Webster, G. D., & Schmitt, D. P. (2009). The dark triad: Facilitating a short-term mating strategy in men. *European Journal of Personality, 23,* 5–18.

Kagan, J. (1999). The role of parents in children's psychological development. *Pediatrics, 104,* 164–167.

Kaplan, H. S., & Gangestad, S. W. (2005). Life history theory and evolutionary psychology. In D. M. Buss (Ed.), *Handbook of evolutionary psychology* (pp. 68–95). New Jersey: Wiley.

Karremans, J. C., Frankenhuis, W. E., Arons, S. (2010). Blind men prefer a low waist-to-hip ratio. *Evolution and Human Behavior, 31,* 182–186.

Kasser, T., & Sharma, Y. S. (1999). Reproductive freedom, educational equality, and females' preference for resource-acquisition characteristics in mates. *Psychological Science, 10,* 374–377.

Katz, J. N. (1995). *The invention of heterosexuality.* New York: Dutton Press.

Katzmarzyk, P. T., & Leonard, W. R. (1998). Climatic Influences on human body size and proportions: Ecological adaptations and secular trends. *American Journal of Physical Anthropology, 106,* 483–503.

Kearney-Cooke, A. (1998). Gender differences and self-esteem. *The Journal of Gender-Specific Medicine, 2,* 46–52.

Keller, H. (1990). Evolutionary approaches. In J. W. Berry, Y. H. Poortinga, & J. Pandey (Eds.), *Handbook of cross-cultural psychology, 2nd Edition* (Vol. 1, pp. 215–255). Boston: Allyn and Bacon.

Kelly, S. A., Panhuis, T. M., & Stoehr, A. M. (2012). Phenotypic plasticity: Molecular mechanisms and adaptive significance. *Comprehensive Physiology, 2,* 1417-1439.

Kenrick, D. T., Nieuweboer, S., & Buunk, A. P. (2010). Universal mechanisms and cultural diversity: Replacing the Blank Slate with a coloring book. In M. Schaller et al. (Eds.), *Evolution, culture, and the human mind* (pp. 257–272). New York: Psychology Press.

Kirkpatrick, L. A. (2011). The role of evolutionary psychology within an interdisciplinary science of religion. *Religion, 41,* 329–339.

Kitzinger, C. (1994). Should psychologists study sex differences?. *Feminism & Psychology, 4,* 501–506.

Kling, K., Hyde, J., Showers, C., & Buswell, B. (1999). Gender differences in self-esteem: A meta-analysis. *Psychological Bulletin, 125,* 470–500.

Konrad, A. M., Ritchie Jr, J. E., Lieb, P., & Corrigall, E. (2000). Sex differences and similarities in job attribute preferences: A meta-analysis. *Psychological Bulletin, 126,* 593–641.

Krueger, R. F., & Johnson, W. (2008). Behavioral genetics and personality. In, *Handbook of personality: Theory and research* (pp. 287–310).

Lancaster, J. B. (1994). Human sexuality, life histories, and evolutionary ecology. In A. S. Rossi (Ed.), *Sexuality across the life course* (pp. 39–62). Chicago: University of Chicago Press.

Lazarus, J. (2002). Human sex ratios: Adaptations and mechanisms, problems and prospects. In I. C. W. Hardy (Ed.), *Sex ratios: Concepts and research methods* (pp. 287–311). Cambridge: Cambridge University Press.

Li, N. P., Bailey, J. M., Kenrick, D. T., & Linsenmeier, J. A. W. (2002). The necessities and luxuries of mate preferences: Testing the tradeoffs. *Journal of Personality and Social Psychology, 82,* 947–955.

Lippa, R. A. (2005). *Gender, nature, and nurture.* LEA Inc.

Lippa, R. A. (2007). The preferred traits of mates in a cross-national study of heterosexual and homosexual men and women: An examination of biological and cultural influences. *Archives of Sexual Behavior, 36,* 193–208.

Lippa, R. A. (2009). Sex differences in sex drive, sociosexuality, and height across 53 nations: Testing evolutionary and social structural theories. *Archives of Sexual Behavior, 38,* 631–651.

Lippa, R. A. (2010). Sex differences in personality traits and gender-related occupational preferences across 53 nations: Testing evolutionary and social-environmental theories. *Archives of Sexual Behavior, 39,* 619–636.

Lippa, R. A., Collaer, M. L., & Peters, M. (2010). Sex differences in mental rotation and line angle judgments are positively associated with gender equality and economic development across 53 nations. *Archives of Sexual Behavior, 39,* 990–997.

Lombardo, W. K., Cretser, G. A., & Roesch, S. C. (2001). For crying out loud—The differences persist into the '90s. *Sex Roles, 45,* 529–547.

Low, B. S. (1990). Marriage systems and pathogen stress in human societies. *American Zoologist, 30,* 325–340.

Low, B. S. (2000). *Why sex matters*. Princeton: Princeton University Press.

Lucas, R. E., & Gohm, C. L. (2000). Age and sex differences in subjective well-being across cultures. In E. Diener & E. M. Suh (Eds.), *Culture and subjective well-being* (pp. 291–317). Cambridge: MIT Press.

Lukaszewski, A. W., Roney, J. R., Mills, M. E., & Bernard, L. C. (2013). At the interface of social cognition and psychometrics: Manipulating the sex of the reference class modulates sex differences in personality traits. *Journal of Research in Personality, 47,* 953–957.

Lynn, R., & Martin, T. (1997). Gender differences in extraversion, neuroticism, and psychoticism in 37 countries. *Journal of Social Psychology, 137,* 369–373.

Maccoby, E. E. (2000). Perspectives on gender development. *International Journal of Behavioral Development, 24,* 398–406.

MacKinnon, C. A. (1982). Feminism, Marxism, method, and the state: An agenda for theory. *Signs, 7,* 515–544.

Magdol, L., Moffitt, T. E., Caspi, A., Newman, D. L., Fagan, J., & Silva, P. A. (1997). Gender differences in partner violence in a birth cohort of 21-year-olds: Bridging the gap between clinical and epidemiological approaches. *Journal of Consulting and Clinical Psychology, 65,* 68–78.

Marcinkowska, U., Kozlov, M. V., Cai, H., Contreras-Garduno, J., Dixson, B. J., Oana, G. A., Kaminski, G., Li, N. P., Lyons, M. T., Onyishi, I. E., Prasai, K., Pazhoohi, F., Prokop, P., Rosales Cardozo, S. L., Sydney, N., Yong, J. C., & Rantala, M. J. (2014). Cross-cultural variation in men's preference for sexual dimorphism in women's faces. *Biology Letters, 10*(4), 20130850. doi:10.1098/rsbl.2013.0850.

Marlowe, F. W., & Berbesque, J. C. (2012). The human operational sex ratio: effects of marriage, concealed ovulation, and menopause on mate competition. *Journal of Human Evolution, 63,* 834–842.

Massimini, F., & Delle Fave, A. (2000). Individual development in a bio-cultural perspective. *American Psychologist, 55,* 24–33.

McCarthy, M. M., & Arnold, A. P. (2011). Reframing sexual differentiation of the brain. *Nature Neuroscience, 14,* 677–683.

McCrae, R. R. (2002). NEO-PI-R data from 36 cultures: Further intercultural comparisons. In R. R. McCrae & J. Allik (Eds.), *The five-factor model of personality across cultures* (pp. 105–125). New York: Kluwer Academic/Plenum Publishers.

McCrae, R. R., & Terracciano, A. (2005). Universal features of personality traits from the observer's perspective: Data from 50 cultures. *Journal of Personality and Social Psychology, 88,* 547–561.

McCullough, M. E., Carter, E. C., DeWall, C. N., & Corrales, C. M. (2012). Religious cognition down-regulates sexually selected, characteristically male behaviors in men, but not in women. *Evolution and Human Behavior, 33,* 562–568.

McHoskey, J. W. (2001). Machiavellianism and sexuality: On the moderating role of biological sex. *Personality and Individual Differences, 31,* 779–789.

Mealey, L. (1985). The relationship between social status and biological success: A case study of the Mormon religious hierarchy. *Ethology & Sociobiology, 6,* 249–257.

Mealey, L. (2000). *Sex differences: Developmental and evolutionary strategies*. San Diego: Academic.

Mesoudi, A., Whiten, A., & Laland, K. N. (2006). Towards a unified science of cultural evolution. *Behavioral and Brain Sciences, 29,* 329–383.

Micttunen, J., Veijola, J., Lauronen, E., Kantojärvi, L., & Joukamaa, M. (2007). Sex differences in Cloninger's temperament dimensions—a meta-analysis. *Comprehensive Psychiatry, 48,* 161–169.

Miller, D. I., & Halpern, D. F. (2014). The new science of cognitive sex differences. *Trends in Cognitive Sciences, 18,* 37–45.

Moore, F. R., Coetzee, V., Contreras-Garduno, J., Debruine, L. M., Kleisner, K., Krams, I., Marcinkowska, U., Nord, A., & Perrett, D. I., Rantala, M. J., Schaum, N., Suzuki, T. N. (2013). Cross-cultural variation in women's preferences for cues to sex- and stress-hormones in the male face. *Biology Letters, 9,* 1–4.

Napier, J. L., Thorisdottir, H., & Jost, J. T. (2010). The joy of sexism? A multinational investigation of hostile and benevolent justifications for gender inequality and their relations to subjective well-being. *Sex Roles, 62,* 405–419.

Nesse, R., & Williams, G. C. (1994). *Why we get sick: The new science of Darwinian medicine*. New York: Vintage Books.

Nettle, D. (2002). Women's height, reproductive success and the evolution of sexual dimorphism in modern humans. *Proceedings of the Royal Society of London Series B-Biological Sciences, 269,* 1919–1923.

Nettle, D. (2007). Empathizing and systemizing: What are they, and what do they contribute to our understanding of psychological sex differences? *British Journal of Psychology, 98,* 237–255.

Nettle, D. (2009). Ecological influences on human behavioural diversity: A review of recent findings. *Trends in Ecology & Evolution, 24,* 618–624.

Ngun, T. C., Ghahramani, N., Sanchez, F. J., Bocklandt, S., Vilain, E. (2011). The genetics of sex differences in brain and behavior. *Frontiers in Neuroendocrinology, 32,* 227–246.

Nolen-Hoeksema, S. (2001). Gender differences in depression. *Current Directions in Psychological Science, 10,* 173–176.

Nordenstroìm, A., Servin, A., Bohlin, G., Larsson, A., & Wedell, A. (2002). Sex-typed toy play behavior correlates with the degree of prenatal androgen exposure assessed by CYP21 genotype in girls with congenital adrenal hyperplasia. *Journal of Clinical Endocrinology & Metabolism, 87,* 5119–5124.

Norenzayan, A., & Heine, S. J. (2005). Psychological universals: What are they and how can we know? *Psychological bulletin, 131,* 763–874.

Pasco, J. A., Nicholson, G. C., Brennan, S. L., & Kotowicz, M. A. (2012). Prevalence of obesity and the relationship between the body mass index and body fat: Cross-sectional, population-based data. *PloS One, 7,* e29580.

Pasterski, V. L., Geffner, M. E., Brain, C., Hindmarsh, P., Brook, C., & Hines, M. (2005). Prenatal hormones and postnatal socialization by parents as determinants of male-typical toy play in girls with congenital adrenal hyperplasia. *Child Development, 76,* 264–278.

Pedersen, F. A. (1991). Secular trends in human sex ratios: Their influence on individual and family behavior. *Human Nature, 2,* 271–291.

Petersen, J. L., & Hyde, J. S. (2010). A meta-analytic review of research on gender differences in sexuality, 1993–2007. *Psychological Bulletin, 136,* 21–38.

Pigliucci, M. (2001). *Phenotypic plasticity: Beyond nature and nurture*. New York: Johns Hopkins University Press.

Pirlott, A., & Schmitt, D. P. (2014). Gendered sexual culture. In A. Cohen (Ed.), *New directions in the psychology of culture* (pp. 191–216). Washington, DC: American Psychological Association Books.

Plomin, R. (2008). *Behavioral genetics*. New York: Macmillan.

Puts, D. A. (2010). Beauty and the beast: Mechanisms of sexual selection in humans. *Evolution and Human Behavior, 31,* 157–175.

Quinlan, R. J. (2007). Human parental effort and environmental risk. *Proceedings of the Royal Society B: Biological Sciences, 274,* 121–125.

Ruble, D. N., & Martin, C. L. (1998). Gender development. In W. Damon & N. Eisenberg (Eds.), *Handbook of child psychology* (5th ed., pp. 933–1016). Hoboken: Wiley.

Rudman, L. A., Fetterolf, J. C., & Sanchez, D. T. (2013). What motivates the sexual double standard? More support for male versus female control theory. *Personality and Social Psychology Bulletin, 39,* 250–263.

Ruff, C. (2002). Variation in human body size and shape. *Annual Review of Anthropology, 31,* 211–232.

Ruigrok, A. N., Salimi-Khorshidi, G., Lai, M. C., Baron-Cohen, S., Lombardo, M. V., Tait, R. J., & Suckling, J. (2013). A meta-analysis of sex differences in human brain structure. *Neuroscience & Biobehavioral Reviews, 39,* 34–50.

Santiago-Menendez, M., & Campbell, A. (2013). Sadness and anger: Boys, girls, and crying in adolescence. *Psychology of Men & Masculinity, 14,* 400–410.

Schaller, M., & Murray, D. R. (2008). Pathogens, personality, and culture: Disease prevalence predicts worldwide variability in sociosexuality, extraversion, and openness to experience. *Journal of Personality and Social Psychology, 95,* 212.

Schmitt, D. P. (2005a). Fundamentals of human mating strategies. In D. M. Buss (Ed.), *The evolutionary psychology handbook* (pp. 258–291). New York: Wiley.

Schmitt, D. P. (2005b). Sociosexuality from Argentina to Zimbabwe: A 48-nation study of sex, culture, and strategies of human mating. *Behavioral and Brain Sciences, 28,* 247–275.

Schmitt, D. P. (2011). Psychological adaptation and human fertility patterns: Some evidence of human mating strategies as evoked sexual culture. In A. Booth & A. C. Crouter (Eds.), *Romance and sex in adolescence and emerging adulthood: Risks and opportunities* (pp. 161–170). Mahwah: Lawrence Erlbaum Associates.

Schmitt, D. P. (2012). When the difference is in the details: A critique of Zentner and Mitura (2012), Stepping out of the caveman's shadow: Nations' gender gap predicts degree of sex differentiation in mate preferences. *Evolutionary Psychology, 10,* 720–726.

Schmitt, D. P. (2014). Evaluating evidence of mate preference adaptations: How do we really know what *Homo sapiens sapiens* really want? In V. A. Weekes-Shackelford & T. K. Shackelford (Eds.), *Evolutionary perspectives on human sexual psychology and behavior* (pp. 3–39). New York: Springer.

Schmitt, D. P., & Buss, D. M. (2000). Sexual dimensions of person description: Beyond or subsumed by the Big Five? *Journal of Research in Personality, 34,* 141–177.

Schmitt, D. P., & Rohde, P. A. (2013). The Human Polygyny Index and its ecological correlates: Testing sexual selection and life history theory at the cross-national level. *Social Science Quarterly, 94,* 1159–1184.

Schmitt, D. P., Alcalay, L., Allensworth, M., Allik, J., Ault, L., Austers, I., et al. (2003). Are men universally more dismissing than women? Gender differences in romantic attachment across 62 cultural regions. *Personal Relationships, 10,* 307–331.

Schmitt, D. P., Alcalay, L., Allik, J., Angleiter, A., Ault, L., Austers, I., et al. (2004). Patterns and universals of mate poaching across 53 nations: The effects of sex, culture, and personality on romantically attracting another person's partner. *Journal of Personality and Social Psychology, 86,* 560–584.

Schmitt, D. P., Realo, A., Voracek, M., & Alik, J. (2008). Why can't a man be more like a woman? Sex differences in Big Five personality traits across 55 cultures. *Journal of Personality and Social Psychology, 94,* 168–192.

Schmitt, D. P., Youn, G., Bond, B., Brooks, S., Frye, H., Johnson, S., Klesman, J., Peplinski, C., Sampias, J., Sherrill, M., & Stoka, C. (2009). When will I feel love? The effects of personality, culture, and gender on the psychological tendency to love. *Journal of Research in Personality, 43,* 830–846.

Schmitt, D. P., Alcalay, L., Allik, J., Alves, I. C. B., Anderson, C. A., Angelini, A. L., Asendorpf, J. B., et al. (2014). *Psychological sex differences across cultures: Findings from the International Sexuality Description Project-2*. Manuscript in preparation.

Schneider, H. J., Pickel, J., & Stalla, G. K. (2006). Typical female 2nd–4th finger length (2D:4D) ratios in male-to-female transsexuals—possible implications for prenatal androgen exposure. *Psychoneuroendocrinology, 31,* 265–269.

Schwartz, S. H., & Rubel, T. (2005). Sex differences in value priorities: Cross-cultural and multimethod studies. *Journal of Personality and Social Psychology, 89,* 1010–1028.

Schwartz, S. H., & Rubel-Lifschitz, T. (2009). Cross-national variation in the size of sex differences in values: Effects of gender equality. *Journal of Personality and Social Psychology, 97,* 171–187.

Sela, Y., Shackelford, T. K., & Liddle, J. R. (2014). When religion makes it worse: Religiously-motivated violence as a sexual selection weapon. In J. Slone & J. Van Slyke (Eds.), *The attraction to religion* (pp. 1–30). London: Bloomsbury Academic.

Sidanius, J., & Pratto, F. (1999). *Social dominance: An intergroup theory of social hierarchy and oppression*. New York: Cambridge University Press.

Sidanius, J., Levin, S., Liu, J., & Pratto, F. (2000). Social dominance orientation, anti-egalitarianism and the political psychology of gender: An extension and cross-cultural replication. *European Journal of Social Psychology, 30,* 41–67.

Silverman, I., Phillips, K., & Silverman, L. K. (1996). Homogeneity of effect sizes for sex across spatial tests and cultures: Implications for hormonal theories. *Brain and Cognition, 31,* 90–94.

Silverman, I., Choi, J., & Peters, M. (2007). The hunter-gatherer theory of sex differences in spatial abilities: Data from 40 countries. *Archives of Sexual Behavior, 36,* 261–268.

Simpson, J. A., & Gangestad, S. W. (1991). Individual differences in sociosexuality: Evidence for convergent and discriminant validity. *Journal of Personality and Social Psychology, 60,* 870–883.

Sprecher, S., Sullivan, Q., & Hatfield, E. (1994). Mate selection preferences: Gender differences examined in a national sample. *Journal of Personality and Social Psychology, 66,* 1074–1080.

Stearns, S. C., & Koella, J. C. (1986). The evolution of phenotypic plasticity in life-history traits: Predictions of reaction norms for age and size at maturity. *Evolution, 40,* 893–913.

Stevens, J. S., & Hamann, S. (2012). Sex differences in brain activation to emotional stimuli: A meta-analysis of neuroimaging studies. *Neuropsychologia, 50,* 1578–1593.

Straus, M. A. (1979). Measuring intrafamily conflict and violence. The Conflict Tactics (CT) Scales. *Journal of Marriage and the Family, 51,* 75–88.

Straus, M. A. (2008). Dominance and symmetry in partner violence by male and female university students in 32 nations. *Children and Youth Services Review, 30,* 252–275.

Su, R., Rounds, J., & Armstrong, P. I. (2009). Men and things, women and people: A meta-analysis of sex differences in interests. *Psychological Bulletin, 135,* 859–884.

Thornhill, R., & Gangestad, S. W. (2008). *The evolutionary biology of human female sexuality*. New York: Oxford University Press.

Tooby, J., & Cosmides, L. (1990). The past explains the present: Emotional adaptations and the structure of ancestral environments. *Ethology and Sociobiology, 11,* 375–424.

Tooby, J. & Cosmides, L. (1992). Psychological foundations of culture. In J. Barkow, L. Cosmides, & J. Tooby (Eds.), *The adapted mind* (pp. 19–136). New York: Oxford University Press.

Touraille, P. (2013). Human sex differences in height: Evolution due to gender hierarchy? In M. Ah-King (Ed.), *Challenging popular myths of sex, gender and biology* (pp. 65–75). New York: Springer International Publishing.

Trabzuni, D., Ramasamy, A., Imran, S., Walker, R., Smith, C., Weale, M. E., & North American Brain Expression Consortium. (2013). Widespread sex differences in gene expression and splicing in the adult human brain. *Nature communications, 4,* 2771.

Trivers, R. (1972). Parental investment and sexual selection. In B. Campbell (Ed.), *Sexual selection and the descent of man: 1871–1971* (pp. 136–179). Chicago: Aldine

Udry, J. R. (2000). Biological limits of gender construction. *American Sociological Review, 65,* 443–457.

Udry, J. R., Morris, N. M., & Kovenock, J. (1995). Androgen effects on women's gendered behaviour. *Journal of Biosocial Science, 27,* 359–359.

Van Damme, R., Entin, P., Vanhooydonck, B., & Herrel, A. (2008). Causes of sexual dimorphism in performance traits: A comparative approach. *Evolutionary Ecology Research, 10,* 229–250.

van Hemert, D. A., van de Vijver, F. J. R., & Vingerhoets, A. J. J. M. (2011). Culture and crying: Prevalences and gender differences. *Cross-Cultural Research: The Journal of Comparative Social Science, 45,* 399–431.

Vandermassen, G. (2011). Evolution and rape: A feminist Darwinian perspective. *Sex Roles, 64,* 732–747.

Van de Velde, S., Bracke, P., & Levecque, K. (2010). Gender differences in depression in 23 European countries. Cross-national variation in the gender gap in depression. *Social Science & Medicine, 71,* 305–313.

Via, S., Gomulkiewicz, R., De Jong, G., Scheiner, S. M., Schlichting, C. D., & Van Tienderen, P. H. (1995). Adaptive phenotypic plasticity: Consensus and controversy. *Trends in Ecology & Evolution, 10,* 212–217.

Vingerhoets, A. J. J. M., & Scheirs, J. (2000). Sex differences in crying: Empirical findings and possible explanations. In A. H. Fischer (Ed.), *Gender and emotion: Social psychological perspectives* (pp. 143–165). Cambridge: Cambridge University Press.

Voyer, D., Voyer, S., & Bryden, M. P. (1995). Magnitude of sex differences in spatial abilities: A meta-analysis and consideration of critical variables. *Psychological Bulletin, 117,* 250–270.

Voyer, D., Postma, A., Brake, B., & Imperato-McGinley, J. (2007). Gender differences in object location memory: A meta-analysis. *Psychonomic Bulletin & Review, 14,* 23–38.

Weeden, J., & Kurzban, R. (2013). What predicts religiosity? A multinational analysis of reproductive and cooperative morals. *Evolution and Human Behavior, 34,* 440–445.

Weisberg, Y. J., DeYoung, C. G., & Hirsh, J. B. (2011). Gender differences in personality across the ten aspects of the Big Five. *Frontiers in Psychology, 2,* 1–11.

Wheatley, J. R., Apicella, C. A., Burriss, R. P., Cárdenas, R. A., Bailey, D. H., Welling, L. L., & Puts, D. A. (2014). Women's faces and voices are cues to reproductive potential in industrial and forager societies. *Evolution and Human Behavior, 35,* 264–271.

Whissell, C. M. (1996). Predicting the size and direction of sex differences in measures of emotion and personality. *Genetic, Social, and General Psychology Monographs, 122,* 253–284.

Whitman, D. W., & Agrawal, A. A. (2009). What is phenotypic plasticity and why is it important? *Phenotypic Plasticity of Insects, 10,* 1–63.

Williams, J. E. & Best, D. L. (1990). *Sex and psyche: Gender and self viewed cross-culturally.* Newbury Park: Sage.

Winegard, B. M., Winegard, B. M., & Deaner, R. O. (2014). Misrepresentations of evolutionary psychology in sex and gender textbooks. *Evolutionary Psychology, 12,* 474–508.

Wood, W., & Eagly, A. H. (2002). A cross-cultural analysis of the behavior of women and men: Implications for the origins of sex differences. *Psychological Bulletin, 128,* 699–727.

Wood, W. & Eagly, A. H. (2012). Biosocial construction of sex differences and similarities in behavior. In M. P. Zanna & J. M. Olson (Eds.), *Advances in experimental social psychology.* San Diego: Academic.

Zentner, M., & Mitura, K. (2012). Stepping out of the caveman's shadow: Nations' gender gap predicts degree of sex differentiation in mate preferences. *Psychological Science, 23,* 1176–1185.

Chapter 12
Sperm Competition and the Evolution of Human Sexuality

Michael N. Pham and Todd K. Shackelford

Sperm Competition and the Evolution of Human Sexuality

Sperm competition occurs when a female copulates with two or more males within a sufficiently brief time period, with the result that the sperm of different males compete to fertilize ova (Parker 1970). Humans and most birds are socially monogamous: Males and females form long-term pair bonds and occasionally pursue extra-pair copulations (EPCs). Among socially monogamous species, sperm competition most commonly occurs when females pursue EPCs (Smith 1984). A male whose regular partner pursues EPCs is at risk of cuckoldry—the unwitting investment of resources into offspring to whom he is genetically unrelated. The costs of cuckoldry may have driven the evolution of male sperm competition tactics—strategic adjustments in psychology, behavior, and physiology that increase sperm competition success. Because males have finite resources for survival and reproduction, males *judiciously* deploy sperm competition tactics: Males attend to specific sperm competition cues and adjust accordingly their sperm competition tactics.

Do Humans Have an Evolutionary History of Sperm Competition?

Cuckoldry is likely to have recurred over human evolution. Current estimates document nonzero rates of discrepant social and genetic fatherhood (Bellis et al. 2005; Anderson 2006; Voracek et al. 2008; Wolf et al. 2012), and a meta-analysis of 32 published studies documented that 3.1% of children are genetically unrelated to their social father (Voracek et al. 2008). Anderson (2006) provided evidence that 29.8% of men with *low* paternity confidence (e.g., those disputing paternity), compared to 1.7% of men with *high* paternity confidence are genetically unrelated to

M. N. Pham (✉) · T. K. Shackelford
Department of Psychology, Oakland University, Rochester, MI 48309, USA
e-mail: mnpham@oakland.edu

T. K. Shackelford, R. D. Hansen (eds.), *The Evolution of Sexuality*,
Evolutionary Psychology, DOI 10.1007/978-3-319-09384-0_12

their child. These results suggest that men's *perceived* cuckoldry risk may reasonably predict their *actual* cuckoldry risk.

Male sexual jealousy provides evidence that cuckoldry recurred over human evolution. Jealousy motivates men to minimize the occurrence of their partner's EPCs, and the jealousy is so strong, in fact, that it is the leading cause of partner-killing across cultures (Buss 2006; Daly and Wilson 1988). Male sexual jealousy could not have evolved without the evolutionary recurrence of cuckoldry (Buss 2013).

Double-Mating

Human sperm competition occurs when a woman "double-mates" or copulates with two or more men within about five days (sperm remain alive in the female reproductive tract for about five days; Baker and Bellis 1995). Using data collected from a nationwide survey of British women, Baker and Bellis (1995) reported the percentage of women who ever double-mated as a function of their sexual experience, operationalized as lifetime number of copulations. For women reporting fewer than 50 lifetime copulations, 17.5 % reported having double-mated at least once. This percentage increases steeply with sexual experience: 71.8 % of women reporting more than 1000 lifetime copulations reported double-mating at least once. Gallup et al. (2006) found that 13.4 % of a sample of American college women copulated with two men within a 24-h period at least once, and 8.3 % copulated with two men simultaneously at least once.

The temporal window for human sperm competition may extend beyond the temporal window for sperm viability. For example, dead sperm may impede cervical pathways for sperm of subsequent ejaculates (Baker and Bellis 1995). Additionally, research has yet to investigate the extent to which nonsperm substances in semen (e.g., spermicidals; Baker and Bellis 1995) might influence the outcome of human sperm competition or the duration that these substances remain potent in the female reproductive tract. For example, seminal fluid in the female reproductive tract may trigger ovulation (reviewed in Gallup and Burch 2006; Gallup et al. 2012).

Additionally, ancestral women's double-mating must sometimes have occurred such that competing sperm were present during the periods of conception risk. Indirect evidence corroborates this premise: Modern women are more likely to pursue EPCs during peak fertility (Gangestad et al. 2005; Pillsworth and Haselton 2006), in-pair copulation (IPC) frequency remains largely uniform across each fertility cycle (except during menses, when IPC frequency decreases precipitously; reviewed in Brewis and Meyer 2005), and men pursue IPCs at a frequency which "tops up" their in-pair partner's reproductive tract with a continuous population of viable sperm (Baker and Bellis 1993a).

Relative Testis Size

Relative testis size—the ratio of testis weight to body weight—is a reliable proxy both for sperm count and for sperm competition level: A species' level of sperm competition typically correlate with its relative testis size (reviewed in Simmons and Fitzpatrick 2012). Additionally, experimentally increasing sperm competition level also increases relative testis size (Hosken and Ward 2001). Although relative testis size in humans is smaller than in highly polyandrous primates, it is larger than in relatively monogamous primates (Short 1981), indicating that sperm competition may have over human evolution. In the remainder of this chapter, we examine evidence of physiological, psychological, behavioral, and anatomical adaptations to the human sperm competition from a comparative evolutionary perspective.

Ejaculate Adjustment

Sperm are "tickets" for the "lottery prize" of fertilizing ova (Parker 1970; Wedell et al. 2002). Males at greater sperm competition risk ejaculate more sperm to increase the probability that their sperm—and not rival sperm—fertilize ova (Wedell et al. 2002). For example, among many avian species, males at greater sperm competition risk ejaculate more sperm at the next copulation (Nicholls et al. 2001; Pizzari 2003). In humans, men at greater "objective" sperm competition risk—the proportion of time they spend apart from their regular partner since their last IPC—ejaculate more sperm at their next IPC (Baker and Bellis 1993a). Males also may adjust other semen parameters in response to the sperm competition risk. In humans, men produce masturbatory ejaculates containing a greater percentage of motile sperm when viewing pornography depicting one woman interacting with two men (i.e., indexing sperm competition), than when viewing pornography depicting three women interacting (i.e., indexing the absence of sperm competition; Kilgallon and Simmons 2005).

Whether humans can adjust their ejaculate in a manner predicted by sperm competition theory remains debated. Critics have cited the decline in semen quality during recent decades as evidence against human sperm competition (e.g., Auger et al. 1995; Bostofte et al. 1982). Future research should employ experimental methodologies (e.g., priming thoughts of partner infidelity) using larger samples sizes.

IPC Interest

Males may possess a sperm competition psychology—a set of information-processing mechanisms that motivate males to judiciously deploy sperm competition tactics. These mechanisms are activated when males perceive sperm competition cues (e.g., female attractiveness: Cornwallis and O'Connor 2009; presence of rival males: Nicholls et al. 2001) and produce outputs that motivate them to deploy sperm competition tactics (Goetz et al. 2007).

Among many nonhuman species, males attend to the presence of sexual rivals and adjust accordingly their sperm competition tactics (Candolin and Reynolds 2002; Gage and Barnard 1996; Nicholls et al. 2001). In humans, men at greater objective sperm competition risk report greater IPC interest, and greater distress (e.g., anger, upset, frustration, and persistence) if their regular partner denies their IPC request (Shackelford et al. 2002, 2007), but *only* among men who perceive that their partner spends more time with male friends (i.e., potential sexual rivals; Pham and Shackelford 2013a). Men assess a greater risk of sperm competition from their regular partner's absence because it is during her absence that they cannot account for her sexual behavior with other men. Additionally, men experimentally primed with thoughts of partner infidelity reported distress if their regular partner denies their IPC request (Starratt et al. 2013). This line of research suggests that men possess a sperm competition psychology that regulates their IPC interest in accordance with their sperm competition risk.

Forced IPC

Men at greater sperm competition risk experience greater distress if their partner denies an IPC request (Pham and Shackelford 2013a; Shackelford et al. 2002, 2007; Starratt et al. 2013), which suggests that men might sometimes pursue IPCs *forcefully*. That is, men at greater sperm competition risk may be motivated to submit their sperm—sometimes forcefully—into competition with rival sperm that may be, or will be, in their regular partner's reproductive tract. In humans, women whose regular partner is unattractive report lower IPC interest and greater EPC interest during the high-fertility phase of their ovulatory cycle (Gangestad et al. 2005). Thus, forced IPCs may be a sperm competition tactic that results from sexual conflict—when male mating strategies and female mating strategies are at odds (Shackelford and Goetz 2012).

Among socially monogamous avian species, forced IPCs occur predictably (and immediately) following events that index cuckoldry risk (e.g., intrusion of rival males, female absence, partner-observed female EPCs; Barash 1997; Cheng et al. 1983; reviewed in Goetz and Shackelford 2006). In humans, men at greater sperm competition risk are more likely to perform forced IPCs (reviewed in Goetz et al. 2008). For example, men who perform a forced IPC also are more likely to accuse their regular partner of EPCs around the time they perform the forced IPC (Finkelhor and Yllo 1985; Russell 1982). According to both men's reports and women's reports, men who are more sexually coercive toward their regular partner also are more likely to verbally impugn her sexual fidelity (Starratt et al. 2008). Among women who are physically abused, those whose regular partner performed forced IPCs also report that he is more sexually jealous (Frieze 1983; Gage and

Hutchinson 2006). Men who report sexually coercing their regular partner also report greater suspicion of her EPC, and women who report a greater likelihood of pursuing EPCs also report that their regular partner is more sexually coercive (Goetz and Shackelford 2006, 2009). Among men convicted of physically assaulting their regular partner, only those who performed forced IPCs—relative to those who did not perform forced IPCs—experienced greater cuckoldry risk events prior to the assault (Camilleri and Quinsey 2009). Men at greater objective sperm competition risk are more likely to perform forced IPCs, but only those men who perceive a greater likelihood of their partner's EPC (McKibbin et al. 2011). Forced IPCs provide insight into the behaviors males occasionally perform to avoid cuckoldry: Men may even kill their partner to avoid cuckoldry (Buss 2006; Shackelford et al. 2003).

Frequent IPCs

Frequent IPC may function as a sperm competition tactic. A male who performs frequent IPCs maintains large numbers of viable sperm in his partner's reproductive tract across her fertility cycle, thereby increasing his chances of success in sperm competition. In many avian species, males at greater sperm competition risk perform frequent IPCs. For example, male Montagu harriers (*Circus pygagus*) experimentally exposed to a rival male—relative to no rival males—performed more frequent IPCs (Mougeot et al. 2001). Thus, males may strategically adjust IPC frequency according to sperm competition risk.

In humans, frequent IPCs also may function as a sperm competition tactic. For example, a man performs frequent IPCs if his regular partner is more attractive (Kaighobadi and Shackelford 2008), if she has more male friends and male coworkers (i.e., potential sexual rivals; Pham et al. in press), and if he performs more mate guarding behaviors (Shackelford et al. 2006). Additionally, deployment of frequent IPCs as a sperm competition tactic may explain why a man (but not a woman) reports continued IPC interest over the duration of a romantic relationship (Klusmann 2002, 2006).

Frequent IPCs may serve other functions, as well. For example, male storks (*Ciconia Ciconia*) perform frequent IPCs to advertise their physical quality to their regular partner (Tortosa and Redondo 1992). Frequent IPCs also may function to solve male adaptive problems associated with concealed female fertility status: Women, unlike females of many nonhuman species, do not display overt cues to their fertility status. Frequent IPCs may function as a male strategy to maintain large numbers of viable sperm in his partner's reproductive tract across her fertility cycle, ensuring that his sperm are present to fertilize her ova. Although men may be able to detect fertility cues (Haselton and Gildersleeve 2011), frequent IPCs may offer "insurance" against cuckoldry. This may also explain why IPC frequency decreases during menses—a time during which ova cannot be fertilized (reviewed in Brewis and Meyer 2005).

Semen Displacement

Among several nonhuman species, males have anatomical, behavioral, and physiological adaptations to displace rival semen from a female's reproductive tract. For example, male tree crickets (*Truljalia hibinosis*) can remove nearly 90 % of a rival's ejaculate from a female's reproductive tract during copulation (Ono et al. 1989).

Humans also may have adaptations to displace semen. Using artificial human penises, artificial female reproductive tracts, and semen-like fluid, Gallup et al. (2003) provided evidence that the human penis may be able to displace semen from the female reproductive tract during copulatory thrusting. Goetz et al. (2005) found that men at greater recurrent sperm competition risk perform more semen-displacing copulatory behaviors (e.g., deeper and more thrusts), and Gallup et al. found that men performed more semen-displacing copulatory behaviors during copulations that followed allegations of partner infidelity, both my men's reports and women's reports. Additionally, men experience post-ejaculatory events that prevent them from displacing their own semen, including decreased copulatory interest with the same woman (reviewed in Gallup and Burch 2004), but *not* with novel women (i.e., Coolidge Effect; O'Donohue and Plaud 1991).

The human penis may displace semen, in general, and not sperm, in particular. Only sperm that do not travel beyond the reproductive tract—sperm that have little chance of fertilizing ova—could be displaced by the penis. However, human semen includes substances that can influence fertilizing success, and the human penis may displace such nonsperm substances. For example, there is some evidence that the posterior portion of human ejaculates contain spermicides that can reduce the fertilizing success of subsequent ejaculates (Baker and Bellis 1995). Thus, men can increase their chances of success in sperm competition by displacing spermicidal substances left by a rival male in a female's reproductive tract.

Human penile foreskin may affect semen displacement. Circumcised penises have more pronounced coronal ridges—the region where semen collects during semen-displacing copulatory behaviors—which might afford greater semen displacement (Gallup and Burch 2004). Additionally, sperm of one man may "piggyback" under the foreskin of another man's penis (Gallup and Burch 2004). If an uncircumcised man (Male A) copulates with a woman (Female A) whom recently copulated with a different man (Male B), then Male A may inadvertently capture underneath his foreskin semen of Male B. Then, if Male A copulates with a different woman (Female B), he may inadvertently deposit semen of Male B into Female B. An unlikely but possible outcome of this phenomenon is that a woman conceives by a man with whom she has never copulated. This hypothesis has not yet been empirically tested.

Sexual Arousal

Male sexual arousal may proximately cause the deployment of sperm competition tactics. Men's pornography preferences provide insight into male sexual arousal because pornography is produced largely to facilitate or enhance male sexual arousal (Mosher 1988). Therefore, the popularity and prevalence of a specific form of pornography suggest the extent to which that form of pornography is sexually arousing. Pound (2002) documented that pornography depicting two men interacting with one woman ("MMF"; i.e., indexing sperm competition) is more prevalent than pornography depicting two women interacting with one man ("MFF"; i.e., indexing the absence of sperm competition). These findings are consistent with men's reports of their preferences in pornography (Pound 2002; cf. Hald 2006). McKibbin et al. (2013) found that the frequency of images on adult DVD covers depicting MMF pornography (i.e., indexing sperm competition) predicted the DVD's sales rank, whereas the frequency of images on adult DVD covers depicting MFF pornography (i.e., indexing the absence of sperm competition) did *not* predict the DVD's sales rank. Consistent with findings from Pound (2002) and from McKibbin et al. (2013), men produce more competitive ejaculates when viewing pornography that indexes sperm competition than when viewing pornography that indexes the absence of sperm competition (Kilgallon and Simmons 2005).

An alternative explanation is that men prefer MMF pornography because they are super-stimulated by witnessing simultaneous, multiple, sexual acts (e.g., oral sex, vaginal sex). However, Pound (2002) and McKibbin et al. (2013) found that MMF pornography was more arousing than pornography depicting multiple males and multiple females—scenes which likely contained the most frequent representation of simultaneous, multiple, sexual acts. Thus, men likely prefer MMF pornography because of adaptations to sperm competition, and not because men are hypersexual.

Men's sexual arousal causes behavioral and physiological adjustments that may function as sperm competition tactics. For example, men who are more sexually aroused express greater copulatory interest (Ariely and Loewenstein 2006). Additionally, men produce less competitive ejaculates following masturbatory (less sexually arousing) events relative to copulatory (more sexually arousing) events (Zavos 1985; Zavos and Goodpasture 1989). Furthermore, men who are experimentally sexually aroused report a greater likelihood of performing sexually aggressive behaviors (Ariely and Loewenstein 2006). As men are sexually aroused when viewing sperm competition cues, and because their sexual arousal causes deployment of sperm competition tactics, assessing men's sexual arousal provides insight into human sperm competition psychology.

Oral Sex

Here, we define oral sex as oral stimulation of genitals. Individuals report performing oral sex for many reasons. Cornell and Halpern-Felsher (2006) surveyed 425 young men and women who reported that they perform oral sex to retain virginity (because oral sex is sometimes not perceived as "real" sex; Sanders and Reinisch 1999), to increase their sexual reputation, to sexually satisfy their partner, and to avoid the risk of pregnancy and diseases associated with penile–vaginal sex.

There is much evidence suggesting a human evolutionary history of oral sex. First, oral sex occurs in many cultures (e.g., Guadamuz et al. 2010; Iwawaki and Wilson 1983; Lurie et al. 1995), including some preindustrial cultures (Hewlett and Hewlett 2010), indicating that oral sex is not a culture-specific practice. Second, oral sex is depicted frequently in pornography, and pornography appeals to evolved mechanisms: Humans do not possess adaptations to experience sexual arousal in response to viewing computer images, but to viewing "real-life" humans with whom they can copulate (Malamuth 1996). Third, oral sex is depicted in ancestral cave paintings (Angulo and García 2005). Fourth, oral sex occurs across many species (Nishimura et al. 1991; Palagi et al. 2003; Soini 1987), and behaviors that occur across species indicate the possibility of convergent adaptations.

The relatively unsanitary conditions of ancestral environments may have discouraged ancestral humans from performing oral sex (Gallup et al. 2012). Humans possess "pathogen-avoidance mechanisms" (e.g., experiencing disgust in response to smelling feces; Oaten et al. 2009), and ancestral humans—relative to modern humans—lived in relatively unsanitary conditions. However, ancestral humans—relative to modern humans—were likely far less concerned with sanitary practices. In fact, research on the immune system suggests that not only were ancestral humans recurrently exposed to various "unsanitary" microbes, but that this exposure selected for a robust immune system (Rook 2007). We do not argue that oral sex is an adaptation to strengthen the immune system. Rather, we argue that ancestral humans may have been less concerned about the sanitary conditions surrounding oral sex, and that oral sex recurred over human evolutionary history. Next, we review how oral sex performed by men on women (cunnilingus) may be related to sperm competition.

Mate Retention Mate retention behaviors reduce the risk of a regular partner's EPC (Buss 1988; Buss and Shackelford 1997). Men may perform oral sex on their regular partner to minimize their sperm competition risk by reducing her motivation to pursue EPCs. Women who receive cunnilingus are more sexually satisfied (Richters et al. 2006), and women who are more satisfied with their regular partner are less likely to pursue EPCs (Santtila et al. 2007). Pham and Shackelford (2013b) found that men who report greater interest in and spend more time performing cunnilingus on their regular partner also report performing more mate retention behaviors.

Infidelity-Detection Men may perform cunnilingus to assess sperm competition risk. Because a portion of semen remains in the reproductive tract following insemination (Baker and Bellis 1993b), men may assess the likelihood of their partner's EPC by smelling and tasting the presence of rival semen that may be in or near her genitals (Pham and Shackelford 2013c; Thornhill 2006). The infidelity-detection hypothesis may explain why men typically perform cunnilingus before (and not after) they ejaculate (Halpern and Sherman 1979): Men's own semen might "mask" the odor of rival semen. However, it is possible that men are simply repulsed by their own semen. Indirect evidence indicates that humans can smell the semen of others. For example, fertility clinicians record the odors they smell from semen as a part of semen quality analysis (e.g., Mauras et al. 2005). Pham and Shackelford

(2013b) found that men who perceive their regular partner to be more attractive—a sperm competition cue—also report greater interest in and spend more time performing cunnilingus on her.

Fertility-Detection Males may perform cunnilingus to assess the health and fertility status of females. Males may sniff and lick (i.e., oral sex) female genitals to gather scent cues to her fertility status. For example, male cotton-top tamarins (*Saguinus Oedipus*) that smell a female's scent marks produced at high-fertility—relative to at low-fertility—experience more frequent penile erections and perform more mounting behaviors (Ziegler et al. 1993). Gathering scent cues to fertility status also may explain why males of several nonhuman species more frequently perform cunnilingus on high-fertility females than on low-fertility females (Dunbar 1977; Johnston 1974; Kiddy et al. 1978; Murphy 1973; Nishimura et al. 1991; Palagi et al. 2003; Sankar and Archunan 2004; Soini 1987). In humans, men may be able to detect a female's fertility status through scent cues. For example, men rate female genital odors that are produced during high-fertility (relative to low-fertility) as more pleasant smelling (Cerda-Molina et al. 2013; Doty et al. 1975).

Sperm Retention Men may perform cunnilingus to promote female orgasm and consequent sperm retention. Female orgasm may promote conception via uterine contractions, which pull sperm further into the reproductive tract (Fox et al. 1970; Wildt et al. 1998; Zervomanolakis et al. 2007, 2009), and women who orgasm temporally near their partner's ejaculation retain more sperm in their reproductive tract (Baker and Bellis 1993b). Additionally, men at greater objective sperm competition risk report greater interest in their regular partner's orgasm (McKibbin et al. 2010). During a sexual encounter, women who receive cunnilingus—relative to those who do not—are more likely to orgasm (Richters et al. 2006). Among men who performed cunnilingus on their regular partner, those at greater sperm competition risk are more likely to perform cunnilingus until she experiences orgasm (Pham et al. 2013a). However, these researchers also found that men whose regular partner experiences cunnilingus-assisted orgasm ejaculated *outside* the time window of greater sperm retention described by Baker and Bellis (1993b). Baker and Bellis (1993b) found greater sperm retention from copulatory female orgasm, but did not specify the means by which orgasm was achieved—that is, whether it was achieved by self-stimulation, by partner-stimulation, or by, for example, digital or oral stimulation. The means by which female orgasm is induced may influence consequent sperm retention (reviewed in King and Belsky 2012).

Sexual Arousal Cunnilingus may be related to male sexual arousal and, therefore, to other sperm competition tactics. Male Indian flying foxes (*Pteropus giganteus*) that spend more time performing cunnilingus on a female also spend more time copulating with her (Maruthupandian and Marimuthu 2013), and Maruthupandian and Marimuthu have interpreted this relationship with respect to the sperm competition theory. Men typically perform oral sex *before* they copulate and ejaculate (Halpern and Sherman 1979), suggesting that oral sex may influence sexual arousal and consequent sperm competition tactics (e.g., copulatory thrusting, ejaculate adjustment). Pham et al. (2013b) found that men who spend more time performing oral sex on their regular partner also spend more time copulating with her, perform more semen-displacing copulatory behaviors, and report greater sexual arousal (e.g., more forceful ejaculation, greater orgasm intensity). Consistent with findings from Pham et al. (2013b), Cerda-Molina et al. (2013) found that men who smell vaginal odors produced at high-fertility (relative to low-fertility) also experience a surge in testosterone and report greater copulatory interest.

The proposed functions of cunnilingus are not necessarily mutually exclusive. Cerda-Molina et al. (2013) found that men report greater sexual arousal and copulatory interest when smelling vaginal fluids produced at high-fertility relative to low-fertility, supporting both the fertility-detection function and the sexual arousal function of cunnilingus. All the proposed functions of cunnilingus support the broader hypothesis that cunnilingus minimizes cuckoldry risk for the in-pair male.

Research on the function of oral sex is preliminary, and several directions exist for future research. Here, we reviewed how cunnilingus benefits *males*. However, because male mating strategies and female mating strategies sometimes are at odds (i.e., there is sexual conflict; Shackelford and Goetz, 2012), future research should explore how cunnilingus benefits *females*. Additionally, oral sex performed by females on males (fellatio) is common in humans, though less common among nonhuman species. Tan et al. (2009) documented that female short-nosed fruit bats *(Cynopterus sphinx)* that spend more time performing fellatio on a male also spend more time copulating with him, and Koelman et al. (2000) found that a woman is less likely to experience preeclampsia if her regular partner ejaculates into her mouth—and especially if she ingests his semen—prior to conception. We suggest that the researchers employ the sperm competition theory to guide future work on oral sex.

Remaining Issues in Human Sperm Competition

Male Precedence Male precedence refers to the order in which a male copulates with a female—relative to other males—and how that order influences sperm competition success. In some species (e.g., *Cyrtodiopsis whitei*), a male that copulates with a female *before* rival males copulate with her fertilizes the largest proportion of her ova ("first male precedence"; Lorch et al. 1993). In other species (e.g., *Drosophila melanogaster*), a male that copulates with a female *after* rival males copulate with her fertilizes the largest proportion of her ova ("last male precedence"; Wilson et al. 1997).

Human sperm precedence has received little attention, perhaps in part because well-designed studies present ethical concerns (e.g., two men inseminating a woman and determining subsequent offspring paternity). Indirect evidence suggests that men may exhibit last male precedence (Gallup and Burch 2004). Some evidence is consistent with the possibility that men have anatomical (Gallup et al. 2003) and physiological (Gallup and Burch 2004) adaptations to displace rival semen—but not their own semen. Additionally, nonsperm substances left in the female reproductive tract from one ejaculate may create a more hospitable environment for subsequent ejaculates (e.g., immunosuppressants that weaken the female immune response to facilitate sperm longevity; reviewed in Gallup et al. 2012).

Among many nonhuman species, females actively manipulate male precedence. In *Drosophila melanogaster*, males receive fewer fertilizing advantages from last male precedence when inseminating an older female (relative to a younger female), suggesting that an older female is less effective at manipulating sperm in her reproductive tract (Mack et al. 2003). In humans, women may adjust their sexual behavior to manipulate male precedence. The IPC proclivity model proposes that women manipulate the time between their IPCs and their EPCs. Gallup et al. (2006) found that 64.1 % of women delayed indefinitely an IPC following an EPC and suggested that women may be actively avoiding sperm competition. However, if women do not delay an *EPC following an IPC*, then not only do women actively promote sperm competition, but they also actively ensure that an extra-pair partner secures a competitive advantage associated with last male precedence (e.g., semen displacement). Either possibility (of female avoidance of sperm competition or of female manipulation of male precedence) provides further evidence of an evolutionary history of sperm competition in humans.

Sperm Competition Intensity Sperm competition *risk* refers to the likelihood that a female has copulated or will copulate with two or more males during a given time period. Sperm competition *intensity* refers to the number of males that a female copulates with during a given time period. Research on sperm competition intensity is common in species in which females are highly promiscuous and relatively indiscriminate about with whom they copulate. The distinction between risk and intensity is important when investigating sperm competition strategies. For example, subdominant red junglefowl (*Gallus gallus*) males ejaculate the largest number of sperm in the presence of one rival male (i.e., low-intensity sperm competition), a moderate number of sperm when rivals are absent (i.e., zero-intensity sperm competition), and the lowest number of sperm when two or more rivals are present (i.e., high-intensity sperm competition), whereas dominant males ejaculate more sperm as the number of rivals increases (Pizzari et al. 2003).

Human sperm competition intensity has received little research attention, perhaps because the contexts that generate high-intensity sperm competition (e.g., orgies, gang rapes) are less common than the contexts that generate low-intensity sperm competition (e.g., female EPCs, Smith 1984). For example, Gallup et al. (2006) found that 3.1 % of college women copulated with three or more men simultaneously at least once. Indirect evidence suggests that men are sensitive to sperm competition intensity. For example, men prefer short-term mating with a woman who has several sexual partners (i.e., high-intensity sperm competition) than with a woman who is in a committed relationship (i.e., low-intensity sperm competition; Shackelford et al. 2006). Pound (2002) found that men prefer viewing pornography depicting one woman interacting with two men (high-intensity sperm competition) over pornography depicting one man interacting with one women (low-intensity sperm competition). As men are "witnessing" a woman copulating with another man, both conditions present an equally high *risk* of sperm competition, but the two conditions differ in the number of men and, therefore, the *intensity* of sperm competition. This line of research suggests that the sperm competition intensity models in humans mirror sperm competition intensity models in *dominant* male *Gallus gallus* (Pizzari et al. 2003): Men facing greater sperm competition intensity may deploy more sperm competition tactics (e.g., produce more competitive ejaculates; experience greater sexual arousal; Kilgallon and Simmons 2005; McKibbin et al. 2013; Pound 2002).

An evolutionary history of forced copulations by multiple males (i.e., "gang rape") may have generated contexts for high-intensity sperm competition. Gang rapes have been documented in several nonhuman species. For example, in Lesser Snow Geese (*Anser caerulescens caerulescens*), a female is often gang raped during her regular partner's absence, and her regular partner is absent because he is likely participating in a gang rape of a different female (Mineau and Cooke 1979).

There is evidence suggesting that rapists more frequently target women at high-fertility than women at low-fertility (i.e., the period during which sperm can fertilize ova; Gottschall and Gottschall 2003). Women at high-fertility dress more provocatively (Haselton et al. 2007) and walk more sexually suggestively (Guéguen 2012), and individuals generally agree on the "ease of rape" when judging the walk of featureless avatars (Gunns et al. 2002). However, other evidence suggests that women take less sexual risks at high-fertility (e.g., Chavanne and Gallup 1998).

One-third of rape complaints and one-fifth of confirmed rape cases in Chicago are gang rapes (Ullman 1999). Porter and Alison (2001) conducted a content analysis of archival data and found that, out of 39 cases of gang rapes in the USA and Norway, seventeen involved two men, ten involved three men, seven involved four men, two involved five men, and three involved six men. Gang rapes are especially prevalent in certain populations. For example, men in urban townships in the former Transkei region of South Africa invite their male friends to rape a woman, and under many circumstances (e.g., the victim supposedly deserved the rape; the victim did not vocally refuse the rape; the rapists did not employ outright violence, but instead employed trickery) this behavior is often legitimized (Wood 2005). These men may recruit their male friends to forcefully copulate with their regular partner if they suspect or know of their partner's EPCs (Wood 2005). Among those who serve in the military, 5 % of women report experiencing gang rape during their military service (Sadler et al. 2005). There are numerous case studies documenting the occurrence of gang rape perpetrated by college fraternity men (Sanday 2007). This research suggests that gang rape may be more prevalent under contexts that promote strong male coalitions (e.g., friendships; Gottschall 2004; Wood et al. 2005)

Gang rape occurs in preindustrial environments (e.g., modern hunter-gatherer tribes), providing further evidence that gang rape may have recurred over human evolutionary history. Gang rape of women by men in the Xingu tribes of South America is so prevalent that the behavior is *ritualized* (McCallum 1994). Among the Yanomamo of South America, men gang rape women who are captured from a defeated, neighboring tribe (Sanday 1981). Gang rape may have been a recurrent context selecting over evolutionary history for adaptations to high-intensity sperm competition in humans.

Conclusion

Since the pioneering studies by Robin Baker and Mark Bellis, research on human sperm competition has been profitably guided by a comparative evolutionary perspective. In this article, we reviewed evidence that human males and males of many nonhuman animals attend to similar sperm competition cues (e.g., female attractiveness, regular partner's absence, presence and number of rival males) and deploy similar sperm competition tactics (e.g., ejaculate adjustment, forced IPCs, frequent IPCs). In particular, many adaptations to sperm competition in humans have analogs in birds. These cross-species similarities may be a consequence of convergent evolution, because humans and most birds have similar mating systems (i.e., social monogamy) that require solutions to similar adaptive problems (e.g., cuckoldry). The substantial body of research documenting physiological, anatomical, behavioral, and psychological adaptations to sperm competition in birds and many other nonhuman animals provides evidence that sperm competition was a recurrent feature of the evolutionary histories of these animals. Research on humans also provides evidence that sperm competition was a recurrent feature of human evolutionary history.

Acknowledgments We would like to thank Jennifer Vonk for her helpful comments and suggestions. This chapter is based on Pham and Shackelford (in press).

References

Anderson, K. (2006). How well does paternity confidence match actual paternity? *Current Anthropology, 47,* 513–520.

Angulo, J., & García, M. (2005). *Sex in stone: Sexuality, reproduction, and eroticism in the Paleolithic epoch*. Madrid: Luzán, 5.

Ariely, D., & Loewenstein, G. (2006). The heat of the moment: The effect of sexual arousal on sexual decision making. *Journal of Behavioral Decision Making, 19,* 87–98.

Auger, J., Kunstmann, J. M., Czyglik, F., & Jouannet, P. (1995). Decline in semen quality among fertile men in Paris during the past 20 years. *New England Journal of Medicine, 332,* 281–285.

Baker, R. R., & Bellis, M. A. (1993a). Human sperm competition: Ejaculate adjustment by males and the function of masturbation. *Animal Behaviour, 46,* 861–885.

Baker, R. R., & Bellis, M. A. (1993b). Human sperm competition: Ejaculate manipulation by females and a function for the female orgasm. *Animal Behaviour, 46,* 887–909.

Baker, R. R., & Bellis, M. A. (1995). *Human sperm competition: Copulation, masturbation and infidelity*. London: Chapman & Hall.

Barash, D. P. (1997). Sociobiology of rape in mallards (*Anas platyrhynchos*): Response to the mated male. *Science, 197,* 788–789.

Bellis, M. A., Hughes, K., Hughes, S., & Ashton, J. R. (2005). Measuring paternal discrepancy and its public health consequences. *Journal of Epidemiology and Community Health, 59,* 749–754.

Bostofte, E., Serup, J., & Rebbe, H. (1982). Has the fertility of Danish men declined through the years in terms of semen quality? A comparison of semen qualities between 1952 and 1972. *International Journal of Fertility, 28,* 91–95.

Brewis, A., & Meyer, M. (2005). Demographic evidence that human ovulation is undetectable (at least in pair bonds). *Current Anthropology, 46,* 465–471.

Buss, D. M. (1988). From vigilance to violence: Tactics of mate retention in American undergraduates. *Ethology and Sociobiology, 9,* 291–317.
Buss, D. M. (2006). *The murderer next door: Why the mind is designed to kill.* New York: Penguin.
Buss, D. M. (2013). Sexual Jealousy. *Psychological Topics, 22,* 155–182.
Buss, D. M., & Shackelford, T. K. (1997). From vigilance to violence: Mate retention tactics in married couples. *Journal of Personality and Social Psychology, 72,* 346–361.
Camilleri, J. A., & Quinsey, V. L. (2009). Testing the cuckoldry risk hypothesis of partner sexual coercion in community and forensic samples. *Evolutionary Psychology, 7,* 64–178.
Candolin, U., & Reynolds, J. D. (2002). Adjustments of ejaculation rates in response to risk of sperm competition in a fish, the bitterling (*Rhodeus sericeus*). *Proceedings of the Royal Society of London, Series B: Biological Sciences, 269,* 1549–1553.
Cerda-Molina, A. L., Hernández-López, L., de la O, C. E., Chavira-Ramírez, R., & Mondragón-Ceballos, R. (2013). Changes in men's salivary testosterone and cortisol levels, and in sexual desire after smelling female axillary and vulvar scents. *Frontiers in Endocrinology, 4,* 1–9.
Chavanne, T. J., & Gallup, G. G. Jr. (1998). Variation in risk taking behavior among female college students as a function of the menstrual cycle. *Evolution and Human Behavior, 19,* 27–31.
Cheng, K. M., Burns, J. T., & McKinney, F. (1983). Forced copulation in captive mallards III. Sperm competition. *The Auk, 100,* 302–310.
Cornell, J. L., & Halpern-Felsher, B. L. (2006). Adolescents tell us why teens have oral sex. *Journal of Adolescent Health, 38,* 299–301.
Cornwallis, C. K., & O'Connor, E. A. (2009). Sperm: Seminal fluid interactions and the adjustment of sperm quality in relation to female attractiveness. *Proceedings of the Royal Society B: Biological Sciences, 276,* 3467–3475.
Daly, M., & Wilson, M. (1988). *Homicide.* Hawthorne: Aldine.
Doty, R. L., Ford, M., Preti, G., & Huggins, G. R. (1975). Changes in the intensity and pleasantness of human vaginal odors during the menstrual cycle. *Science, 190,* 1316–1318.
Dunbar, I. F. (1977). Olfactory preferences in dogs: The response of male and female beagles to conspecific odors. *Behavioral Biology, 20,* 471–481.
Finkelhor, D., & Yllo, K. (1985). *License to rape: Sexual abuse of wives.* New York: Holt, Rinehart, & Winston.
Fox, C. A., Wolff, H. S., & Baker, J. A. (1970). Measurement of intra-vaginal and intra-uterine pressures during human coitus by radio-telemetry. *Journal of Reproduction and Fertility, 22,* 243–251.
Frieze, I. H. (1983). Investigating the causes and consequences of marital rape. *Signs, 8,* 532–553.
Gage, A. R., & Barnard, C. J. (1996). Male crickets increase sperm number in relation to competition and female size. *Behavioral Ecology and Sociobiology, 38,* 349–353.
Gage, A. J., & Hutchinson, P. L. (2006). Power, control, and intimate partner sexual violence in Haiti. *Archives of Sexual Behavior, 35,* 11–24.
Gallup, G. G. Jr., & Burch, R. L. (2004). Semen displacement as a sperm competition strategy in humans. *Evolutionary Psychology, 2,* 12–23.
Gallup, G. G. Jr., & Burch, R. L. (2006). The semen displacement hypothesis: Semen hydraulics and the intra-pair copulation proclivity model of female infidelity. In S. Platek & T. Shackelford (Eds.), *Female infidelity and paternal uncertainty: Evolutionary perspectives on male anti-cuckoldry tactics.* Cambridge: Cambridge University Press.
Gallup, G. G. Jr., Burch, R. L., Zappieri, M. L., Parvez, R. A., Stockwell, M. L., & Davis, J. A. (2003). The human penis as a semen displacement device. *Evolution and Human Behavior, 24,* 277–289.
Gallup, G. G. Jr., Burch, R. L., & Berens Mitchell, T. J. (2006). Semen displacement as a sperm competition strategy: Multiple mating, self-semen displacement, and timing of in-pair copulations. *Human Nature, 17,* 253–264.
Gallup, G. G. Jr., Burch, R. L., & Petricone, L. R. (2012). Sexual conflict, infidelity, and vaginal/semen chemistry. In T. K. Shackelford & A. T. Goetz (Eds.), *The Oxford handbook of sexual conflict in humans* (pp. 217–231). New York: Oxford University Press.

Gangestad, S. W., Thornhill, R., & Garver-Apgar, C. E. (2005). Women's sexual interests across the ovulatory cycle depend on primary partner developmental instability. *Proceedings of the Royal Society B: Biological Sciences, 272,* 2023–2027.

Goetz, A. T., & Shackelford, T. K. (2006). Sexual coercion and forced in-pair copulation as sperm competition tactics in humans. *Human Nature, 17,* 265–282.

Goetz, A. T., & Shackelford, T. K. (2009). Sexual coercion in intimate relationships: A comparative analysis of the effects of women's infidelity and men's dominance and control. *Archives of Sexual Behavior, 38,* 226–234.

Goetz, A. T., Shackelford, T. K., Weekes-Shackelford, V. A., Euler, H. A., Hoier, S., Schmitt, D. P., & LaMunyon, C. W. (2005). Mate retention, semen displacement, and human sperm competition: A preliminary investigation of tactics to prevent and correct female infidelity. *Personality and Individual Differences, 38,* 749–763.

Goetz, A. T., Shackelford, T. K., Platek, S. M., Starratt, V. G., & McKibbin, W. F. (2007). Sperm competition in humans: Implications for male sexual psychology, physiology, anatomy, and behavior. *Annual Review of Sex Research, 18,* 1–22.

Goetz, A. T., Shackelford, T. K., & Camilleri, J. A. (2008). Proximate and ultimate explanations are required for a comprehensive understanding of partner rape. *Aggression and Violent Behavior, 13,* 119–123.

Gottschall, J. (2004). Explaining wartime rape. *Journal of Sex Research, 41,* 129–136.

Gottschall, J. A., & Gottschall, T. A. (2003). Are per-incident rape-pregnancy rates higher than per-incident consensual pregnancy rates? *Human Nature, 14,* 1–20.

Guadamuz, T. E., Kunawararak, P., Beyrer, C., Pumpaisanchai, J., Wei, C., & Celentano, D. D. (2010). HIV prevalence, sexual and behavioral correlates among Shan, Hill tribe, and Thai male sex workers in Northern Thailand. *AIDS Care, 22,* 597–605.

Guéguen, N. (2012). Gait and menstrual cycle: Ovulating women use sexier gaits and walk slowly ahead of men. *Gait and Posture, 35,* 621–624.

Gunns, R. E., Johnston, L., & Hudson, S. M. (2002). Victim selection and kinematics: A point-light investigation of vulnerability to attack. *Journal of Nonverbal Behavior, 26,* 129–158.

Hald, G. M. (2006). Gender differences in pornography consumption among young heterosexual Danish adults. *Archives of Sexual Behavior, 35,* 577–585.

Halpern, J., & Sherman, M. A. (1979). *Afterplay: A key intimacy.* New York: Pocket Books.

Haselton, M. G., & Gildersleeve, K. (2011). Can men detect ovulation? *Current Directions in Psychological Science, 20,* 87–92.

Haselton, M. G., Mortezaie, M., Pillsworth, E. G., Bleske-Rechek, A., & Frederick, D. A. (2007). Ovulatory shifts in human female ornamentation: Near ovulation, women dress to impress. *Hormones and Behavior, 51,* 40–45.

Hewlett, B. S., & Hewlett, B. L. (2010). Sex and searching for children among Aka foragers and Ngandu farmers of central Africa. *African Study Monographs, 31,* 107–125.

Hosken, D. J., & Ward, P. I. (2001). Experimental evidence for testis size evolution via sperm competition. *Ecology Letters, 4,* 10–13.

Iwawaki, S., & Wilson, G. D. (1983). Sex fantasies in Japan. *Personality and Individual Differences, 4,* 543–545.

Johnston, R. E. (1974). Sexual attraction function of golden hamster vaginal secretion. *Behavioral Biology, 12,* 111–117.

Kaighobadi, F., & Shackelford, T. K. (2008). Female attractiveness mediates the relationship between in-pair copulation frequency and men's mate retention behaviors. *Personality and Individual Differences, 45,* 293–295.

Kiddy, C. A., Mitchell, D. S., Bolt, D. J., & Hawk, H. W. (1978). Detection of estrus-related odors in cows by trained dogs. *Biology of Reproduction, 19,* 389–395.

Kilgallon, S. J., & Simmons, L. W. (2005). Image content influences men's semen quality. *Biology Letters, 1,* 253–255.

King, R., & Belsky, J. (2012). A typological approach to testing the evolutionary functions of human female orgasm. *Archives of Sexual Behavior, 41,* 1145–1160.

Klusmann, D. (2002). Sexual motivation and the duration of partnership. *Archives of Sexual Behavior, 31,* 275–287.

Klusmann, D. (2006). Sperm competition and female procurement of male resources. *Human Nature, 17,* 283–300.

Koelman, C. A., Coumans, A. B., Nijman, H. W., Doxiadis, I. I., Dekker, G. A., & Claas, F. H. (2000). Correlation between oral sex and a low incidence of preeclampsia: A role for soluble HLA in seminal fluid? *Journal of Reproductive Immunology, 46,* 155–166.

Lorch, P. D., Wilkinson, G. S., & Reillo, P. R. (1993). Copulation duration and sperm precedence in the stalk-eyed fly *Cyrtodiopsis whitei* (Diptera: Diopsidae). *Behavioral Ecology and Sociobiology, 32,* 303–311.

Lurie, P., Eugenia, M., Fernandes, L., & Hughes, V. (1995). Socioeconomic status and risk of HIV-1, syphilis and hepatitis B infection among sex workers in São Paulo State, Brazil. *AIDS (London, England), 9,* 31–37.

Mack, P. D., Priest, N. K., & Promislow, D. E. (2003). Female age and sperm competition: Last-male precedence declines as female age increases. *Proceedings of the Royal Society of London. Series B: Biological Sciences, 270,* 159–165.

Malamuth, N. M. (1996). Sexually explicit media, gender differences, and evolutionary theory. *Journal of Communication, 46,* 8–31.

Maruthupandian, J., & Marimuthu, G. (2013). Cunnilingus apparently increases duration of copulation in the Indian flying fox, *Pteropus giganteus. PloS One, 8,* e59743.

Mauras, N., Bell, J., Snow, B. G., & Winslow, K. L. (2005). Sperm analysis in growth hormone-deficient adolescents previously treated with an aromatase inhibitor: Comparison with normal controls. *Fertility and Sterility, 84,* 239–242.

McCallum, C. (1994). Ritual and the origin of sexuality in the Alto Xingu. In P. Harvey & P. Gow (Eds.), *Sex and violence: Issues in representation and experience* (pp. 90–114). New York: Routledge.

McKibbin, W. F., Bates, V. M., Shackelford, T. K., LaMunyon, C. W., & Hafen, C. A. (2010). Risk of sperm competition moderates the relationship between men's satisfaction with their partner and men's interest in their partner's copulatory orgasm. *Personality and Individual Differences, 49,* 961–966.

McKibbin, W. F., Starratt, V. G., Shackelford, T. K., & Goetz, A. T. (2011). Perceived risk of female infidelity moderates the relationship between objective risk of female infidelity and sexual coercion in Humans (*Homo sapiens*). *Journal of Comparative Psychology, 125,* 370–373.

McKibbin, W. F., Pham, M. N., & Shackelford, T. K. (2013). Human sperm competition in post-industrial ecologies: Sperm competition cues predict pornographic DVD sales rank. *Behavioral Ecology, 24,* 819–823.

Mineau, P., & Cooke, F. (1979). Rape in the lesser snow goose. *Behaviour, 70,* 280–291.

Mougeot, F., Arroyo, B. E., & Bretagnolle, V. (2001). Decoy presentations as a means to manipulate the risk of extrapair copulation: An experimental study in a semicolonial raptor, the Montagu's harrier (*Circus pygargus*). *Behavioral Ecology, 12,* 1–7.

Mosher, D. L. (1988). Pornography defined: Sexual involvement theory, narrative context, and goodness-of-fit. *Journal of Psychology and Human Sexuality, 1,* 67–85.

Murphy, M. R. (1973). Effects of female hamster vaginal discharge on the behavior of male hamsters. *Behavioral Biology, 9,* 367–375.

Nicholls, E. H., Burke, T., & Birkhead, T. R. (2001). Ejaculate allocation by male sand martins, Riparia riparia. *Proceedings of the Royal Society of London. Series B: Biological Sciences, 268,* 1265–1270.

Nishimura, K., Utsumi, K., Okano, T., & Iritani, A. (1991). Separation of mounting-inducing pheromones of vaginal mucus from estrual heifers. *Journal of Animal Science, 69,* 3343–3347.

O'Donohue, W., & Plaud, J. J. (1991). The long-term habituation of sexual arousal in the human male. *Journal of Behavior Therapy and Experimental Psychiatry, 22,* 87–96.

Oaten, M., Stevenson, R. J., & Case, T. I. (2009). Disgust as a disease-avoidance mechanism. *Psychological Bulletin, 135,* 303–321.

Ono, T., Siva-Jothy, M. T., & Kato, A. (1989). Removal and subsequent ingestion of rivals' semen during copulation in a tree cricket. *Physiological Entomology, 14,* 195–202.

Palagi, E., Telara, S., & Tarli, S. B. (2003). Sniffing behavior in *Lemur catta*: Seasonality, sex, and rank. *International Journal of Primatology, 24,* 335–350.

Parker, G. G. (1970). Sperm competition and its evolutionary consequences in the insects. *Biological Review, 45,* 525–567.

Pham, M. N., & Shackelford, T. K. (in press). Human sperm competition: A comparative evolutionary analysis. *Animal Behavior and Cognition.*

Pham, M. N., & Shackelford, T. K. (2013a). The relationship between objective sperm competition risk and men's copulatory interest is moderated by partner's time spent with other men. *Human Nature, 24,* 476–485.

Pham, M. N., & Shackelford, T. K. (2013b). Oral sex as mate retention behavior. *Personality and Individual Differences, 55,* 185–188.

Pham, M. N., & Shackelford, T. K. (2013c). Oral sex as infidelity-detection. *Personality and Individual Differences, 54,* 792–795.

Pham, M. N., Shackelford, T. K., Holden, C. J., Zeigler-Hill, V., Hummel, A., & Memering, S. L. (in press). Partner attractiveness moderates the relationship between number of sexual rivals and in-pair copulation frequency in humans (*Homo sapiens*). *Journal of Comparative Psychology.*

Pham, M. N., Shackelford, T. K., Sela, Y., & Welling, L. L. M. (2013a). Is cunnilingus-assisted orgasm a male sperm-retention strategy? *Evolutionary Psychology, 11,* 405–414.

Pham, M. N., Shackelford, T. K., Welling, L. L. M., Ehrke, A. D., Sela, Y., & Goetz, A. T. (2013b). Oral sex, semen displacement, and sexual arousal: Testing the ejaculate adjustment hypothesis. *Evolutionary Psychology, 11,* 1130–1139.

Pillsworth, E. G., & Haselton, M. G. (2006). Male sexual attractiveness predicts differential ovulatory shifts in female extra-pair attraction and male mate retention. *Evolution and Human Behavior, 27,* 247–258.

Pizzari, T., Cornwallis, C. K., Løvlie, H., Jakobsson, S., & Birkhead, T. R. (2003). Sophisticated sperm allocation in male fowl. *Nature, 426,* 70–74.

Porter, L. E., & Alison, L. J. (2001). A partially ordered scale of influence in violent group behavior: An example from gang rape. *Small Group Research, 32,* 475–497.

Pound, N. (2002). Male interest in visual cues of sperm competition risk. *Evolution and Human Behavior, 23,* 443–446.

Richters, J., de Visser, R., Rissel, C., & Smith, A. (2006). Sexual practices at last heterosexual encounter and occurrence of orgasm in a national survey. *Journal of Sex Research, 43,* 217–226.

Rook, G. A. (2007). The hygiene hypothesis and the increasing prevalence of chronic inflammatory disorders. *Transactions of the Royal Society of Tropical Medicine and Hygiene, 101,* 1072–1074.

Russell, D. E. H. (1982). *Rape in marriage*. New York: Macmillan.

Sadler, A. G., Booth, B. M., & Doebbeling, B. N. (2005). Gang and multiple rapes during military service: Health consequences and health care. *Journal of the American Medical Women's Association, 60,* 33–41.

Sanday, P. (2007). *Fraternity gang rape: Sex, Brotherhood, and Privilege on Campus*. New York: NYU Press.

Sanday, P. R. (1981). The socio-cultural context of rape: A cross-cultural study. *Journal of Social Issues, 37,* 5–27.

Sanders, S. A., & Reinisch, J. M. (1999). Would you say you had sex if…? *The Journal of the American Medical Association, 281,* 275–277.

Sankar, R., & Archunan, G. (2004). Flehmen response in bull: Role of vaginal mucus and other body fluids of bovine with special reference to estrus. *Behavioural Processes, 67,* 81–86.

Santtila, P., Wager, I., Katarina, W., Harlaar, N., Jern, P., Johansson, A., & Sandnabba, N. K. (2007). Discrepancies between sexual desire and sexual activity: Gender differences and associations with relationship satisfaction. *Journal of Sex and Marital Therapy, 34,* 29–42.

Shackelford, T. K., & Goetz, A. T. (Eds.). (2012). *The Oxford handbook of sexual conflict in humans*. New York: Oxford University Press.

Shackelford, T. K., LeBlanc, G. J., Weekes-Shackelford, V. A., Bleske-Rechek, A. L., Euler, H. A., & Hoier, S. (2002). Psychological adaptation to human sperm competition. *Evolution and Human Behavior, 23,* 123–138.

Shackelford, T. K., Buss, D. M., & Weekes-Shackelford, V. A. (2003). Wife-killings committed in the context of a "lovers triangle". *Basic and Applied Social Psychology, 25,* 127–133.

Shackelford, T. K., Goetz, A. T., LaMunyon, C. W., Quintus, B. J., & Weekes-Shackelford, V. A. (2004). Sex differences in sexual psychology produce sex similar preferences for a short-term mate. *Archives of Sexual Behavior, 33,* 405–412.

Shackelford, T. K., Goetz, A. T., Guta, F. E., & Schmitt, D. P. (2006). Mate guarding and frequent in-pair copulation in humans. *Human Nature, 17,* 239–252.

Shackelford, T. K., Goetz, A. T., McKibbin, W. F., & Starratt, V. G. (2007). Absence makes the adaptations grow fonder: Proportion of time apart from partner, male sexual psychology, and sperm competition in humans (Homo sapiens). *Journal of Comparative Psychology, 121,* 214–220.

Short, R. V. (1981). Sexual selection in man and the great apes. In C. E. Graham (Ed.), *Reproductive biology of the great apes* (pp. 319–341). New York: Academic.

Simmons, L. W., & Fitzpatrick, J. L. (2012). Sperm wars and the evolution of male fertility. *Reproduction (Cambridge, England), 144,* 519–534.

Smith, R. L. (1984). Human sperm competition. In R. L. Smith (Ed.), *Sperm competition and the evolution of animal mating systems* (pp. 601–659). New York: Academic.

Soini, P. (1987). Sociosexual behavior of a free-ranging *Cebuella pygmaea* (Callitrichidae, platyrrhini) troop during postpartum estrus of its reproductive female. *American Journal of Primatology, 13,* 223–230.

Starratt, V. G., Goetz, A. T., Shackelford, T. K., & Stewart-Williams, S. (2008). Men's partner-directed insults and sexual coercion in intimate relationships. *Journal of Family Violence, 23,* 315–323.

Starratt, V. G., McKibbin, W. F., & Shackelford, T. K. (2013). Experimental manipulation of psychological mechanisms responsive to female infidelity. *Personality and Individual Differences., 55,* 59–62.

Tan, M., Jones, G., Zhu, G., Ye, J., Hong, T., Zhou, S., et al. (2009). Fellatio by fruit bats prolongs copulation time. *PloS One, 4,* e7595.

Thornhill, R. (2006). Foreward: Human sperm competition and women's dual sexuality. In T. K. Shackelford & N. Pound (Eds.), *Sperm competition in humans: Classic and contemporary readings (v–xix).* New York: Springer.

Tortosa, F. S., & Redondo, T. (1992). Frequent copulations despite low sperm competition in white storks (*Ciconia ciconia*). *Behaviour, 121,* 287–315.

Ullman, S. E. (1999). A comparison of gang and individual rape incidents. *Violence and Victims, 14,* 123–133.

Voracek, M., Haubner, T., & Fisher, M. L. (2008). Recent decline in nonpaternity rates: A cross-temporal meta-analysis 1, 2. *Psychological reports, 103,* 799–811.

Wildt, L., Kissler, S., Licht, P., & Becker, W. (1998). Sperm transport in the human female genital tract and its modulation by oxytocin as assessed by hysterosalpingoscintigraphy, hysterotonography, electrohysterography and Doppler sonography. *Human Reproduction Update, 4,* 655–666.

Wilson, N., Tubman, S. C., Eady, P. E., & Robertson, G. W. (1997). Female genotype affects male success in sperm competition. *Proceedings of the Royal Society of London. Series B: Biological Sciences, 264,* 1491–1495.

Wedell, N., Gage, M. J., & Parker, G. A. (2002). Sperm competition, male prudence and sperm-limited females. *Trends in. Ecology and Evolution, 17,* 313–320.

Wolf, M., Musch, J., Enczmann, J., & Fischer, J. (2012). Estimating the prevalence of nonpaternity in Germany. *Human Nature, 23,* 208–217.

Wood, K. (2005). Contextualising group rape in post-apartheid South Africa. *Culture. Health and Sexuality, 7,* 303–317.

Zavos, P. M. (1985). Seminal parameters of ejaculates collected from oligospermic and normospermic patients via masturbation and at intercourse with the use of a Silastic seminal fluid collection device. *Fertility and Sterility, 44,* 517–520.

Zavos, P. M., & Goodpasture, J. C. (1989). Clinical improvements of specific seminal deficiencies via intercourse with a seminal collection device versus masturbation. *Fertility and Sterility, 51,* 190–193.

Zervomanolakis, I., Ott, H. W., Hadziomerovic, D., Mattle, V., Seeber, B. E., Virgolini, I., Heute, D., et al. (2007). Physiology of upward transport in the human female genital tract. *Annals of the New York Academy of Sciences, 1101,* 1–20.

Zervomanolakis, I., Ott, H. W., Müller, J., Seeber, B. E., Friess, S. C., Mattle, V., & Wildt, L. (2009). Uterine mechanisms of ipsilateral directed spermatozoa transport: Evidence for a contribution of the utero-ovarian countercurrent system. *European Journal of Obstetrics and Gynecology and Reproductive Biology, 144,* S45–S49.

Ziegler, T. E., Epple, G., Snowdon, C. T., Porter, T. A., Belcher, A. M., & Küderling, I. (1993). Detection of the chemical signals of ovulation in the cotton-top tamarin. *Saguinus oedipus. Animal Behaviour, 45,* 313–322.

Index

T. K. Shackelford, R. D. Hansen (eds.), *The Evolution of Sexuality*,
Evolutionary Psychology, DOI 10.1007/978-3-319-09384-0

CPSIA information can be obtained
at www.ICGtesting.com
Printed in the USA
LVHW03*1813101018
593128LV00009B/153/P

9 783319 093833